ETHNOBOTANY AND
THE SEARCH FOR NEW DRUGS

D1131211

The Ciba Foundation is an international scientific and educational charity (Registered Charity No. 313574). It was established in 1947 by the Swiss chemical and pharmaceutical company of CIBA Limited—now Ciba-Geigy Limited. The Foundation operates independently in London under English trust law.

The Ciba Foundation exists to promote international cooperation in biological, medical and chemical research. It organizes about eight international multidisciplinary symposia each year on topics that seem ready for discussion by a small group of research workers. The papers and discussions are published in the Ciba Foundation symposium series. The Foundation also holds many shorter meetings (not published), organized by the Foundation itself or by outside scientific organizations. The staff always welcome suggestions for future meetings.

The Foundation's house at 41 Portland Place, London W1N 4BN, provides facilities for meetings of all kinds. Its Media Resource Service supplies information to journalists on all scientific and technological topics. The library, open five days a week to any graduate in science or medicine, also provides information on scientific meetings throughout the world and answers general enquiries on biomedical and chemical subjects. Scientists from any part of the world may stay in the house during working visits to London.

Ciba Foundation Symposium 185

ETHNOBOTANY AND THE SEARCH FOR NEW DRUGS

1994

JOHN WILEY & SONS

Chichester · New York · Brisbane · Toronto · Singapore

Published in 1994 by John Wiley & Sons Ltd
Baffins Lane, Chichester
West Sussex PO19 1UD, England

Reprinted August 1995

Other Wiley Editorial Offices

John Wiley & Sons, Inc., 605 Third Avenue,
New York, NY 10158–0012, USA

Jacaranda Wiley Ltd, G.P.O. Box 859, Brisbane,
Queensland 4001, Australia

John Wiley & Sons (Canada) Ltd, 22 Worcester Road,
Rexdale, Ontario M9W 1L1, Canada

John Wiley & Sons (SEA) Pte Ltd, 37 Jalan Pemimpin #05-04,
Block B, Union Industrial Building, Singapore 2057

Suggested series entry for library catalogues:
Ciba Foundation Symposia

Ciba Foundation Symposium 186
x + 280 pages, 17 figures, 24 tables

Library of Congress Cataloging-in-Publication Data
Ethnobotany and the search for new drugs / [G. T. Prance, Derek J.
Chadwick (organizer), and Joan Marsh].
    p. cm.—(Ciba Foundation symposium;185)
    "Symposium on Ethnobotany and the Search for New Drugs, held at
the Hotel Praia Centro, Fortaleza, Brazil, 30 November–2 December
1993."
    Includes bibliographical references and index.
    ISBN 0 471 95024 6
    1. Materia medica, Vegetable—Congresses.   2. Ethnobotany—
Congresses. 3. Medicinal plants—Congresses.   I. Prance, Ghillean
T., 1937– . II. Chadwick, Derek.   III. Marsh, Joan.
IV. Symposium on Ethnobotany and the Search for New Drugs (1993:
Fortaleza, Brazil) V. Series.
RS164.E84 1994
615´.32—dc20   94-28300
                                                          CIP

British Library Cataloguing in Publication Data
A catalogue record for this book is
available from the British Library

ISBN 0 471 95024 6

Phototypeset by Dobbie Typesetting Limited, Tavistock, Devon
Printed and bound in Great Britain by Antony Rowe Ltd, Chippenham,
Wiltshire

Contents

Symposium on Ethnobotany and the search for new drugs, held at The Hotel Praia Centro, Fortaleza, Brazil, 30 November–2 December 1993

This symposium is based on a proposal made by Professor Paul Cox

Editors: Derek J. Chadwick (Organizer) and Joan Marsh

G. T. Prance Introduction 1

M. J. Balick Ethnobotany, drug development and biodiversity conservation—exploring the linkages 4
Discussion 18

P. A. Cox The ethnobotanical approach to drug discovery: strengths and limitations 25
Discussion 36

N. R. Farnsworth Ethnopharmacology and drug development 42
Discussion 51

W. H. Lewis and **M. P. Elvin-Lewis** Basic, quantitative and experimental research phases of future ethnobotany with reference to the medicinal plants of South America 60
Discussion 72

E. Elisabetsky and **D. A. Posey** Ethnopharmacological search for antiviral compounds: treatment of gastrointestinal disorders by Kayapó medical specialists 77
Discussion 90

A. A. Craveiro, M. I. L. Machado, J. W. Alencar and **F. J. A. Matos** Natural product chemistry in north-eastern Brazil 95
Discussion 102

R. E. Schultes Amazonian ethnobotany and the search for new drugs* 106
Discussion 112

*Professor Schultes was unable to attend the symposium; his paper was circulated and discussed in his absence.

M. M. Iwu African medicinal plants in the search for new drugs based on ethnobotanical leads 116
Discussion 126

X. Lozoya Two decades of Mexican ethnobotany and research on plant-derived drugs 130
Discussion 140

S. K. Jain Ethnobotany and research on medicinal plants in India 153
Discussion 164

P. G. Xiao Ethnopharmacological investigation of Chinese medicinal plants 169
Discussion 173

G. M. Cragg, M. R. Boyd, J. H. Cardellina II, D. J. Newman, K. M. Snader and **T. G. McCloud** Ethnobotany and drug discovery: the experience of the US National Cancer Institute 178
Discussion 190

S. R. King and **M. S. Tempesta** From shaman to human clinical trials: the role of industry in ethnobotany, conservation and community reciprocity 197
Discussion 206

J. H. Barton Ethnobotany and intellectual property rights 214
Discussion 221

G. J. Martin Conservation and ethnobotanical exploration 229
Discussion 239

B. Berlin and **E. A. Berlin** Anthropological issues in medical ethnobotany 240
Discussion 259

G. T. Prance Conclusions 266

Index of contributors 269

Subject index 270

Participants

G. Albers-Schönberg Natural Products Chemistry, Merck Sharp & Dohme Research Laboratories, Rahway, NJ 07065, USA

W. Balée Department of Anthropology, Tulane University, 1021 Audubon Street, New Orleans, LA 70118, USA

M. J. Balick Institute of Economic Botany, The New York Botanical Garden, Bronx, NY 10458-5126, USA

J. H. Barton Law & High Technology Program, Stanford University, Stanford, CA 94305-8610, USA

B. Berlin Department of Anthropology, University of Georgia, Athens, GA 30602-1619, USA

L. Bohlin Department of Pharmacognosy, Biomedical Center, University of Uppsala, Box 579, S-751 23 Uppsala, Sweden

P. A. Cox Department of Botany & Range Science, Brigham Young University, Provo, UT 84602, USA

G. M. Cragg Natural Products Branch, National Cancer Institute, Frederick Cancer Research & Development Center, Fairview Center 206, Frederick, MD 21702-1201, USA

A. A. Craveiro Laboratorio de Produtos Naturais, Universidade Federal do Ceará, Campus do Pici, CP 12200, 60.021-970 Fortaleza, Ceará, Brazil

E. Dagne Department of Chemistry, Addis Ababa University, PO Box 1176, Addis Ababa, Ethiopia

E. Elisabetsky Laboratorio Etnofarmacologia, Instituto de Biociências, Universidade Federal do Rio Grande do Sul, Rua Sarmento Leite 500, 90.050-170 Porto Alegre RS, Brazil

M. P. Elvin-Lewis   Department of Biology, Washington University, Box 1137, One Brookings Drive, St Louis, MO 63130-4899, USA

N. R. Farnsworth   Program for Collaborative Research in the Pharmaceutical Sciences (M/C 877), University of Illinois at Chicago, 833 South Wood Street, Chicago, IL 60612, USA

M. M. Iwu   Phytotherapy Research Laboratory, Department of Pharmacognosy, University of Nigeria, Nsukka, Nigeria

S. K. Jain   National Botanical Research Institute, Lucknow 226 001, India

S. R. King   Ethnobotany and Conservation, Shaman Pharmaceuticals Inc., 213 East Grand Avenue, South San Francisco, CA 94080-4812, USA

R. P. L. Lemos   (*Ciba Foundation Bursar*) Instituto do Meio Ambiente de Alagoas, Av. Major Cicero de Gois Monteiro 2197, Mutange, 57.017-320 Maceió AL, Brazil

W. H. Lewis   Department of Biology, Washington University, Box 1137, One Brookings Drive, St Louis, MO 63130-4899, USA

X. Lozoya   Instituto Mexicano del Seguro Social, Centro Medico Nacional XXI, Jefatura de Servicios de Investigación Médica, Ed. Academia Nacional de Medicina 4° piso, Avenida Cuahutemoc 330, Mexico DF-CP 06725, Mexico

J. D. McChesney   Research Institute of Pharmaceutical Sciences, School of Pharmacy, University of Mississippi, University, MS 38677, USA

G. J. Martin   The WWF/UNESCO/Kew People and Plants Initiative, Division of Ecological Sciences, Man and the Biosphere Program, UNESCO, 7, Place de Fontenoy, F-75352 Paris, Cedex 07 SP, France

F. J. A. Matos   Horto de Plantas Medicinais, Laboratorio de Produtos Naturais, Universidade Federal do Ceará, Campus do Pici, CP 12200, 60.021-970 Fortaleza, Ceará, Brazil

A. B. de Oliveira   Universidade Federal de Minas Gerais, Faculdade de Farmácia, Av. Olegário Maciel 2360—Cidade Jardim, 30.180-112 Belo Horizonte MG, Brazil

H. H. Peter   Department of Biotechnology/Microbial Chemistry, PH 2.255, K-681.2.42, Ciba-Geigy AG, CH-4002 Basel, Switzerland

C. Picheansoonthon   (*Ciba Foundation Bursar*) Department of Pharmaceutical Botany and Pharmacognosy, Khon Kaen University, 123 Friendship Highway, Amphoe Muang, Khon Kaen 40002, Thailand

D. Posey   Instituto Etnobiologica da Amazonia, Conj Maguary, Ala 3, Casa 1a, Icoracy-Belém, Para, Brazil

G. T. Prance   (*Chairman*) Royal Botanic Gardens, Kew, Richmond TW9 3AB, UK

S. M. Rubin   Conservation International, Suite 1000, 1015 18th Street NW, Washington DC 20036, USA

G. Schwartsmann   Servico de Oncologia, Hospital de Clinicas de Porto Alegre, Rua Ramiro Barcelos 2350/sala 2030, Porto Alegre RS, Brazil

P. G. Xiao   Institute of Medicinal Plant Development, Chinese Academy of Medical Sciences, Haidian District, Dong Beiwang, Beijing 100094, China

Introduction

Professor G. T. Prance

Royal Botanic Gardens, Kew, Richmond TW9 3AB, UK

Many drugs that are on the market have come to us from folk use and use of plants by indigenous cultures. These drugs are being used in some way in modern medicine, but not necessarily for the same purpose as they were used by the native cultures. Very often, something has been used by local peoples because it is biologically active, but a more appropriate use for it in Western medicine is for something different. Folk medicine has been a pointer towards many of the drugs that we use and it is certainly a very useful indicator of biologically active substances. For ethnobotanists, the recent accelerated work on ethnobotany because of the renewed interest of the pharmaceutical and medical world is most welcome. It has helped the science of ethnobotany, not just medical ethnobotany but ethnobotany in general, to advance. One of the reasons the study of ethnobotany has accelerated is that we are aware that if we don't do some of this work now, it will be too late because of the unfortunate acculturation of tribal peoples around the world.

We also need to bear in mind the vital importance of conservation of culture, as well as biodiversity. To my mind, biodiversity gets too much emphasis compared with cultural conservation. We should be asking ourselves: how can our work in ethnobotany help to maintain cultural identity? Several papers in this symposium point out the importance of cultural identity, the importance of tribal customs and of their religions.

This leads on to the issue of ethics. What have indigenous peoples gained from the use of their knowledge by Western culture? I am afraid the answer to date is very little. What have developing countries gained from the use of their genetic material in medicines? This is a hot issue in Brazil. A report from a Manaus newspaper last week, which was also in a national newspaper, had the headline 'Contraband of Amazonian plants to the exterior'—based on absolutely no facts whatsoever. The report says that the New York Botanical Garden is taking plant material for analysis in the National Institutes of Health cancer programme, when they have deliberately avoided doing any work in Brazil. Because of the danger of misunderstanding here, the Garden is extremely careful not to damage its relationship in Brazil so that it can continue the important biodiversity and taxonomic work being done in this country. We need to appreciate the complexities presented by work in medical ethnobotany.

We are working in an area that can easily be misunderstood. We need to leave this symposium with a good sense of the ethics of our future work.

The other interesting development in ethnobotany and ethnomedicine is that we can achieve so much more today because of the amazing improvements in screening techniques through *in vitro* bioassays. So many analyses of plants in the past have not found the active component because the assays were not sensitive enough. A plant that was tested negative 20 years ago would not necessarily give a negative result today.

A problem that I hope we will discuss is the fact that the product used as a medicine by local peoples is usually not what is tested in the laboratory. Most of the effective brews I have seen and experienced are mixtures of green plants in the field. We tend to collect the individual plants, dry them, take them back and then see what chemicals they contain. Something we really have to get over in our ethnobotanical search for medicine is: how can we analyse what the indigenous peoples are really taking? They take elaborate chemical mixtures; chemical reactions occur within the tea that an indigenous person brews, and that is very different from a single compound isolated from a dried plant.

Another problem is the varied chemistry within individual plant species, depending on the ecology, the soil and the climate. We were working at Kew, for example, on a member of the Labiateae for the essential oil that it contained. We were delighted with the amount of the oil that the plant had shown in the field. This plant grew very well in our greenhouses. A chemist harvested it, thinking there would be a lot more of this product for analysis and found that there was none. The seeds were from the same plant that had given a good yield. We persisted with this and after three years of cultivation, suddenly it started producing the desired product in large quantities, probably owing to a change in some local ecological conditions in the growing area or even in the outside climate that we don't know. We often miss things because of this sort of variation. Plants are very sensitive to local conditions; they don't always produce the same chemicals consistently.

Another interesting area is the different concepts of disease between native peoples and practitioners of Western medicine. We often go into our studies thinking of disease in Western terms. We will produce much better results if we learn to understand the concepts and the spiritual side of the medicine of the people with whom we are working. We should not be afraid to discuss that aspect of it here. I believe that ethnobotany includes the knowledge of local folk cultures, as well as of indigenous tribes. For example, I have learned a great deal from the *caboclo* of Amazonia who have taken up many indigenous beliefs and many plant uses. I hope that in our discussions we will not refer only to indigenous peoples, but include the peasant *campesiño*, *caboclo*, *ribereño*, *mestizo*, or whatever they are called in different parts of the world.

As a plant taxonomist, I am very aware of the difficulties of identification of the plants studied. Some failures to repeat experiments are often due to poor taxonomy. Ensuring that our work is backed up by adequate taxonomy is another important issue that needs to be considered.

Finally, a danger associated with the discovery of a new medicine is that the plant species itself can become threatened by overexploitation. We saw this danger coming when taxol was discovered in the Pacific yew, until it could be synthesized from a precursor in commoner yew species. When I was in Cameroon, I saw the *Prunus africana* being eliminated from the forest, as it has been from the forest of most of Africa, because of the commercial exploitation of the bark for medicine for prostate problems. This could be an ideal product for sustainable harvesting, but people are often too greedy, so the entire bark is taken from the trees and the trees die. Promoting sustainable use of the different plants in which useful products are discovered is vital. Here in Brazil, various species of *Pilocarpus* are now endangered through their overexploitation for pilocarpine for the treatment of glaucoma. Fortunately, I think this will end because the company developing the drug is now setting plantations. Nevertheless, some species of *Pilocarpus* are on the brink of extinction.

These are some issues that I hope will come out in our general discussions and in some of your papers.

Ethnobotany, drug development and biodiversity conservation—exploring the linkages

Michael J. Balick

Institute of Economic Botany, The New York Botanical Garden, Bronx, NY 10458, USA

Abstract. Numerous ethnobotanical studies aimed at identifying new pharmaceutical products have been initiated in recent times. Ethnobotany has once again become a recognized tool in the search for new pharmaceuticals. Initiatives by governmental agencies and the private sector have helped spark this renewal. Many of these projects are interdisciplinary efforts involving scientists from the fields of anthropology, botany, medicine, pharmacology and chemistry. The Belize Ethnobotany Project links pharmaceutical prospecting with the conservation of traditional medical systems and biological resources. It illustrates the concept of the 'ethno-biomedical reserve' and provides an opportunity for pharmaceutical and herbal industries to contribute to the conservation effort. Terra Nova Rainforest Reserve is an ethno-biomedical reserve in Belize that was given legal status in June of 1993. Too often the exploitation of wild harvested resources has led to their severe degradation. There is a need for increased efforts to develop technologies to sustain their extraction.

1994 Ethnobotany and the search for new drugs. Wiley, Chichester (Ciba Foundation Symposium 185) p 4–24

In exploring the link between ethnobotany, drug development and the conservation of biodiversity, I would like to address several relevant topics. This paper will report on work done by the New York Botanical Garden staff in collaboration with the National Cancer Institute (NCI). I would like to broaden the scope of the 'drug development' aspect of this paper to include health care in general, whether delivered via pharmaceutical products, herbal formulations or single plant species. As I shall discuss, the conservation of biodiversity can best be achieved through establishing a linkage with the health-care delivery system in its broadest sense.

Ethnobotany and screening for activity against cancer
or the human immunodeficiency virus

I previously reported *in vitro* data received from the NCI, comparing the activity of selected medicinal plants against the human immunodeficiency virus (HIV) with that of plants collected randomly (Balick 1990). These results showed that such activity was greater in the medicinal plants. The plant extracts were subsequently dereplicated at the NCI and retested in the HIV screens. Dereplication involves removal of certain components of the crude plant extract such as tannins and polysaccharides. After dereplication, the percentage of plants showing anti-HIV activity falls dramatically and the overall percentage is virtually identical using either random or ethno-directed collection methodology (Table 1). Thus, as Gordon Cragg and his colleagues at NCI have shown with a much larger data set (Cragg et al 1994, this volume), general ethnobotanical collection does not appear to be advantageous in developing leads for HIV treatment.

The question arises as to whether potentially significant compounds are discarded in the dereplication process. Dereplication removes compounds that are known to have immunostimulatory activity. At least one major research programme (King & Tempesta 1994, this volume) shows that potent antiviral compounds can be isolated from tannins.

Another comment may be relevant to an evaluation of the ethno-directed approach involving anticancer therapies. In the field, many plants are collected and documented as being used against 'cancer', after their identification by a field guide or traditional healer. As Xavier Lozoya has pointed out (personal communication), ethnobotanists are usually limited in their powers of medical observation and often fail to include in their notes comments on the type of cancer for which the plants are used, as well as other vital information, such as method of preparation and dosage. From our work in Belize, we have found that several plants are used to treat 'cancer', including *Acalypha arvensis* and *Phlebodium decumanum*. During my first few months of field work in Belize, I recorded these plants as purported to have 'anticancer' activity. However, working with a naturopathic physician and observing the uses of these plants in patients, we realized that the word 'cancer' amongst healers in Belize and

TABLE 1   Percentage of plants showing activity in an *in vitro* anti-HIV screen

	Collection method	
	Random collections (61 samples)	Ethnobotanical collections (73 samples)
Initial screen	1.6	15
After dereplication	1.6	1.3

elsewhere in Central America is really the local word for 'an ailment characterized by severe, weeping open wounds that are chronic, spreading and difficult to heal' (Arvigo & Balick 1994). In order to establish a linkage between ethno-directed sampling and anticancer activity, one must have an intimate understanding of the disease concepts of the culture whose pharmacopoeia is under examination. For the most part, there are few traditional therapies for cancer as defined in the Western sense. It appears that much of the information that has been gathered previously on Central American plants with purported 'anticancer' activity is confused, unclear or has been biased in some way through the collection process. In order to evaluate the ethno-directed approach rigorously and fairly, we need to integrate botany, pharmacology, medicine and traditional beliefs more closely.

Finally, what would the ethno-directed approach yield if the search for anticancer and anti-HIV agents were focused on screening systems and therapeutic treatments other than those involving cytotoxicity? For example, would the plants collected through the ethno-directed approach show significantly greater immunostimulatory activity than those collected randomly? Given present financial and technological limitations, this question may not be answered for many years.

The link between medicinal plants, drug development and conservation

It is usually assumed that the discovery of a new plant-derived drug will ultimately be of value to conservation efforts, especially in rain forest regions. This idea is based on the profit potential and economic impact, as well as the feeling that governments and people will somehow place a greater value on a resource if it or its derivatives can produce a product with a multinational market. The economic value and the potential of medicinal plants to support conservation efforts can be viewed from three perspectives: regional traditional medicine, the international herbal industry and the international pharmaceutical industry (Table 2). The distribution of economic benefits varies greatly. In traditional medical systems, they accrue to professional collectors who sell the plants to traditional healers or to the healers themselves. The local and international herbal industries benefit a broad range of people and institutions, including collectors, wholesalers, brokers and companies that produce and sell herbal formulations. In the international pharmaceutical industry, most of the economic benefit goes to those at the upper end of the economic stratum, at the corporate level, as well as to those involved in wholesale and retail sales.

It is estimated that the international herbal industry is about 10 times the size of the US herbal industry, the value of which is about $1300 million annually (M. Blumenthal, personal communication). Thus, the market value of traditional medical products, which are used by thousands of millions of people around the world, is thousands of millions of dollars each year. Whether or not this

TABLE 2 **The economic value of plant medicines and their potential value for conservation**

Sector	Distribution of economic benefit	Amount of taxes collected	Pitfalls
International pharmaceutical industry	Upper end of economic system	Substantial	Overharvest Synthesis (if no provision for benefits included) Plantations established outside area of discovery
National and international herbal industry	Full spectrum of economic system	Medium	Overharvest Plantations established outside area of discovery
Regional traditional medicine	Lower end of economic system	Small	Overharvest (sustainability)

In each sector the market value of medicinal plants is thousands of millions of US$. The conservation potential of plant use in each sector ranges from low to high.

is comparable to the $80 000–90 000 million of global retail sales of pharmaceutical products has not been calculated, to my knowledge. However, it can be argued that commerce in traditional plant medicines, consisting primarily of local activity such as previously described, constitutes a significant economic force. If it is assumed that 3000 million people use traditional medicines from plants for their primary health care, and each person utilizes $2.50–$5.00 worth annually (whether harvested, bartered or purchased), then the annual value of these plants is $7500–$15 000 million, which is comparable to that of the two other sectors of the global pharmacopoeia.

An interesting perspective emerges when the tax yields to government are compared. Obviously, in traditional medical systems, taxes are neither assessed nor paid. The international herbal industry is subject to taxes such as those levied at the point of sale and on corporate profits. Governments benefit most from taxes through commerce for therapies produced and sold by the international pharmaceutical industry.

Those who promote the linkage between conservation and the search for new pharmaceutical products often fail to point out that the time from collection of a plant in the forest to sale on the pharmacist's shelves is 8–12 years and that programmes initiated today must be viewed as having long-term benefits, at best. An exception to this are agreements such as that between Merck, Sharp and Dohme and INBio, the National Biodiversity Institute of Costa Rica.

This agreement provides a substantial 'up-front' payment from Merck for infrastructure development at INBio and for the national parks system in Costa Rica. Hopefully, this will be a model for such North/South collaborations in the future. In traditional medicine and the herbal industry, the yields are immediate and the economic impact on the individual, community and region can be significant.

The potential for strengthening conservation efforts ranges from low to high, depending on whether the extraction of the resource can be sustainably managed over the long term or is simply exploited for short-term benefits by collectors and an industry that has little interest in ensuring a reliable supply in the future. Conservation potential is minimal if the end products are derived from synthetic processes or from plantations developed outside the original area of collection. To address this issue, the NCI's Developmental Therapeutics Program seeks to ensure that the primary country of origin of the plant will have the first opportunity to produce the plant, should commercially valuable products arise as a result of their programme (Cragg et al 1994, this volume).

Table 2 also summarizes the pitfalls inherent in each approach, including overharvesting, synthesis with no provision for benefits, issues of land tenure and the establishment of plantations outside the original range of the species. In any attempt to plan for the maximum conservation potential of a discovery, these must be kept in mind.

Harvest itself is not without pitfalls. One of the primary concerns about extraction is sustainability. A case in point is the extraction of a drug used in the treatment of glaucoma, pilocarpine. The source of pilocarpine is several species of trees in the genus *Pilocarpus*, which occur naturally in the north-east of Brazil, *P. pinnatifolius, P. microphylla* and *P. jaborandi*. Leaves have been harvested from the trees for many decades, usually under contract from chemical companies. Limited attempts at sustainable management were undertaken in the 1980s but, for the most part, harvest continued in a destructive fashion. Extinction—of the population in many areas—has been the fate of these plants. Finally, over the last few years, cultivated plantations of *Pilocarpus* species have been developed, which will reduce the value of the remaining wild stands, as well as eliminate any incentive there was for conserving them.

Development of a forest-based traditional medicine industry

As discussed above, one of the primary dilemmas in the development of a programme for extraction of non-timber forest products has been the long history of over-collecting of the resources, with a resultant decline in these resources, as well as the export of raw materials for processing to centres and countries far from their origin. Rattan is a classic example of this overexploitation, with people in producing countries who are closest to the resource receiving the smallest percentage of the profits involved in its manufacture into high quality furniture. In Central America, as elsewhere, locally developed brands of commercialized traditional medicines are now being marketed. I have seen these brands in Costa Rica, Honduras and Belize during my recent travels there. A key difference in these types of endeavours is that the value-added component of the product is added in the country of origin of the raw material. As these product brands develop, and as new brands and products appear owing to the success of the original endeavours, greater demand for ingredients from rain forest species will result. This could contribute to preservation of tropical forest ecosystems, but only *if* people carefully manage the production or extraction of the plant species that are primary ingredients in these unrelated products. In addition, it is expected that small farmers will cultivate some of the native species, for sale to both local herbalists and commercial companies. To address the latter possibility, our work in Belize has included a project with the Belize College of Agriculture, Central Farms, to learn how to propagate and grow over 24 different plants currently utilized in traditional medicine in that country. Hugh O'Brien, Professor of Horticulture at the College has coordinated this effort, which has included the genera *Achras, Aristolochia, Brosimum, Bursera, Cedrela, Croton, Jatropha, Myroxylon, Neurolaena, Piscidia, Psidium, Senna, Simarouba, Smilax, Stachytarpheta* and *Swietenia.*

The Belize ethnobotany project

This project was initiated in 1988, as a collaborative endeavour between the Ix Chel Tropical Research Foundation, a Belizean non-governmental organization, and the Institute of Economic Botany of The New York Botanical Garden. The principal purpose of the project has been to conduct an inventory of the ethnobotanical diversity of Belize, a country with significant tracts of intact forest as well as nine different cultural and ethnic groups present within its borders. The project has carried out dozens of expeditions to various locales and has collected some 2700 plant specimens, over 50% of which have ethnobotanical descriptions. The specimens have been deposited at the Belize College of Agriculture and Forestry Department Herbaria, as well as in The New York Botanical Garden and US National Herbarium. A database has been set up at The New York Botanical Garden with planned distribution to several computer facilities within Belize. The project has benefited from traditional knowledge graciously provided by over two dozen traditional healers, of Mopan, Yucatec and Kekchi Maya, Ladino, Garifuna, Creole, East Indian and Mennonite descent.

The ethnobotanical inventory has been combined with a herbarium and literature-based production of a list of the flowering plants and ferns of Belize, with annotations on their common names and uses. This list, which has been produced using the input of over 100 taxonomic specialists, will help provide an idea of the comprehensiveness of the ethnobotanical survey.

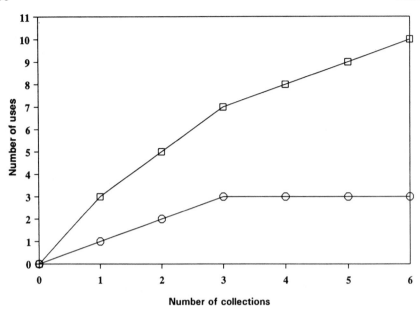

FIG. 1.   Multiple use curves for *Vitex gaumeri* (○) and *Neurolaena lobata* (□).

The concept of the multiple use curve

In ethnobotany, a key issue is sample size. When carrying out an ethnobotanical study of a particular group, it is important to determine the number of collections that must be made and the number of people to be contacted before one has a reasonable certainty that the information on a specific plant is relatively complete. Many ethnobotanical studies depend on one or several collections as the basis for their information and conclusions. The adequacy of such numbers of collections or interviews can be assessed using a mathematical relationship based on the concept of the species–area curve (Campbell et al 1986). The resulting 'multiple use curve' shows the relationship between the number of different uses of a particular species and the number of healers interviewed and voucher collections obtained.

Figure 1 shows that for *Vitex gaumeri* the curve reaches asymptote at three different uses. *Yax nik*, as this species is commonly called, is an important medicinal plant utilized for the treatment of leishmaniasis, as well as other skin sores. *Neurolaena lobata* is used as an analgesic, an antimalarial agent and a parasiticide, also to treat nausea, swelling, fever and other conditions. This plant, with its multiple uses, conforms to the concept of a 'powerful plant' discussed elsewhere (Balick 1990). Each individual healer has general uses for the plant, as well as employing it for a specialized purpose. A comparison of the two curves illustrates the difficulty of drawing a general conclusion about the number of specimens that must be collected before a 'complete' idea of a plant's value can be obtained. In the first, the available information is probably attained after three healers have been interviewed, while in the second, asymptote is not reached after interviews with six healers (Fig. 1). Obviously, the conclusions to be obtained from these data are limited by the small sample size. The six individuals we worked with during the specific collections documented in Fig. 1 are all well-known traditional healers or bush-masters in Belize. The curve would look somewhat different if a country-wide survey of traditional healers were made or an entire community was interviewed for its 'generalist' knowledge. But, it seems fair to conclude that for certain classes of plants, such as medicinals, multiple collections and/or interviews must be made to obtain a complete picture of the knowledge pertaining to an individual taxon. Perhaps with other classes of plants, such as those used as construction materials, the curves would reach asymptote after one or two collections. Application of such multiple use curves to ethnobotanical data sets could add a new dimension to both the planning and the analysis of a particular study.

Figure 2 illustrates the multiple use curve in two different formats. The solid line sorts the data such that each subsequent interview identifies the maximum number of new uses, which is compatible with the protocol for developing a species–area curve. The dotted line represents a chronological presentation of

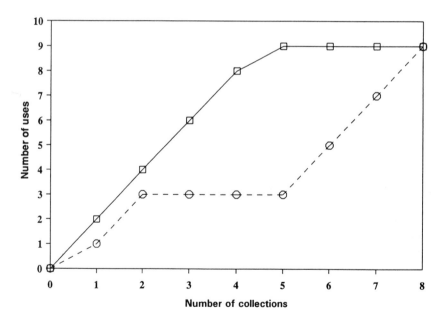

FIG. 2. Multiple use curve for *Piper jaquemontianum* in 'best case' (□) and 'chronological' (○) format.

the data, revealing that if the interview ceased after five collections, it would appear that asymptote was reached at that point. However, healers six, seven and eight all added new information to the curve. Thus, depending on how it is presented, the information in a multiple use curve could be misinterpreted. In this way, the philosophy of the multiple use curve differs somewhat from that of the species–area curve. To avoid confusion, I suggest that the chronological approach be utilized. In addition, the small sample size of ethnobotanical data makes their presentation quite different from the thousands of data points that are common in ecological plots and further supports the use of the chronological approach. The most productive application of the multiple use curve seems to be when relatively large amounts of information (multiple collections) are available for a few interesting taxa.

Valuation studies

One method of ascertaining the value of non-timber forest products in the tropical forest is to make an inventory within a clearly defined area and estimate the economic value of the species found there. Peters et al (1989) were the first to point out the commercial value of non-timber forest products found within a hectare of forest in the Peruvian Amazon. They included medicinal plants in their inventory and, at the suggestion of the authors, this aspect was evaluated in Belize. From two separate plots, of 30- and 50-year-old forest, respectively, total biomasses of 308.6 and 1433.6 kg (dry weight) of medicines whose value could be judged in local markets were collected (Tables 3 and 4). Local herbal pharmacists and healers purchased and processed medicinal plants from herb-gatherers and small farmers for an average price of US$2.80 per kg. Multiplying the quantity of medicine found per hectare by this price suggests that harvesting the medicinal plants from a hectare would yield the collector between $864 and $4014 gross revenue. Subtracting the costs required to harvest, process and ship the plants, the net revenue from clearing a hectare was calculated to be $564 and $3054 for the two plots. Details of the study can be found in Balick & Mendelsohn (1992).

Not enough information is available to understand the life cycles and regeneration time needed for each species, therefore we cannot comment on the frequency and extent of collection involved in sustainable harvesting. However, taking the current age of the forest in each plot as the rotation time, we estimated the present value of harvesting plants sustainably using the standard Faustman formula:

$$V = \frac{R}{(1 - e^{-rt})}$$

where R is the net revenue from a single harvest, r is the real interest rate and t is the length of the rotation in years. For a 30-year rotation in plot 1, this

TABLE 3   Medicinal plants harvested from a 30-year-old valley forest plot in Cayo, Belize

Common name	Scientific name	Use[a]
Bejuco verde	Agonandra racemosa (DC.) Standl.	Sedative, laxative, analgesic; to treat 'gastritis'
Calawalla	Phlebodium decumanum (Willd.) J. Smith	To treat ulcers, pain, 'gastritis', chronic indigestion, high blood pressure, 'cancer'
China root	Smilax lanceolata L.	As a blood tonic; to treat fatigue, 'anaemia', acid stomach, rheumatism and skin conditions
Cocomecca	Dioscorea sp.	To treat urinary tract ailments, bladder infection, stoppage of urine, kidney sluggishness and malfunction; to loosen mucus in coughs and colds; as a febrifuge and blood tonic
Contribo	Aristolochia trilobata L.	To treat influenza, colds, constipation, fevers, stomach ache, indigestion, 'gastritis' and parasites

[a]Uses listed are based on disease concepts recognized in Belize, primarily of Maya origin, that may or may not have equivalent states in Western medicine. For example, kidney sluggishness is not a condition commonly recognized by Western-trained physicians, but is a common complaint among people in this region.

TABLE 4   Medicinal plants harvested from a 50-year-old ridge forest plot in Cayo, Belize

Common name	Scientific name	Use[a]
Negrito	Simaruba glauca DC.	To treat dysentery & diarrhoea, dysmenorrhoea, skin conditions; as a stomach and bowel tonic
Gumbolimbo	Bursera simaruba (L.) Sarg.	As an antipruritic and diuretic; to treat stomach cramps and kidney infections
China root	Smilax lanceolata L.	As a blood tonic; to treat fatigue, 'anaemia', acid stomach, rheumatism and skin conditions
Cocomecca	Dioscorea sp.	To treat urinary tract ailments, bladder infections, stoppage of urine, kidney sluggishness and malfunction; to loosen mucus in coughs and colds; as a febrifuge and blood tonic

[a]See Table 3.

gives the present value of medicine as $726 per hectare. For plot 2, with a 50-year rotation, the estimated value is $3327 per hectare. These calculations assume a 5% interest rate.

These estimates of the value of using tropical forests for the harvest of medicinal plants compared favourably with alternative land uses in the region, such as *milpa* (corn, bean and squash cultivation) in Guatemalan rain forest, which yielded $288 per hectare. We also identified commercial products such as allspice, copal, chicle and construction materials in the plots which could be harvested and added to the total value. Thus, this study suggested that protection of at least some areas of rain forest as extractive reserves for medicinal plants appears to be economically justified. It seems that a periodic harvest strategy is a realistic and sustainable method of utilizing the forest. From our evaluation of forest similar to the second, 50-year-old forest plot analysed, it would appear that one could harvest and clear one hectare per year of a 50 ha plot indefinitely, assuming that all of the species found in each plot would regenerate at similar rates. More than likely, however, some species, such as *Bursera simaruba*, would become more dominant in the ecosystem while others, such as *Dioscorea*, could become rare.

The analysis used in this study is based on current market values. The estimates of the worth of the forest could change according to local market forces. For example, if knowledge about tropical herbal medicines became even more widespread and their collection increased, prices for specific medicines would fall. Similarly, if more consumers became aware of the potential of some of these medicines or if the cost of commercially produced pharmaceuticals became too great, demand for herbal medicines could increase, substantially driving up prices. Finally, destruction of the tropical forest habitats of many of these important plants would increase their scarcity, driving up local prices. This scenario has already been observed in Belize with some species. It seems that the value of tropical forest for the harvest of non-timber forest products will increase relative to other land uses over time, as these forests become more scarce.

Establishment of an ethno-biomedical forest reserve

The concept of the extractive reserve as a tool for conservation has received a great deal of attention over the last few years. Many of these reserves involve tracts of forest where non-timber forest products can be harvested by local individuals or groups who then, it is argued, have a stake in the preservation of the biological integrity of the ecosystem. Products such as rubber, Brazil nuts, copal resin, plant oils, fruits, fibre, construction materials, foliage and house plants for the florist trade, and a wealth of other items have been selected for harvest and marketing from extractive reserves in the Amazon, Central America, Asia and Africa. Numerous perspectives on these resources, both positive and negative, have been highlighted recently (Browder 1992, Ryan 1991).

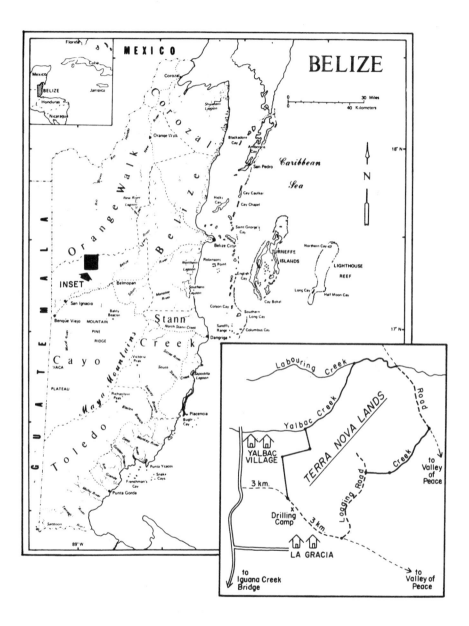

FIG. 3.   Map of Belize showing location of Terra Nova Rain Forest Reserve, with inset providing details of the 2429 ha property.

In June 1993, a 2429 ha parcel of lowland tropical forest was given 'forest reserve' status by the government of Belize, to serve as a location for the extraction of medicinal plants, as well as for teaching and apprenticeship. This particular forest, in the Yalbak region of Belize (Fig. 3), contains diverse medicinal plant species. Also within its borders are many different animals, including jaguar, tapir, peccary, howler monkeys, other mammals, birds and reptiles.

The area has been named 'Terra Nova Rain Forest Reserve'. The Belize Association of Traditional Healers has submitted a management plan to operate the reserve. Terra Nova is proposed to have three primary components contributing to its programme and upkeep. Foremost would be the activities of the traditional healers, their apprentices and local students. Trails would be cut, useful plants identified and labelled, and the regulated harvest of materials undertaken. Seedlings of medicinal plants 'rescued' from nearby areas that have been deforested are being transplanted to Terra Nova, primarily by local students, to enrich certain parts of the reserve where the native flora has been degraded. A major concern will be to prevent the overharvest of its economically important components during the initial burst of enthusiasm for the raw materials.

The second component of the proposed programme for the reserve will be ethnobotanical and ecological research, designed to identify the plant resources it contains and develop appropriate technologies for their sustainable extraction. Dr David Campbell from Grinnell College (Iowa, USA) and his students have constructed ecological transects in selected parts of the reserve to serve as long-term study sites. Some of these plots are areas where extraction would take place; in others, changes in the native vegetation would be monitored. Ethnobotanical inventories have been started to catalogue economically important plants in the reserve, as well as in the surrounding Cayo District. Other scientists would be invited to participate in these studies and in archaeological work on some of the sites within Terra Nova.

A third component of Terra Nova's proposed programme is ecological tourism. Belize is host to hundreds of thousands of such tourists annually, who visit its forests, archaeological sites and coastal areas. When facilities and infrastructure are developed, interested visitors would be invited to Terra Nova to enjoy nature walks and to participate in seminars and classes with traditional healers in a forest setting.

A unique feature of this reserve is its focus on the extraction of medicinal plants used locally as part of the primary health-care network. We propose to call this type of extractive reserve an 'ethno-biomedical forest reserve', a term intended to convey a sense of the interaction among people, plants and animals and the health-care system in the region.

It will be many years before this first ethno-biomedical forest reserve can be judged as a success or failure. A great deal of work must go into developing the management plan and finding the financial and human resources to implement it. Land-use pressures surrounding the reserve, specifically logging and agriculture, as well as sociological and political factors could endanger the long-term existence of the reserve. However, in Belize there is a great deal of optimism about this reserve, in view of its innovative nature, and much support for it amongst the people.

Conclusion

What is the potential contribution of ethnobotany to pharmaceutical development and medicine? As a labour-intensive, long-term endeavour involving highly trained botanists, the most valuable contribution is probably in programmes that are small and tightly focused. High-throughput screening systems in large companies and government laboratories require vast numbers of extracts each month. Given the paucity of knowledge about the chemical composition and full medicinal potential of 99.5% of the plant kingdom, the chance of success of these operations makes it worth using the random approach. The ethno-directed approach is most likely to succeed when it is focused on disease systems actually treated by healers with specific plants. Examples of this might include plants used to treat hepatitis, diabetes, diarrhoea, gastrointestinal problems, skin infections, fungi, wounds, etc. This concept is being tested by the private sector and the results over the next decade will enable a proper evaluation of the ethno-directed approach.

The contributions of medicine to ethnobotany, for the most part, have yet to be realized. In the same way that contemporary ethnobotanists recognize that knowledge of the language and culture of the people one is working with are prerequisites for a full understanding of the uses of plants, we also must combine medical and ethnobotanical skills to obtain a proper understanding of plant use in terms of both Western science and traditional beliefs. This will involve either an interdisciplinary approach with larger teams that include botanists and medical personnel, or an expanded academic curriculum to train those interested in specializing in the medical aspects of ethnobotany.

Finally, I would like to conclude by noting that biodiversity and cultural conservation should be two of the most important objectives of the contemporary ethnobotanist; establishing alliances with those involved in the search for new drugs seems to be an important vehicle for helping to achieve these goals.

Acknowledgements

I am grateful for the collaboration and support of many agencies in Belize, including the Belize Center for Environmental Studies, Belize Zoo and Conservation Center, Belize Association of Traditional Healers, Belize Forestry Department, Belize College of

Agriculture, Ix Chel Tropical Research Foundation and the US Agency for International Development. I should like to thank my colleagues Rosita Arvigo, David Campbell, Sarah Laird, Rob Mendelsohn, Scott Mori, Lon Nicolait, Hugh O'Brien, Polo Romero, Jennie W. Sheldon, Gregory Shropshire and Jay Walker. This paper reports, in part, on research supported by the following sources: US Agency for International Development, US National Cancer Institute, Metropolitan Life Insurance Foundation, The Healing Forest Conservancy, The Overbrook Foundation, The Nathan Cummings Foundation, The Rex Foundation, The Noble Foundation and The Philecology Trust, as well as private gifts.

References

Arvigo R, Balick M J 1993 Rainforest remedies: 100 healing herbs of Belize. Lotus Press, Wilmot, WI

Balick MJ 1990 Ethnobotany and the identification of therapeutic agents from the rainforest. In: Bioactive compounds from plants. Wiley, Chichester (Ciba Found Symp 154) p 22–39

Balick M, Mendelsohn R 1992 Assessing the economic value of traditional medicines from tropical rain forests. Conserv Biol 6:128–130

Browder JO 1992 Social and economic constraints on the development of market-oriented extractive reserves in Amazon rain forests. Adv Econ Bot 9:33–41

Campbell DG, Daly DC, Prance GT, Maciel UN 1986 Quantitative ecological inventory on terra firme and várzea tropical forest on the Rio Xingu, Brazilian Amazon. Brittonia 38:369–393

Cragg GM, Boyd MR, Cardellina JH II, Newman DJ, Snader KM, McCloud TG 1994 Ethnobotany and drug discovery: the experience of the US National Cancer Institute. In: Ethnobotany and the search for new drugs. Wiley, Chichester (Ciba Found Symp 185) p 178–196

King SR, Tempesta MS 1994 From shaman to human clinical trials: the role of industry in ethnobotany, conservation and community reciprocity. In: Ethnobotany and the search for new drugs. Wiley, Chichester (Ciba Found Symp 185) p 197–213

Peters CP, Gentry AH, Mendelsohn RO 1989 Valuation of an Amazonian rainforest. Nature 339:655–656

Ryan JC 1991 Goods from the woods. Forest Trees People 14:23–40

DISCUSSION

Dagne: Could you make clear to us the difference between a botanist and an ethnobotanist? Is it possible to consider some of the traditional healers as ethnobotanists?

Balick: Many of us are trained in systematic botany in a paradigm that was developed over the last 30–40 years. The paradigm has changed in the last 10–15 years. Contemporary ethnobotany is much more interdisciplinary, integrating ideas from many disciplines with different perspectives. There still needs to be a person who identifies the plants, so that one knows what one is working with.

With regard to healers as ethnobotanists, Paul Cox views all the people he works with as his teachers and in this sense they are the ultimate ethnobotanists; in a large measure, I agree with that concept.

Prance: I was rather distressed when I went back to one of the sites in the forest where we had done some quantitative analysis. We had slashed trees to identify them, as you always do in field work because field people depend very much on bark characters. I was distressed to see the enormous effect that those slashes had had on introducing disease into the trees. If a simple slash to identify a tree does that, what is the effect of taking off a large portion of the bark, as is done in the harvesting of *Bursera*.

Balick : That's a very interesting question. Many people talk about sustainable harvest, but the two or three year investment required to wound the trees and observe the effects, in most cases, has never been made. *Bursera* grows its bark back quite quickly. In the case of the other 12 or 15 species that we are looking at, no one knows. No one has ever taken the time to make and observe those slashes; usually the trees are just cut down or girdled completely. This is a key area of interest for the healers who want to preserve as much of this reserve as intact as possible. At their suggestion we have started this particular area of research.

Farnsworth: You showed that a species of *Aristolochia* was widely used (Table 3). Several years ago a pharmaceutical company in Germany determined that *Aristolochia indica* was mutagenic and carcinogenic. This has been repeated several times. A bulletin from the World Health Organization was sent to member states advising them that *Aristolochia*-containing products should be withdrawn from the market. As far as I know, every *Aristolochia* species has the carcinogen (aristolochic acid) in it. It would be of value for ethnobotanists to note this in order to help people avoid exposure to *Aristolochia* species.

Balick: Part of our project is the publication of the healer's manual, 'Rainforest remedies—100 healing herbs of Belize', which is written by Dr Rosita Arvigo and myself (1993). This describes the plants most utilized by traditional healers in Belize. Largely from NAPRALERT, we have been able to put in some pharmacological data. For each of the 5–6 species in widespread use for which toxicity has been demonstrated, this is spelled out on the page, for example that pregnant women should not take this or that people should not use it internally, it's just for external use. *Aristolochia* is a very commonly used species in Belize, and in Mexico and Central America where it is known as *contribo*. Our manual advises people of the toxicity of aristolochic acid.

Farnsworth: With regard to the lowering of the hit rate after dereplication, there is a long road in drug discovery from the ethnobotanical/ethnomedical aspect to the market place. I don't know of any major drug company that is interested in pursuing polysaccharides or polyphenolic compounds or tannins as drugs, for the simple reason that most polysaccharides break down to glucose on oral administration, and in ethnomedicine that's the usual route of administration. Polyphenolic (tannin) compounds bind to protein, so if they are absorbed, a large proportion becomes bound to serum and is not bioavailable. In addition, the molecules are very difficult to work with. So I'm reasonably sure that some plant extracts work in traditional medicine for diarrhoeal diseases because of their polyphenolic (tannin) content, since these are not well absorbed and are very astringent.

Iwu: This is the core of the ethnobotanical approach: the fact that poly-saccharides and polyphenols are not used. Such compounds leach over time into the circulatory system and are gradually released from the proteins or macromolecules. This is a completely different pharmaceutical approach which we will have to look into. Dereplication might be a suitable method for our current state of knowledge, but once we understand how these compounds work, these assays may no longer be appropriate. It is important that we test the whole extract and record the results of such tests. NCI researchers are finding that the early screening they did for cancer using non-human solid tumours was not really a very good indicator of activity. We should not have a strict *a priori* rule about which types of compounds are responsible for bioactivity.

King: It remains to be seen whether tannins can become an appropriate product for pharmaceutical development. The question of bioavailability of tannins applies to oral consumption, not topical application. Many tannin-containing plants are used for a variety of purposes that may well have an application in the pharmaceutical industry and clearly have a wide application in traditional medicine.

Elisabetsky: If I wanted to study the chemistry of very polar compounds, in Brazil there's no place to go: we lack a tradition of working with such compounds. The majority of traditional medicines are prepared in water. Couldn't tannins and sugars carry the active principle through the body to a site where it is liberated? We don't look at those things, so we have yet another weak link in the process of transforming traditional medicine into a cross-culturally effective one. As far as I know, the Japanese are the ones who are really doing good polar chemistry and they're coming out with all sorts of new things. Is this being developed in the chemical environment in the USA?

Farnsworth: I personally believe that a traditional healer, 90% of the time, uses a boiling water extract or a hot water extract. Yet in many cases the active compounds turn out to be completely insoluble in water. The reason most likely is that there are saponins in the extract or other materials that affect surface tension and solubilize or bring into the aqueous phase water-insoluble compounds. Once you take them away from that solubilizing agent you have a different situation. The biggest problem in drug development with plants is answering a very simple question: what kind of extract should we test? Every time you make two extracts, you double the cost. If you partition the extract, you make at least two fractions that must be tested. In a drug company that has a capacity to screen 1000 or more extracts a week, they usually go with one extract; they figure, 'win some, lose some'.

Lewis: Mike, you mentioned the scarcity or lack of apprentices to the elderly healers. Why is this the case in Belize?

Balick: Quite simply, young people are more interested in modern culture—television, carbonated drinks, vehicles, etc—than in learning about life in the forest. It is the march of civilization that accords our particular society a higher value than a traditional society—the imposition of our values on those people. Virtually everyone we work with has no apprentice. There is a change—two years ago the Belize Association of Traditional Healers was formed, this is a professional organization. In the past, left over from the initial contact of the native people with Europeans 500 years ago, there was a great deal of persecution and paranoia about letting anyone know that you knew about medicinal plants and that you practised with them. The authorities that took over the New World persecuted traditional healers. The Spanish conquistadores burned many of the books of the Maya, for example, the great codices, of which only two remain today, so teaching shifted from a written to an oral tradition. But with the formation of associations of traditional healers, and I know there are many in Africa, these people can develop a greater sense of worth in their skills and a sense of social and even economic empowerment.

The whole idea of the medicinal plant reserve is that the Association will be teaching young people in the reserve. Our book is being used in the college curricula and at the village level. These are small but necessary steps—in many places, it seems to be the same story.

Martin: I am interested in the historical process which has created a cultural diversity in the use of plants. You mentioned that first there was a single culture, the Maya, with a written knowledge of plants. Then the Spanish suppressed this and it turned into oral knowledge. Then people from many different cultures arrived, bringing their own knowledge; this has produced a diversity of uses. Now, you are recording the current knowledge in written form. Do you think this will lead to homogenization of what people know about plants? Will what you have found in the Maya community now be transferred to other ethnic communities and, if so, what effect will this have on the diversity of knowledge?

Balick: There are very few cultural boundaries because Belize is such a small and diverse country. Knowledge has been flowing back and forth for 500 years, since even before the initial contact or conquest. I think our data will show that there's probably a 70–80% overlap in the knowledge or concepts among these various groups.

Posey: I question the validity of using schools and written materials to transmit the richness of this information. If you look at traditional societies, it is schools that have separated the apprentice from the master. If the process of learning is something that begins very early and continues for a very long time, we may be transmitting some kind of information through schools and books, but we are never going to translate all that knowledge. We have, somehow, to deal with this conflict.

Balick: I agree fully with that. At the most, we've gathered 5–10% of what exists in seven years. In view of the magnitude of the task, I don't think there's hope for more than 10–15% to be put on paper. An apprenticeship is a long-term commitment to a person over 20–40 years; we are not seeing very many of those happening right now. I rely on people like you to help guide such efforts to stimulate interest in apprenticeships. But as far as providing primary health care is concerned, I think the books, the schools, and the nurseries where some of these rare plants can be grown, will all help.

The students are now rescuing seedlings of trees from land being used for hydroelectric projects and housing projects, and putting them in nurseries. These seedlings will not bear bark for 30–40 years; the students are planting these trees in areas that they know will be protected for their grandchildren and their children. To me, this is an important change of attitude that's only happened in the last 2–3 years. Professional associations help foster this sort of thing. I think college curricula can also play a role in this sort of re-shaping of attitude.

Elvin-Lewis: The unique types of ethnomedicine that you have described may not have empirical verification except within the practising lifetime of the healer, because they are unique to the particular types of treatments that you have studied. Traditional medical practices are always in various phases of dynamic evolution. As such, there are inherent limitations in assigning a value to them, except in the context of time or of their origin, such as West Asian (Indian) or African. Healers who have moved to a new region may not necessarily have access to identical plants to those they have used previously, so it is necessary for them to select plants with similar activities. It is this value judgement that makes them successful in their profession or not.

Balick: The Maya crossed the Bering Straits between 12 000 and 25 000 years ago and therefore had 200–300 generations in the New World. In contrast, we have also worked with a healer called Mr Ramcharan, who came from Calcutta 60–70 years ago. People like Mr Ramcharan are looking for Ayurvedic plants or substitutes. What we see amongst these people is a continual experimentation, especially when they don't have the plant that their grandfather or his grandfather showed them, they only remember the plant from another country or continent. For example, some depend heavily on the doctrine of signatures to guide their experimentation.

Elvin-Lewis: Another suggestion is that when you are dealing with data regarding the value of certain plants, you should order them. We have, for example, ordered the number of tooth-blackening plants among the Jívaro. We find that we started uncovering some logic behind why a plant is used, how often it's used, why it's used more than another plant. Data acquired without these types of qualifications may not be absolute, because although a plant is described as being used, it may be a second or third choice of relatively little value.

Balick: We have collected over 3000 herbarium sheets in this area from amongst these nine different cultures, most strongly in 4–5 of the cultures. The ordering will be possible when we take a broad view and look for patterns, hopefully using some anthropological programmes and even anthropologists to help us.

Jain: One criticism of reports of ethnobotanical uses for plants is always credibility and the time frame of the studies. When a person starts working on the ethnobotany of a particular region or of a particular ethnic group, the question arises, how long should the field work go on? How many times should you go to the area? How many people should you interview? Have you tried to determine the optimum period of work in a particular area?

Balick: Our situation is a little different. I work with Dr Rosita Arvigo, who is herself a traditional healer and a naprapathic physician. She works in partnership with the healers as part of their practice. So we don't spend much time taking surveys or asking what people use; we spend a lot of time observing people actually using the plants or going with the people as they make their morning rounds to collect the materials they are going to use that day. In a way, this is a richer approach than making a simple ethnobotanical inventory or general collecting. My conclusions changed dramatically in response to what I learned: what I thought I knew in Year 1 of the study is very different from my conclusions in Year 7, after our team has worked so closely with about two dozen healers.

Jain: Does this mean that the information that has been collected comes from one person?

Balick: No. In each village there is a healer. Instead of working with the villagers, in general we work with the healer, although we do involve the villagers and record general knowledge about a plant. Having worked in about 24 villages now with their healers, I would say that limited numbers of people are the source of this information. But from the perspective of utility of the traditional medicine, it's probably a deeper set of knowledge.

McChesney: Most economic analyses are done in order either to justify conservation or to promote conservation, etc. In the particular case you have cited, the value derived, particularly for the older forest, was rather astounding. However, supply and demand can create a very great change in such a situation. Have you tried to factor in such considerations? I suspect there's a reasonable economic balance in supply and demand at this point in time, which sets the baseline value of the collected material. Increased labour involvement and increased supply may drive down the market value substantially. Has anyone looked at those kinds of projections as a way of anticipating the issues that you mentioned with regard to plantation cultivation of important commercial plants?

Balick: These are secondary forests; you actually find less medicine in the primary forests, which in Belize are virtually non-existent because of the long period of habitation there. As these forests are destroyed (and these Central American forests are being destroyed probably more quickly than those of some of the Amazonian regions), the supplies of medicinal plants are drying up. The demand continues, supply is less and the price increases. At the same time, cultivation, management and re-planting some of these plants may counterbalance that. Supply and demand are key issues. You mentioned the astounding magnitude of the dollar values. When we first did the analysis, we put the value-added component in, the value of actually producing a tincture in that area. The result was that the $3300 became up to 10 times that much, which was even more unbelievable. But to me that would be the equivalent of valuing the crude rubber used in a jaguar car tyre on the basis of its manufacture and retail sale. This is a very important point: if value-added activities are performed in these reserves or in the local areas, this will help overcome some of the problems of supply and demand and economic issues that are affecting extractive reserves today.

Iwu: From a Third World perspective, there is a difficulty in differentiating between the value in terms of what people are willing to pay and the actual worth of the forest. The forest has no commercial value until the ethnobotanists, the pharmaceutical companies, the individual researchers or the traditional medical people give 'market' value to it. In Nigeria, if you take medicine from the forest for preparation of remedies for self medication, you may not pay the owner of the forest for it, so it has no 'economic' value. The cultural backgrounds have to be taken into account when evaluating the worth of a forest. In economic terms, the value of products used in traditional medicine in a cash-flow analysis is zero. But when you consider the total economy, which includes the well-being of the people, you are presented with a completely different picture. How do we put market value on the forest so that local communities will benefit from ethnobotanical studies? Nobody has yet addressed this issue.

Balick: The situation is apparently quite different in the neotropics. In Honduras, Belize, Costa Rica and Nicaragua, there are now industries that are including the value-added component in these forest areas. Tinctures and herbal industries take a few cents worth of herb and turn it into something much more valuable. The analysis I described was predicated on the fact that the herbalists charge 20 Belize dollars or US$10 per sack of herbs that they sell to the healers. This was a very conservative analysis and I was surprised that the numbers were so large. But I agree the African situation is quite different.

Reference

Arvigo R, Balick M 1993 Rainforest remedies: 100 healing herbs of Belize. Lotus Press, Twin Lakes, WI

The ethnobotanical approach to drug discovery: strengths and limitations

Paul Alan Cox

Department of Botany & Range Science, Brigham Young University, Provo, Utah 84602, USA

Abstract. For pharmaceuticals ranging from digitalis to vincristine the ethnobotanical approach to drug discovery has proven successful. The advent of high-throughput, mechanism-based *in vitro* bioassays coupled with candidate plants derived from pain-staking ethnopharmacological research has resulted in the discovery of new pharmaceuticals such as prostratin, a drug candidate for treatment of human immunodeficiency virus, as well as a variety of novel antiinflammatory compounds. Not all Western diseases are equally likely to be recognized by indigenous peoples. Gastrointestinal maladies, inflammation, skin infections and certain viral diseases are likely to be of high saliency to indigenous healers, whereas diseases such as cancer and cardiovascular illness are unlikely to be easily diagnosed by indigenous peoples. Yet indigenous remedies may indicate pharmacological activity for maladies such as schizophrenia, for which the biochemical mechanisms have yet to be discovered. Ethnopharmacological information can be used to provide three levels of resolution in the search for new drugs: (1) as a general indicator of non-specific bioactivity suitable for a panel of broad screens; (2) as an indicator of specific bioactivity suitable for particular high-resolution bioassays; (3) as an indicator of pharmacological activity for which mechanism-based bioassays have yet to be developed.

1994 Ethnobotany and the search for new drugs. Wiley, Chichester (Ciba Foundation Symposium 185) p 25–41

Ethnobotany and drug discovery

At least 25% of the prescription drugs issued in the USA and Canada contain bioactive compounds that are derived from or modelled after plant natural products (Farnsworth 1984). Many of these drugs were discovered by following leads provided from indigenous knowledge systems. Historically, ethnobotanical leads have resulted in three different types of drug discovery: (1) unmodified natural plant products where ethnomedical use suggested clinical efficacy (e.g. digitalis); (2) unmodified natural products of which the therapeutic efficacy was only remotely suggested by indigenous plant use (e.g. vincristine); and (3) modified natural or synthetic substances based on a natural product used in folk medicine (e.g. aspirin).

Historical success of the ethnobotanical approach: Withering and digitalis

In 1775 the English physician William Withering was told by a folk healer that the leaves of *Digitalis purpurea* were useful for treating dropsy, a swelling of the body caused by inadequate pumping action of the heart. Treating his dropsy patients with the leaves, Withering discovered a powerful cardiotonic effect:

'It has a power over the motion of the heart, to a degree yet unobserved in any other medicine' (Withering 1785, p 2).

Yet some of Withering's patients suffered toxic reactions. In an effort to standardize dosage, he began to administer water infusions of the leaves and later ground leaf powder. *Digitalis* as administered by Withering was very efficacious: his data indicate a success rate of 65–80 % (Aronson 1985). Since Withering's day, more than 30 cardiac glycosides, including digitoxin, digoxin and digitoxigenin, have been isolated from *D. purpurea* (Fig. 1). Each year over 1500 kg of pure digoxin and 200 kg of digitoxin are prescribed to thousands of patients with cardiac problems (Lewington 1990). It is now known that *Digitalis* affects the heart by inhibiting an enzyme that powers the transport of potassium and sodium ions across cell membranes, Na^+/K^+ ATPase (Aronson 1985, Post 1974).

Withering's success in translating folk knowledge into new pharmaceuticals is not unique. Of all the plant-derived drugs listed in Samuelsson (1992), at least 50 were derived from ethnobotanical leads (Table 1). This list is far from complete: Farnsworth (1990) estimated that there are 88 ethnobotanically derived drugs. Many of the drugs in Table 1 are from leads that have been known to Western science for many years. For example, Gerard in 1597 indicated pharmacological efficacy of the sea squill, *Drimia maritima*: 'It is given to those that have the dropsie' (Gerard 1597 p 137). Since then, the cardiotonic drug proscillaridin has been derived from *D. maritima*. Many other drugs have similar histories of discovery rooted in folk knowledge.

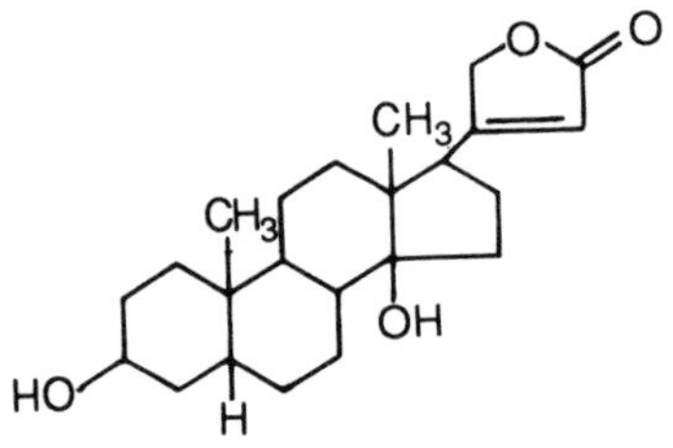

FIG. 1.   Digitoxigin, a cardiotonic drug discovered from *Digitalis purpurea* using the ethnobotanical approach.

TABLE 1 Fifty drugs discovered from ethnobotanical leads

Drug	Medical use	Plant source
Ajmaline	For heart arrhythmia	*Rauvolfia* spp.
Aspirin	Analgesic, antiinflammatory	*Filipendula ulmaria*
Atropine	Pupil dilator	*Atropa belladonna*
Benzoin	Oral disinfectant	*Styrax tonkinensis*
Caffeine	Stimulant	*Camellia sinensis*
Camphor	For rheumatic pain	*Cinnamomum camphora*
Cascara	Purgative	*Rhamnus purshiana*
Cocaine	Ophthalmic anaesthetic	*Erythoxylum coca*
Codeine	Analgesic, antitussive	*Papaver somniferum*
Colchicine	For gout	*Colchicium autumnale*
Demecolcine	For leukaemia, lymphomata	*C. autumnale*
Deserpidine	Antihypertensive	*Rauvolfia canescens*
Dicoumarol	Antithrombotic	*Melilotus officinalis*
Digoxin	For atrial fibrillation	*Digitalis purpurea*
Digitoxin	For atrial fibrillation	*D. purpurea*
Emetine	For amoebic dysentery	*Psychotria ipecacuanha*
Ephedrine	Bronchodilator	*Ephedra sinica*
Eugenol	For toothache	*Syzygium aromaticum*
Gallotanins	Haemorrhoid suppository	*Hamamelis virginia*
Hyoscyamine	Anticholinergic	*Hyoscyamus niger*
Ipecac	Emetic	*Psychotria ipecacuanha*
Ipratropium	Bronchodilator	*H. niger*
Morphine	Analgesic	*Papaver somniferum*
Noscapine	Antitussive	*Papaver somniferum*
Papain	Attenuator of mucus	*Carica papaya*
Papaverine	Antispasmodic	*Papaver somniferum*
Physostigmine	For glaucoma	*Physostigma venenosum*
Picrotoxin	Barbiturate antidote	*Anamirta cocculus*
Pilocarpine	For glaucoma	*Pilocarpus jaborandi*
Podophyllotoxin	For condyloma acuminatum	*Podophyllum peltatum*
Proscillaridin	For cardiac malfunction	*Drimia maritima*
Protoveratrine	Antihypertensive	*Veratrum album*
Pseudoephedrine	For rhinitis	*E. sinica*
Psoralen	For vitiligo	*Psoralea corylifolia*
Quinine	For malaria prophylaxis	*Cinchona pubescens*
Quinidine	For cardiac arrhythmia	*C. pubescens*
Rescinnamine	Antihypertensive	*R. serpentia*
Reserpine	Antihypertensive	*R. serpentia*
Sennoside A,B	Laxative	*Cassia angustifolia*
Scopalamine	For motion sickness	*Datura stramonium*
Sigmasterol	Steroidal precursor	*Physostigma venenosum*
Strophanthin	For congestive heart failure	*Strophanthus gratus*
Tubocurarine	Muscle relaxant	*Chondrodendron tomentosum*
Teniposide	For bladder neoplasms	*Podophyllum peltatum*
Tetrahydro-cannabinol	Antiemetic	*Cannabis sativa*
Theophylline	Diuretic, antiasthmatic	*Camellia sinensis*
Toxiferine	Relaxant in surgery	*Strychnos guianensis*
Vinblastine	For Hodgkin's disease	*Catharanthus roseus*
Vincristine	For paediatric leukaemia	*C. roseus*
Xanthotoxin	For vitiligo	*Ammi majus*

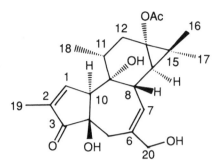

FIG. 2.   Prostratin, a drug candidate for the treatment of human immunodeficiency virus discovered from *Homalanthus nutans* using the ethnobotanical approach.

Ethnobotanical approaches to drug discovery are not merely reversions to some quaint previous time in history as some suggest (see Cox 1990a p 50). The recent advent of new mechanism-specific, high-throughput bioassays has ushered in a new era of targeted ethnobotanical drug discovery programmes.

Recent success of the ethnobotanical approach: NCI and prostratin

In 1984 I entered into collaborative research with the Natural Products Branch of the US National Cancer Institute (NCI). I had began an intensive investigation of Samoan ethnopharmacology and the NCI agreed to assist me in evaluating the pharmacological efficacy of Samoan medicinal plants (Cox et al 1989, Cox 1990b,c). I collected in aqueous ethanol stem wood samples of *Homalanthus nutans* (Euphorbiaceae), which several healers told me was efficacious against yellow fever, a viral disease. The healers carefully prepare hot water infusions that are then drunk by the patients. I supplied lyophilized extracts of my samples to an NCI team led by Michael Boyd. NCI found the extracts to exhibit potent activity in an *in vitro* tetrazolium-based assay designed to detect cytopathic effects of the human immunodeficiency virus, HIV-1, associated with acquired immune deficiency syndrome (AIDS) (Gustafson et al 1992). Bioassay-guided fractionation resulted in the isolation of prostratin, 12-deoxyphorbol 13-acetate (Fig. 2). At non-cytotoxic concentrations prostratin was found to prevent HIV-1 reproduction in lymphocytic and monocytoid target cells. Prostratin also fully protected human cells from lytic effects of HIV-1.

The identification of a phorbol as the active component was of concern because phorbols are known tumour promoters (Evans 1986). Fortunately, prostratin is reportedly not a tumour promoter (Zayed et al 1984) and the NCI team found that it does not induce hyperplasia in mice, even though it stimulates protein kinase C activity. The NCI team concluded that prostratin represents

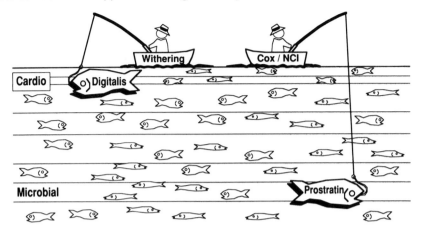

FIG. 3.   The discoveries of digitalis and prostratin portrayed as the results of trolling in the waters of indigenous knowledge.

a non-promoting activator of protein kinase C which strongly inhibits the killing of human host cells *in vitro* by HIV. By these criteria, prostratin is unique (Gustafson et al 1992, p 1984). Currently, NCI considers prostratin as a candidate for drug development.

Strengths of the ethnobotanical approach

Digoxin and prostratin are very different, not only in structure and action, but also in economic importance: digitalis is a drug of known therapeutic and economic value while prostratin has yet to be proven in clinical trials. But the paths leading to the discovery of their possible therapeutic utility illustrate the general sequence of ethnobotanical drug discovery programmes: (1) folk knowledge accumulates concerning possible pharmacological activity of a plant; (2) the plant is used therapeutically by a healer; (3) the healer communicates this knowledge to a Western researcher; (4) the researcher collects and identifies the plant; (5) plant extracts are tested using a bioassay; (6) bioassay-guided fractionation leads to isolation of an active substance; and (7) the structure of the pure substance is determined.

It is as if Withering and I, nearly two centuries apart, had cast a fishing line into similar waters (Fig. 3). Withering landed a large fish of tremendous importance, digitalis, while it remains to be seen if my fish, prostratin, will be useful at all. Yet both of us fished the same waters—indigenous knowledge systems—after having first received a fishing permit—the cooperation of indigenous healers. (It is not known what compensation, if any, Withering offered his healer, but the NCI and Brigham Young University have guaranteed

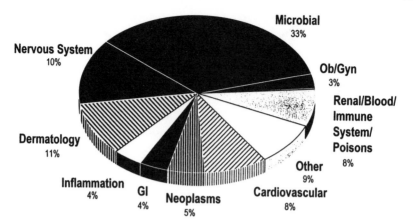

FIG. 4.   Uses of drugs in the US pharmacopoeia, classified according to the treatment categories of Goodman & Gilman (Gilman et al 1991). GI, gastrointestinal; Ob/Gyn, obstetric and gynaecological.

that a significant proportion of any commercial proceeds from prostratin will return to Samoa.) Bioassays might be considered as our hooks and lures, for it would be impossible to catch a fish without them. Finally, both discoveries eventually required teams of chemists and pharmacologists: my fish was landed, identified and evaluated for therapeutic potential by a very competent NCI crew consisting of G. Cragg, M. Suffness, M. Boyd, J. Beutler, J. Cardellina, K. Gustafson, P. Blumberg, J. McMahon and many others.

Analysis of indigenous plant uses

How deep are the waters of indigenous knowledge systems? How many and what type of drugs remain to be discovered through the ethnobotanical approach? Some estimate of the potential catch can be provided by an analysis of maladies for which plants are used by indigenous peoples. My students, A. LeFranca, W. McClatchey, A. Paul, and I reviewed literature accounts of plant uses in 15 widespread geographical areas, including Australia (Cribb & Cribb 1981), Fiji (Spencer 1941), Haiti (Pierre-Noel 1960), India (Rao 1981), Kenya (Morgan & Morgan 1981), Mexico (Alfaro 1980), Nepal (Joshi & Edington 1990), Nicaragua (Dennis 1988), North America (Moerman 1986), Peru (Lewis et al 1987), Rotuma (McClatchey 1993), Saudi Arabia (Abulafatih 1981), Thailand (Anderson 1986), Tonga (George 1989) and West Africa (Oliver-Bever 1986). We classified putative medicinal uses of the plants using Goodman & Gilman's treatment categories (Gilman et al 1991), which include treatment of diseases of the nervous and cardiovascular systems, of neoplasms, of inflammation (including fevers), and a miscellaneous category that included

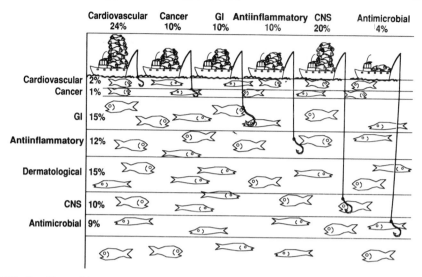

FIG. 5. Uses of indigenous plants by treatment category: the width of each stratum in the water corresponds to the mean percentage plant use by category. Use data are from Australia, Fiji, Haiti, India, Kenya, Mexico, Nepal, Nicaragua, North America, Peru, Rotuma, Saudi Arabia, Thailand, Tonga and West Africa (see text for references). The height of the fish catch on each boat represents the percentage of the corresponding treatment category from the ethnobotanically derived drugs in Table 1.

treatment of renal ailments, dehydration, parasitic diseases, immune disorders and blood diseases, and poisons. We grouped ceremonial and ill-defined uses of indigenous plants as 'other'. This analysis of putative indigenous uses was compared to a baseline of Western drug uses derived from the US pharmacopoeia (US Pharmacopeial Convention 1990) (Fig. 4). We also categorized the 50 ethnobotanically derived drugs (Table 1) according to this scheme. As an independent test of robustness, our data for indigenous uses were compared to Farnsworth's (1990) summary of ethnomedical records in the NAPRALERT database.

There is a striking difference between the mean percentages of disorders treated with indigenous plants and Western drugs (Fig. 5). Indigenous plant remedies are focused more on gastrointestinal (GI) complaints, inflammation, dermatological ailments and obstetric/gynaecological disorders, while the main uses of Western drugs are to treat cardiovascular or nervous system disorders and neoplasms or as antimicrobial remedies. We believe our sample of indigenous uses to be robust; in only three disease categories do the NAPRALERT data (Farnsworth 1990) differ by over three percentage points from our indigenous data, while both data sets agree on the distinctive difference between indigenous and Western remedies. The exception is use of the former in dermatology which is under-represented in NAPRALERT because of the lack

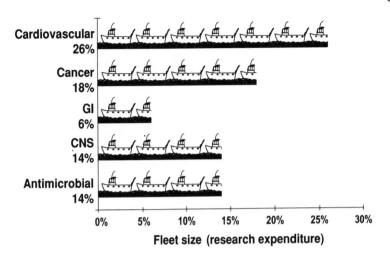

FIG. 6.   Percentage drug research expenditures in the USA portrayed as fleet size. Data from Facts and Comparisons, Inc. (1990).

of a dermatological category in NAPRALERT (N.R. Farnsworth, personal communication).

Why do Western remedies deal with certain types of disease while indigenous remedies focus on others? There are several possible answers.

Perceived peril. Cardiovascular illness, neoplasms, microbial infections and nervous system ailments are the biggest killers in Western cultures. Diarrhoea, complications of maternity and inflammation are regarded as more perilous by indigenous peoples, who do not have the lifestyles or predicted lifespans associated with cardiovascular disease and cancer.

Saliency. Inflammation, dermatological and GI ailments are all rather easily detected by indigenous peoples. Most cancers and cardiovascular disease are difficult to diagnose by indigenous peoples. In fact, few indigenous languages have words for cancer, leukaemia, lymphoma, hypertension, etc.

Toxicity. Indigenous peoples are likely to avoid plant medicines that are highly toxic in low doses. Most cardiovascular and anticancer drugs and those active on the central nervous system have extremely 'narrow' dosage windows and thus would likely be unacceptable to indigenous peoples (Withering's dosage problems with *Digitalis* might be the exception that proves the rule).

Economic incentives. As part of our survey, we analysed the percentage of medical research dollars spent in the USA for each treatment category. This is illustrated in Fig. 6 as different fleet sizes. In the Western world, cardiovascular

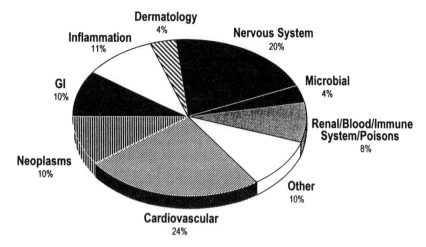

FIG. 7. The relative frequencies of the types of Western diseases treated with drugs derived from ethnobotanical studies.

illness, neoplasms, nervous system disorders and microbial diseases receive 72% of every research dollar (Facts and Comparisons, Inc. 1990); such economic pressures are unique to Western scientists and are not experienced by indigenous healers. Thus, the percentages of drug types previously discovered by the ethnobotanical approach (Table 1, Fig. 7) more closely represent funding categories than indigenous use categories.

On the basis of this analysis, we could predict great success for properly designed ethnobotanical surveys for GI, antiinflammatory, obstetric/gynaecological and dermatological drugs. Using the fishing metaphor, we can say that these are the most common fish in indigenous waters (Fig. 5). Does this mean that no new cardiovascular, anticancer or antimicrobial drugs are likely to be found using the ethnobotanical method? Again, using the fishing metaphor, our analysis indicates the presence of many fish species, including some (such as those for use in dermatology) that we have not really tried to catch (Fig. 5). But are there still new vincristines or new digitoxins out there to be caught? Just how deep are the waters?

From our analysis of previously discovered ethnobotanically derived drugs, it seems the waters are very deep indeed. Of the 50 drugs listed in Table 1, 24% are cardiovascular substances (compared to 2% of indigenous remedies), 20% are used for the nervous system (compared to 10% of indigenous remedies) and 4% are used for neoplasms (compared to only 1% of indigenous remedies) (Fig. 7). 'Seek and ye shall find' seems to be the operative principle.

Ethnobotanical drug discovery: obligations to indigenous peoples

A cynic might ask what returns indigenous peoples have received from the drugs listed in Table 1. Many of these drugs have dramatically improved the health of nearly all peoples around the world; aspirin and quinine, for example, are used throughout the tropics. The truth, however, is that most of these drugs were discovered during the colonial phase of world history, when the rights and aspirations of indigenous peoples were routinely overlooked (Cox 1993, Cox & Elmqvist 1993). Modern ethnobotanists consider indigenous peoples to be colleagues rather than informants and have almost uniformly rejected exploitation as a consequence of joint scientific investigation with indigenous peoples. Yet promising vast financial gain to indigenous peoples, many of whom live in poverty, from such an uncertain enterprise as drug discovery is as cruel as promising cures to those who are desperately ill. Monetization may also conflict with some indigenous views about the sacredness of the creation and the selflessness of healing, views that could inform much current Western debate on both conservation and health care (Cox & Elmqvist 1993).

One solution is to plague indigenous peoples with teams of ethnocentric lawyers who force them to accept Western terms of both equity and equitability, as well as burdensome Western infrastructures to administer such 'equity'. Yet until Western non-governmental organizations, top-heavy with lawyers and anthropologists, are willing to subject themselves and their own budgets to complete control by indigenous peoples (Cox & Elmqvist 1991, 1993), such solutions smack of a new colonialism, merely substituting ethnocentricity and cultural imperialism for the economic and political colonialism of the 19th century. It is, instead, our task to solve issues of indigenous intellectual property rights in terms acceptable to both Western society *and* the particular indigenous culture involved. As a result, each solution, while incorporating some common features, will be unique. *Consent* of the indigenous people, *respect* for their culture and *submission* to indigenous political control are features that should characterize all responsible agreements.

In the case of prostratin in Samoa, research was performed with the explicit oral and written consent of the healers, the villagers and the village chiefs. The aims and intents of all research were carefully explained and consent was obtained before research began. The village authorities negotiated and signed *in advance* of any discovery an agreement concerning the disposition of any commercial proceeds. The NCI, Brigham Young University and several potential pharmaceutical collaborators have agreed to abide by that agreement: the NCI has guaranteed that a major portion of any royalties will return to Samoa; Brigham Young University has guaranteed that 30% of its royalties will return to the village. Similar agreements involving profits related to provision of the plants (in case natural rather than synthetic prostratin is marketed) have also been signed. Yet, as of this writing, it is not known if prostratin will ever make it to market or whether any royalty benefit will accrue to the villagers. To ensure some benefit to the villagers, we have raised private funds to build three schools and two clinics in Falealupo village in exchange for a village promise to protect approximately 12 000 ha of rain forest; all donated funds are controlled and administered by the village chiefs (Cox & Elmqvist 1991, 1993). Conservation of critical resources is perhaps one of the greatest bequests that can be granted to indigenous peoples.

In conclusion, the potential success of the ethnobotanical approach to drug discovery can no longer be questioned: historical and current discoveries attest to its power. While its strength appears to be its great potential for discovering potent new compounds, it may be limited in the type of drugs it is most likely to provide. On the basis of these considerations, hit rates for, for example, antileukaemia drugs may differ little from hit rates from random screening, unless general indications of bioactivity increase the likelihood of finding new anticancer therapies. But the major limitation of the ethnobotanical approach may be that it forces Western scientists to confront the fact that indigenous knowledge systems have effectively discovered bioactivity without the tools of Western science. The peril is that the epistemology of some Western scientists may be threatened by this success. Yet the promise of the ethnobotanical approach is that we may learn not only new pharmacology, but also new ways of caring for the sick and for the planet we share.

References

Abulafatih HA 1981 Medicinal plants in southwestern Saudi Arabia. Econ Bot 41:354–360
Alfaro M 1984 Medicinal plants used in a Totonac community of the Sierra Norte de Puebla: Tuzamapan, Galeana, Puebla, Mexico. J Ethnopharmacol 11:203–221
Anderson EF 1986 Ethnobotany of hill tribes of north Thailand. II. Lahu medicinal plants. Econ Bot 40:442–450
Aronson JK 1985 An account of the foxglove and its medical uses 1785–1985. Oxford University Press, Oxford
Cox PA 1990a Ethnopharmacology and the search for new drugs. In: Bioactive compounds from plants. Wiley, Chichester (Ciba Found Symp 154) p 40–47
Cox PA 1990b Samoan ethnopharmacology. In: Wagner H, Farnsworth NR (eds) Economic and medicinal plant research, vol 4: Plants and traditional medicine. Academic Press, London, p 123–139
Cox PA 1990c Polynesian herbal medicine. In: Cox PA, Banack SA (eds) Islands, plants, and Polynesians. Dioscorides Press, Portland, OR, p 147–169
Cox PA 1993 Saving the ethnopharmacological heritage of Samoa. J Ethnopharmacol 38:181–188
Cox PA, Elmqvist T 1991 Indigenous control: an alternative strategy for the establishment of rainforest preserves. Ambio 20:317–321
Cox PA, Elmqvist T 1993 Ecocolonialism and indigenous knowledge systems: village controlled rain forest preserves in Samoa. Pac Conserv 1:11–25
Cox PA, Sperry LR, Tumonien M, Bohlin L 1989 Pharmacological activity of the Samoan ethnopharmacopoeia. Econ Bot 43:489–497
Cribb AB, Cribb JW 1981 Wild medicine in Australia. Collins, Sydney

Dennis PA 1988 Herbal medicine among the miskito of E. Nicaragua. Econ Bot 42:16–28
Facts and Comparisons, Inc. 1990 Drug facts and comparisons. Lippincott, St Louis, MO
Farnsworth NR 1984 The role of medicinal plants in drug development. In: Krogsgaard-
    Larsen P, Christensen SB, Kofod H (eds) Natural products and drug development.
    Ballière, Tindall, and Cox, London, p 8–98
Farnsworth NR 1990 The role of ethnopharmacology in drug development. In: Bioactive
    compounds from plants. Wiley, Chichester (Ciba Found Symp 154) p 2–21
George LO 1989 An ethnobotanical study of traditional medicine in Tonga. MSc thesis,
    Brigham Young University, Provo, UT, USA
Gerard J 1597 The herball or generall historie of plantes. John Norton, London
Gilman AG, Goodman LS, Rall TW, Murad F 1991 Goodman and Gilman's the
    pharmacological basis of therapeutics, 7th edn. MacMillan, Philadelphia, PA
Gustafson KR, Cardellina JH II, McMahon JB et al 1992 A nonpromoting phorbol from
    the Samoan medicinal plant *Homalanthus nutans* inhibits cell killing by HIV-1. J Med
    Chem 35:1978–1986
Lewington A 1990 Plants for people. The Natural History Museum, London
Lewis WH, Elvin-Lewis M, Gnerre MC 1987 Introduction to the ethnobotanical
    pharmacopoeia of the Amazonian Jívaro in Peru. In: Medicinal and poisonous plants
    of the tropics Symp 5–35. Proc 14th Int Botanical Congress, Berlin. p 96–103
Joshi AR, Edington AM 1990 The use of medicinal plants by two village communities
    in the central development region of Nepal. Econ Bot 44:71–83
McClatchey WC 1993 Studies on the ethnobotany of the island of Rotuma. MSc thesis,
    Brigham Young University, Provo, UT, USA
Moerman DE 1986 Medicinal plants of native America. University of Michigan Museum
    of Anthropology, Ann Arbor, MI, vol 2
Morgan M, Morgan W 1981 Ethnobotany of the Turkana: use of plants by a pastoral
    people and their livestock in Kenya. Econ Bot 35:96–130
Oliver-Bever BEP 1986 Medicinal plants in tropical West Africa. Cambridge University
    Press, Cambridge
Pierre-Noel VV 1960 Les plantes et les legumes d'Haiti qui guerissant. Imprimerie de
    l'Etat, Port-au-Prince
Post RL 1974 A reminiscence about sodium, potassium-ATPase. Ann NY Acad Sci
    242:6–11
Rao RR 1981 Ethnobotany of the Meghalaya: medicinal plants used by the Kahshi and
    Garo Tribes. Econ Bot 35:4–9
Samuelsson G 1992 Drugs of natural origin. Swedish Pharmaceutical Press, Stockholm
Spencer D 1941 Disease, religion, and society in the Fiji Islands. University of Washington
    Press, Seattle, WA
United States Pharmacopeial Convention 1990 USP XXII NF XVII The United States
    pharmacopeia, The National Formulary. The United States Pharmacopeial Convention,
    Rockville, MD
Withering W 1775 An account of the foxglove and some of its medical uses: with practical
    remarks on dropsy and other diseases. M. Swinney, Birmingham

DISCUSSION

Bohlin: The method of administration of drugs is critical, as we all know.
In Samoan traditional medicine, topical application is extremely important. Can
you comment a little more about the way they administer their drugs?

Cox: This is important because if we look only for Western types of administration, we are going to ignore a lot of bioactivity. Most of the plant remedies in Samoa are administered topically; the plant is rubbed on the skin after an initial application of coconut oil. This may indicate something about the chemistry of any bioactive compounds present. For example, we would not expect high molecular weight compounds to be transmitted through the skin. Topical administration also allows a wider dosage window—it's possible to administer a potentially more toxic drug by giving a lower dose over a longer time. In our Western culture, we have topical applications of scopolamine for motion sickness and nicotine patches for people who are trying to give up tobacco. Perhaps ethnobotanical drugs will alter our perceptions of drug administration in the future.

Schwartsmann: You mentioned a plant product, prostratin, that was used by the Samoans for yellow fever, which you brought to the NCI for anticancer screening. Has it been checked in models for yellow fever, whether it is actually effective for the purpose for which it is used?

Cox: It would be great to test this against yellow fever. I said putative, because it's claimed to treat yellow fever but this has not been tested. If anybody has access to a yellow fever model, we would love to test it.

Farnsworth: Yellow fever virus is a retrovirus; that's how it ties in with AIDS.

You set up two medical clinics in Samoa. Are these stocked with herbal drugs or Western drugs?

Cox: Both. The expenditure of funds and the design of the clinics is under the control of the villages and the Government of Western Samoa. They have small clinics in some of the districts, where a nurse practises and there are occasional visits by a doctor, but they also allow indigenous healers to work there.

Posey: What would happen if you found a big 'catch' and it was one of the species prohibited by local communities? It's very nice to talk about sharing of royalties and respecting culture, but in screening for drugs, I've never heard anyone talk about prohibitions the local communities have about the use of some of these plants.

Cox: I deal with that with very explicitly in Ciba Foundation Symposium 154 (Cox 1990). My students and I feel that we have equal obligations to our indigenous colleagues—we call them colleagues and not informants. If they regard certain information as sacred or secret, we don't transmit that information. We only use and screen plants for which they have given explicit informed consent, meaning the indigenous people understand the purpose of our research, our designs and our intentions.

King: Before you hook some of those fish, there has to be a market for them; you didn't address that. Alternatively, for example in dermatology, there is a need but there are products already that meet many of those needs. Recently, there has been commercialized a product derived from *Larrea tridentata* (creosote bush), which is used traditionally by south-west American native people, to treat precancerous skin lesions and dermatological conditions.

Schwartsmann: What do you think about skin tumours? I would anticipate that in indigenous cultures, most of the so-called 'skin tumours' would be in fact chronic infectious or granulomatous diseases rather than malignant tumours.

Cox: Podophyllin comes from the genus *Podophyllum*. It was used by North American Indians to treat skin tumours and is now used to treat testicular neoplasms. But this is unusual; there are very few indigenous cultures that recognize cancer and leukaemia. This is why I believe that indicators of ethnobotanical success for cancer drugs are not indicative of the success of this approach for drugs as a whole.

Elisabetsky: There is a problem concerning the concepts of diseases and how to translate indigenous disease concepts into Western disease concepts and vice versa. Sometimes we publish these numbers and tables; they are useful for us to reason within a specific context, but they might not be generally true. What does neoplasm mean to an indigenous people? Trotter et al (1983) showed that nosological conditions are frequently used by indigenous people as a representative of their epidemiological picture. Gastrointestinal disorders are the most frequent type of disease in Third World countries. Everyone who has been to medical school in Third World countries is taught that if you want to find cures for tropical diseases, first you should find a way to infect the First World with it because there is no investment in drugs unless the disease affects the First World!

Cox: In our analysis of indigenous diseases, 35% were regarded as 'other', meaning that they were not ascribable to any particular Western disease category. We used only categories that were clearly defined.

I agree that it is very difficult to translate indigenous disease categories into Western disease categories, In Latin America, the disease *susto* is very difficult to express in Western terms, but it is very real and people suffer from it. There are many other untranslatable diseases in the South Pacific. I defend these data because they are conservative and they are robust; they are also consistent with Norman Farnsworth's data that we used as a control.

Lewis: I agree with your figure of 35% for 'other' diseases. If one looked at data from any area in the world, about one-third of the diseases would not fall into recognized categories. One disease I hadn't thought of looking at from the point of view of these categories is the 'frightening syndrome'. The Jívaro use quite a few plants to treat this. Is it a problem with the heart and circulation or is it a central nervous system problem? As we learn more about what the indigenous people mean by this term, it will fall into one of the two categories that are underrepresented in your classification. I think indigenous peoples are using more plants for those categories than we are able to discern at this time.

Cox: The Jívaro may have the same problem—wondering where to place what you call central nervous system and cardiovascular disease. As an ethnobotanist, I regard both indigenous and Western disease states as equally real.

The 'other' category may also contain several drugs that affect the immune system. With Professor Bohlin, I have been examining tonics used in Samoa for possible immunostimulatory activity. We found that, in general, immunostimulatory activity is very low. In addition, the 'other' category includes things like hunting medicine and love potions.

Martin: Paul, I like your metaphor of the search for new drugs from plants as a fishing expedition. As Elaine Elisabetsky and Steve King have pointed out, the pharmaceutical industry will be fishing for things that can treat illnesses of the developed world. On the other hand, energy can be put into promoting the capability to do this sort of work in the developing world. There, people will go fishing for other sorts of plants and drugs, those which can treat ailments common in tropical countries. Who do you think will catch more fish? Will we continue to develop the pharmaceutical industry in the developed world, or will there be a sincere effort to build up the capacity and the ability to discover and develop new drugs in developing countries?

Cox: Currently, 600 million people are infected with schistosomiasis. You could in a single stroke improve the lives and help 600 million people if you were able to come up with a cheap and effective plant-derived molluscicide.

I'm looking forward to hearing from our African colleagues later, because they focus very much on diseases of the Third World, including schistosomiasis and malaria. We are doing this work because we wish to project the healers' view of healing into the pharmaceutical industry; this means we must focus as well on illnesses of Third World countries.

Third World diseases are becoming more of a concern in developed countries because of increased travel. If we look at emergent viruses, dengue fever is currently poised to emerge as a serious problem in the United States this decade. We already have hantavirus in the south-western United States.

Iwu: I am also concerned about numbers. Once these get into the literature they get quoted and we could set dangerous precedents. From our knowledge of antimicrobial compounds, there may be many different types in the pharmocopoeia, but that does not reflect the extent of their use. From the Western materia medica, which is something that everybody can verify, the extent of use is different from the availability of drugs in the pharmocopoeia. In terms of the extent of use, cardiovascular drugs may be the most common, but there are few drugs with that activity.

When you go 'fishing' for dermatological preparations, if you limit your horizons to the immediate environment, you get false results. The Europeans, particularly the Body Shop, have done a lot of work on dermatological preparations from Third World countries; they have actually put some of them on the market.

This brings us to the issue of what constitutes a drug? Is it only when it is available as a pill or in an injectable form—as something that is recognized as a drug in Western culture? In your classification of dermatological, are you omitting antifungal agents or antiinflammatory compounds?

Cox: I am defensive of the general pattern we found, because of the remarkable consistency between data from very different geographical areas. Are you disputing, for example, that gastrointestinal disease, particularly diarrhoea, is of important saliency to indigenous people?

Iwu: No, I agree with that.

Cox: There may be some difference in proportions, yet when we consulted statisticians, they were stunned by the degree of consensus between data from different geographical areas. I was expecting to see a difference between tropical and non-tropical areas, and that did not occur. There was a high degree of consensus across different latitudes.

Elisabetsky: Does this geographical consistency arise because you are misinterpreting their uses and somehow biasing the data?

Cox: That's very possible. Because all these data were derived from the literature, they have been filtered through Western investigators, who may have different approaches to understanding indigenous medicine. We have even wondered if the gender of the investigator affects the data. Drugs for obstetric and gynaecological problems that have been derived from ethnobotanical sources are very underrepresented, compared with the number of plants that are ascribed to this category by the indigenous people. Such uses may be more salient to women investigators than to men. I am not claiming that our data should be set in stone, but I do think they are indicative of general patterns.

Dagne: I was surprised that antimalarials are not in the list of medicines used by traditional healers.

Cox: I was surprised as well. Our data suggest that only 2% of the plants used in West Africa are for malaria, whereas in East Africa 9% are. There are two possible reasons: one is that the people there have found a few very active plants and they don't need any more. The other is that the disease is just accepted as unavoidable or possibly, as Nina Etkin suggested, it is treated through diet. I was also surprised that there are so few antiparasitic drugs.

Iwu: I am surprised that there are so many plants to which high medicinal values are ascribed. There are many plants with antiparasitic activity in these cultures, which didn't show in your data.

Farnsworth: In your analysis, you had 50 plant-derived drugs. Our published data indicate at least 119 (Farnsworth et al 1985). Why did you build your analysis on 50 rather than 119 plus any that your group found that mine did not?

Cox: I wanted to come up with an independent list. It was not an attempt to make an exhaustive list, merely a representative list. Also, your list includes a number of drug compounds that are actually inert, like gum acacia (Farnsworth & Soejarto 1985).

Farnsworth: In our paper, gum arabic was not included; we included only plant-derived drugs of known structure (Farnsworth et al 1985).

As Maurice Iwu said, your analysis will enter the literature, and you have used a set of data that is incomplete to make a very profound speculation.

You mentioned correctly that there is no pharmacological category in NAPRALERT for dermatological effects. In South America, there are many reports of insects boring under the skin; we record treatments of this as insecticidal activity not dermatological. Many antiinflammatory compounds are applied externally, which would also have dermatological effects.

Balée: I am curious about the universe of indigenous knowledge systems that you are using, toward the bridge from the indigenous system to the Western system. Is it bounded in terms of indigenous disease categories alone? I get the impression that it's not, because you mentioned love potions.

Cox: Goodman & Gilbert's disease categories are the standard for Western drug disease categories. We tried not to distort the data by forcing them into indigenous disease categories that did not correspond very closely with Western categories, hence we end up with a very large percentage, 35%, of diseases in the 'other' category. I would be interested to do this from an indigenous perspective and look at how our Western drugs are categorized by indigenous healers. I suspect that some of their most important diseases, such as *susto*, would be completely neglected in the Western pharmacopoeia.

Balée: Although not many drugs may come from indigenous knowledge that is outside of the realm of indigenous nosological systems. Some drugs did come from, for example, hunting technology, such as curare and an anticoagulant that was recently discovered in a species of the Brazil nut family, *Cariniana* sp. Are you working with, for example, other possibly bioactive compounds known to indigenous peoples that are not necessarily used in their own therapies?

Cox: It is an important point; our analysis did not include those.

References

Cox PA 1990 Ethnopharmacology and the search for new drugs. In: Bioactive compounds from plants. Wiley, Chichester (Ciba Found Symp 154) p 40–47

Farnsworth NR, Akerele O, Bingel AS, Soejarto DD, Guo Z-G 1985 Medicinal plants in therapy. Bull WHO 63:965–981

Farnsworth NR, Soejarto DP 1985 Potential consequence of plant extinction in the United States on the current and future availability of prescription drugs. Econ Bot 39:231–240

Trotter RT II, Logan MH, Rocha JM, Boneta JL 1983 Ethnography and bioassay: combined methods for a preliminary screen of home remedies for potential pharmacological activity. J Ethnopharmacol 8:113–119

Ethnopharmacology and drug development

Norman R. Farnsworth

Program for Collaborative Research in the Pharmaceutical Sciences, College of Pharmacy, University of Illinois at Chicago, 833 South Wood Street, Chicago, IL 60612, USA

Abstract. The value of ethnomedical information in drug development is based on several factors: accuracy in recording or observing the medical use of the ethnomedical preparation, whether or not the ethnomedical use can be corroborated under scientific conditions in the laboratory, the formal or informal experience of the practitioner who provides the information, the role of the placebo effect and perhaps many others. Published ethnomedical information has many strengths and weaknesses relative to the ability to establish a corresponding biological effect in the laboratory. Many of the publications contain insufficient detail for the laboratory scientist. The ability to correlate ethnomedical reports with corresponding scientific studies could lead to improved selection of plants for further study in the areas of arthritis, cancer, diabetes, epilepsy, hypertension, malaria, pain and fungal and viral infections. These analyses have been accomplished by computer analysis utilizing the NAPRALERT database. This combination of analysing ethnomedical information and published scientific studies on plant extracts (ethnopharmacology) may reduce the number of plants that need to be screened for drug discovery attempts, resulting in a corresponding greater success rate than by random selection and mass bioscreening.

1994 Ethnobotany and the search for new drugs. Wiley, Chichester (Ciba Foundation Symposium 185) p 42–59

The process of drug development, no matter how approached, is generally recognized as a high risk/high pay-off endeavour. According to the experience of the Merck Company, for every 10 000 substances that are evaluated in a battery of biological assays, 20 are selected for animal testing. Of these 20, 10 will be evaluated in humans and only one will be approved by the Food and Drug Administration (FDA) in the United States for sale as a drug. Merck also claims that this process requires about 12 years at a cost of US$231 million (Vagelos 1991). By any measure, drug development along classical lines is a time-consuming and lengthy process. Although not stated in the article, it is reasonable to assume that the substances being developed at Merck are primarily synthetic in origin with a few microbial metabolites being included. Plant-derived secondary metabolites are undoubtedly few in number in the Merck programme, or at least were at the time these figures were made public (Vagelos 1991).

It is not the intent to elaborate on drug development programmes based on synthesis. Suffice it to say that an idea must be developed of a structural type of compound that might have a specific biological activity. The general features of this theoretical model are either searched through the inventory of substances previously synthesized to find relatives of the theoretical molecule or the theoretical molecule and analogues are synthesized. These molecules are then subjected to specific biological assays to test the hypothesis that initiated the programme, but they are almost always entered into all of the high-throughput screens currently being offered by the pharmaceutical firm. Hundreds or even thousands of compounds can be evaluated over one to four days. The results are analysed by means of computers and active substances are assigned priority according to the goals of the programme.

While there is nothing inherently wrong with this approach, only the largest pharmaceutical or chemical firms can afford to initiate such programmes; surely this type of programme in an academic setting is out of the question.

It cannot be denied that higher plants have yielded many useful drugs to alleviate the medical problems facing the world's population. In 1985, we identified 119 secondary metabolites isolated from higher plants that were being used globally as drugs (Farnsworth et al 1985). About 75% of these drugs have the same or related uses as the plant from which they were discovered. Many of these represent 'prototype' drugs that have been used for decades and remain popular today, e.g. atropine, scopolamine, morphine, codeine, tubocurarine, reserpine, pilocarpine, physostigmine (eserine), vinblastine, vincristine and many others. These 119 useful drugs are still obtained commercially, for the most part, by extraction from only about 90 species of plants (Farnsworth et al 1985). With more than 250 000 species of higher plants known to exist on this planet, common sense dictates that many more useful drugs remain to be discovered from this source. But how can plants be identified that will most likely yield new drugs or provide the basic templates for further drug development? The remainder of this paper will focus on how plants can be selected, on the basis of numerous criteria, with the anticipation that a ratio better than 1 : 10 000 will result, if plants are used as starting material rather than the synthetic approach so prevalent today.

Ethnobotany, ethnomedicine, traditional medicine and ethnopharmacology

Ethnobotany is a broad term referring to the study of plants by humans. This includes plants used as foods, medicines, building materials and for any other economic application. Ethnomedicine refers to the use of plants by humans as medicines, but probably should be called ethnobotanical medicine. Traditional medicine is the sum total of all non-mainstream medical practices, usually excluding so-called 'Western' medicine. Generally speaking, I feel that plants used in various forms of traditional medicine have different potential medicinal values, according to the degree of authentication of the medical practice involved. For example, traditional Chinese, Ayurvedic, Unani and Unani-Tibb medical systems and naturopathy are all based on a theory (whether or not one elects to believe the theory), formal education and a written documented history. The systems are periodically revised on the basis of experience and current thought. Information on plants used in these systems is probably more reliable than that from other systems, such as traditional healers, *curanderos*, shamans, 'witch doctors' and herbalists. In these practices, the apprentice system is practised, whereby information is passed on from person to person, father to son, etc. There is usually no formal educational component to these systems and frequently the information is considered highly secretive and is not documented in writing. Plants used in these practices would be less reliable, in my opinion.

In spite of the scepticism of Western-trained physicians and scientists concerning the value of information resulting from any of these systems, all together they are currently serving the primary health-care needs of most of the world's population and this source must not be ignored in any programme of rational drug development starting with plant materials.

Rationale for studying plants used as traditional medicines

It is thought that about 80% of the 5200 million people of the world live in less developed countries. The World Health Organization estimates that about 80% of these people rely almost exclusively on traditional medicine for their primary health-care needs. Since medicinal plants are the 'backbone' of traditional medicine, this means that more than 3300 million people in the less developed countries utilize medicinal plants on a regular basis. Thus, for this segment of the world population, who are generally unable to pay the cost of 'Western' drugs, there is a need to study these plants for safety and efficacy and to develop Galenical products that are standardized and stable. On the other hand, from a developed country's point-of-view, plants are chemical factories that produce a vast array of unusual chemical structures that display a variety of biological activities. Most major pharmaceutical firms are thus interested in acquiring as many of the plants on this planet for their screening programmes as is possible. Only small amounts of plant material are required for the mainly *in vitro* bioassays currently being used: if the extracts show activity, usually less than 500 g are required to identify the active principle(s). The major thrust from an industrial point of view is to find biologically active molecules with the intent of altering the structures in an attempt to enhance activity or decrease toxicity. Such molecules will then be candidates for strong patent positions that are necessary to recoup the costs of development. For the most part, these pharmaceutical firms do not intend to rely on the plants as a source of the marketable drug. The ultimate goal is complete synthesis. Because of this, there is little damage to plants being collected for screening programmes, because only small amounts are required—destruction of the tropical rain forests will not result from this type of small sampling.

A different approach involves a lesser number of plants acquired on the basis of a selection process that takes account of ethnomedical claims for plants. These claims can be identified by an examination of published scientific reports on the medicinal uses of plants or by obtaining information through direct contact with users of medicinal plants or with traditional healers.

Limitations of the ethnomedical approach

For the most part, botanists who conduct field work in areas where use of medicinal plants is a way of life are not trained or do not fully understand the disease state. The information gathered is generally inadequate for the laboratory scientist to evaluate in terms of selecting plants for expensive biological investigations. For example, the literature in ethnobotany, ethnomedicine and ethnopharmacology usually documents the following information: (a) the Latin binomial of the plant used, (b) common or local names of the plant used, (c) plant part(s) used, (d) geographical area where used and (e) a medical use (often vague with few details). Data that are required for assessment of the value of the plant medicine which are usually missing from ethnobotanical writings are: (a) method of preparation of the medicine, (b) dosing (amount and frequency of use), (c) source of information (e.g. traditional healer, the actual user, person who knows someone who uses the plant for medicinal applications), (d) route of administration (oral, external, rectal, vaginal, nasal, ophthalmic) and (e) the specific medical use and/or symptomatology of the disease.

There are numerous examples that might be cited of these deficiencies; the following should make the point. An informant states that a plant is used to allay thirst. Since thirst is one of the many symptoms of diabetes, follow-up questions should be made to see if the user urinates frequently, is susceptible to fungal infections or has any other characteristic symptoms of diabetes. In the case of information concerning use of a plant as a 'contraceptive', rarely does the literature indicate whether the plant is taken by men, women or both. If taken by a woman, questions should be asked about the fertility of the woman (vaginal infections are not uncommon among women living in tropical rain forests, which often result in sterility), the age, sexual activity, normal or abnormal menstrual patterns, etc.

The point is that if ethnomedical information is to be of value in drug discovery, it must be collected in more detail. Otherwise, this approach is no better than random selection followed by targeted biological screening. The approach being used by Shaman Pharmaceuticals Inc. (USA), which involves a team of an ethnobotanist and a medical doctor collecting information in the field directly from users or healers, with follow-up laboratory studies to verify or disprove the ethnomedical claims, is unique in the field of drug discovery.

In order to remedy this deficiency in the type of ethnomedical information found in the literature, scientists trained in botany, who intend to pursue studies in which the information they collect will be of predictive value for drug discovery, should be encouraged to include in their educational programme courses in medical terminology or in pharmacology. As Michael Balick (1994, this volume) has pointed out, there is a dire need to collect good ethnomedical information before the current healers pass on and their information is lost forever. This will require more well-trained ethnobotanists.

Additional applications of ethnomedical information

There is no good way easily to access ethnobotanical or ethnomedical information from the literature. These types of information are poorly indexed in current databases. Original manuscripts, review articles, books and theses abound, but are scattered in hundreds of libraries all over the world. If data from all the published sources could be computerized, appropriate patterns of use could be detected which, even though the data are not as specific as one would desire, could be useful. For example, if a plant was claimed to be used to treat diabetes in one or more countries on several continents, this repetitive type of information would encourage one to believe that the effect was real. We are making an attempt to do this with our two databases, NAPRALERT and MEDFLOR (Medicinal Flora) (see below).

An interesting way to use ethnobotanical information is to examine the Latin binomials of all plants, especially the specific epithets, for those that relate to particular disease states. In the past, botanists who described new species often utilized the Latin name of the disease being treated in their descriptions. Some examples of this are as follows.

Sweet plants. *Lippia dulcis* (*dulcis* = sweet) contains an intensely sweet substance, hernandulcin.

Sedative plants. *Papaver somniferum* (*somniferum* = to sleep) contains morphine, a depressant compound.

Antipyretic plants. *Dichroa febrifuga* (*febrifuga* = lower elevated body temperature) contains the alkaloid febrifugine that has antimalarial properties —alleviating malaria reduces body temperature.

Anthelmintic plants. *Chenopodium anthelminticum* (*anthelminticum* = remove worms from the body) contains several sesquiterpene lactones that have anthelmintic properties, e.g. santonin.

Bitter plants. *Quassia amara* (*amara* = bitter) contains extremely bitter quassinoids (degraded triterpenes). Bitter substances increase the appetite.

Frequently used plants. *Saponaria officinalis* (*officinalis* = 'of the shops', i.e. a commercial item) has been widely mentioned as being used as a folk medicine. The current availability of the Index Kewensis on CD-ROM should make it possible to scan the entire list of known plants to identify specific epithets that relate to disease conditions and thus target candidate plants for further investigation.

In many countries, governments have established official pharmacopoeias. In order for drugs to be entered into these pharmacopoeias, they must comply with the general guidelines including proven therapeutic efficacy and widespread use. Many pharmacopoeias are revised periodically, usually every five years. As newer and 'better' drugs are discovered and find widespread use, they are often included in the pharmacopoeias. If one examines pharmacopoeias of the developed countries prior to the advent of drug development via synthesis, i.e. the late 1940s, one can see that the medical establishment relied heavily on plant drugs and inorganic substances. Thus, one might consider in modern times that plant drugs holding 'official status' in the pharmacopoeias before 1950 were folkloric (ethnomedical) in origin as regards their therapeutic efficacies. On the other hand, these plant drugs must have had some utility as it is not reasonable to assume that all of them acted only through a placebo effect. Thus, one approach to drug development would be to identify those plants holding official status as drugs before 1950 and reexamine them pharmacologically and chemically as sources of new lead compounds.

Precautions in the interpretation of ethnomedical claims

Primitive people often perceived a relationship between the shape, colour or odour of plant materials and features of their diseases. This prompted the use of such plants for afflictions related to these factors. This is known as the doctrine of signatures. Some examples might be a plant that exudes a red sap or resin, or plants that are red in colour, being useful for haemorrhages, for menstrual induction or for increasing blood supply to the body; plants that have a serpentine root system being useful for snakebite; plants with a white latex being able to increase male fertility (ejaculate) or the supply of milk in lactating women; plants having the appearance of a phallus being used for male impotence. Over the millennia, some of the plants chosen might, quite by chance, have been useful for the intended purpose. However, on the basis of scientific experience, one should avoid selecting plants for scientific study that have obvious relationships to the doctrine of signatures.

Even if plants are selected carefully by a proper evaluation of ethnomedical claims, it must be recognized that the placebo is a powerful drug in many cases. Beecher (1955) reviewed the placebo effect in studies that involved severe pain, anxiety and tension, mood changes, cough, sea-sickness and the common cold. His analyses covered 1682 patients of whom 35.2% claimed satisfactory relief

TABLE 1   A summary of placebo effects

Condition	Placebo Agent	Route	No. patients	% Satisfactory relief by placebo
Severe postoperative	Saline	Intravenous	118	21
wound pain	Saline	Subcutaneous	29	31
	Saline	Intravenous	34	26
	Lactose	Oral	176	33
	Saline	Subcutaneous	100	39
Cough	Lactose	Oral	44	40
Drug-induced mood changes	Saline	Subcutaneous	50	30
Pain from angina pectoris	Lactose	Oral	109	34
Headache	Lactose	Oral	199	52
Sea-sickness	Lactose	Oral	33	58
Anxiety and tension	Lactose	Oral	31	30
Experimental cough	Saline	Subcutaneous	1[a]	37
Common cold (acute)	Lactose	Oral	110	35
Common cold (subchronic acute)	Lactose	Oral	48	35
Total patients			1682	35.2

Data from Beecher (1955).
[a] Data based on many experiments in one individual.

of their condition when given placebo medication, usually lactose (Table 1). One can extrapolate these data to ethnomedical claims of efficacy and conclude that at least 35% of the claims could be due to the placebo effect.

Other approaches to drug development using plants as starting material

The ethnomedical approach in drug development has been discussed, but other avenues are available. All of these must be coupled with one or more appropriate bioassays to identify promising leads. Examples are as follows.

Random collection of plants.   The most prominent example of this approach is that followed by the National Cancer Institute (NCI) of the United States of America (Cragg et al 1993). Basically, botanists are contracted to collect plants randomly. Extracts prepared from these plants are tested on cancer cells or cells infected with HIV (human immunodeficiency virus). The NCI used this approach from the mid-1950s until 1982. During this period, NCI claims to have collected about 35 000 species of higher plants for testing. From these, several hundred cytotoxic (*in vitro*) and anticancer (*in vivo*) compounds of known structure were discovered. Of these, 15 showed activity and safety profiles that allowed them to be evaluated in humans. Of these, taxol was recently approved for sale in the United States as an anticancer agent. Camptothecin, discovered at about the same time as taxol, is currently in human trials in the form of several analogues. Published data suggest that the analogues are showing sufficient efficacy that one can expect one or more of them to be approved shortly for human use. If we consider that two drugs have been derived from 35 000 plant species (1:17 500), this is not as effective a method as following the synthetic approach (1:10 000). However, there has been some discussion at the NCI over the number of species actually tested in the early NCI programme (Gordon Cragg, personal communication 1993). The type of record keeping at the time often did not differentiate between synthetic and natural compounds tested and often several plant parts of a single species were tested without proper documentation as to how many actual species were involved. Gordon Cragg believes that the number of species tested most likely was only about 16 000–20 000. If this is true, the actual yield of marketable anticancer drugs per species produced by the early NCI programme would be 1:8000. If all these species had been studied for conditions other than cancer, additional drugs might have been discovered.

Selection of plants based on chemotaxonomy.   If one knows that a certain class of chemical compounds is more promising biologically for an activity of interest to the scientist or pharmaceutical firm than are other classes, one can select plants that are predicted to contain that class using principles of chemotaxonomy, e.g. indole alkaloids in the Apocynaceae, coumarins in the Rutaceae and sesquiterpene lactones in the Compositae.

Selection of plants from a combination of literature reports on plant extract pharmacology and ethnomedical claims associated with the same medicinal use.   One can laboriously scan the literature for plants that have been reported to produce a biological activity of interest and also have an ethnomedical claim for the same use. Further literature searching would reveal whether or not any chemical compounds had been isolated from the plants that were responsible for the desired effect. In our laboratory we have developed a rapid method for doing this. Since 1975 we have been computerizing the global literature on the pharmacological effects of extracts of living organisms (primarily plants, but including marine organisms, fungi and animals), ethnomedical information on the uses of plants, the secondary metabolites that occur in living organisms and the biological effects reported for these secondary metabolites. The database is known as NAPRALERT (NAtural PRoducts ALERT) (Loub et al 1985). More than 110 000 scientific articles have been scanned and their important features computerized. Articles are being added at the rate of approximately

TABLE 2 Candidate plants suitable for investigation in drug discovery because of having both an ethnomedical use and experimental activity suggesting that the use is valid

Drug target	Number of species: active compound unknown	Number of species: active compound known
Arthritis	290	3
Cancer	94	14
Diabetes	108	5
Epilepsy	47	1
Fungal infections	63	2
Hypertension	21	1
Malaria	57	8
Pain	114	2
Viral infections	42	3
Warts	3	1

Data based on NAPRALERT analysis as of 1st October 1993.

7000 per annum. These articles have been selected from a systematic search of the global literature. NAPRALERT is a relational database and thus programs can be written to relate selected aspects in the database with each other to the exclusion of undesired information. Information on more than 47 000 species of organisms can be found in the database at the present time. About 27 000 species of higher plants can be searched in NAPRALERT, of which about 9000 have ethnomedical literature associated with them. Previously, we pointed out that, as of 1989, 300 types of ethnomedical uses were associated with plants in the NAPRALERT database, representing approximately 41 940 computer records (Farnsworth 1990).

The database is especially thorough, with systematic searches dating back to 1900, on natural products in the areas of cancer (chemotherapy, chemoprevention, carcinogenesis), fertility regulation (both male and female), diabetes, malaria and other parasitic diseases, viral diseases, sugar substitutes, molluscicides and anti-HIV agents. As an exercise to demonstrate the power of computer analysis, we selected 10 diseases and searched NAPRALERT for species that had both ethnomedical information and *in vitro* and/or *in vivo* experimental data supporting the ethnomedical claims. Further, for each disease we documented the number of species in which secondary metabolites were known to exist that explained the biological effects (Table 2). These data could provide the basis for selection of plants with a high degree of expectation that they would contain novel bioactive compounds, if subjected to experimental studies.

In a NAPRALERT analysis, the database is first searched for plants whose extracts have shown a desired biological activity, for example, for diabetes. These are sorted and saved in group A. Group B represents all plants that contain secondary metabolites reported to have antidiabetic activity. Group C represents plants that have ethnomedical claims associated with diabetes. After the database merges these, two lists of plants emerge as priorities in drug development. The 'D' plants are those that have an ethnomedical reputation as being used for diabetes, which have been shown experimentally to have antidiabetic activity and which contain no compounds that can explain the antidiabetic effects. The 'E' plants are those which have ethnomedical claims for diabetes, which also have been shown experimentally to have antidiabetic effects and which also contain one or more compounds reported to have antidiabetic effects. Thus, the 'D' plants are those of interest to pharmaceutical firms that want to identify bioactive, hopefully novel, secondary metabolites that can be used as lead compounds in drug development. The 'E' plants are those that could be suitable for use in the less developed countries where all ethnomedical and experimental evidence suggests efficacy and which contain active compounds that could serve as markers for use in quality control procedures in Galenical preparations.

References

Balick MJ 1994 Ethnobotany, drug development and biodiversity conservation—exploring the linkages. In: Ethnobotany and the search for new drugs. Wiley, Chichester (Ciba Found Symp 185) p 4–24

Beecher HK 1955 The powerful placebo. J Am Med Assoc 159:1602–1606

Cragg GM, Boyd MR, Cardellina JH II et al 1993 Role of plants in the National Cancer Institute drug discovery and development program. In: Kinghorn AD, Balandrin MF (eds) Human medicinal agents from plants. American Chemical Society, Washington, DC (Am Chem Soc Symp Ser 534) p 80–95

Farnsworth NR 1990 The role of ethnopharmacology in drug development. In: Bioactive compounds from plants. Wiley, Chichester (Ciba Found Symp 154) p 2–21

Farnsworth NR, Akerele O, Bingel AS, Soejarto DD, Guo ZG 1985 Medicinal plants in therapy. Bull WHO 63:965–981

Loub WD, Farnsworth NR, Soejarto DD, Quinn ML 1985 NAPRALERT: computer handling of natural product research data. J Chem Inf Comput Sci 25:99–103

Vagelos PR 1991 Are prescription drug prices high? Science 252:1080–1084

DISCUSSION

King: Norman, you have told us numerous times that 90% of the 119 plant-derived pharmaceutical products that are on the market today come from natural sources. They are cultivated or harvested from wild plants. This comes back to the conservation aspect that Mike Balick was describing—alternatives to deforestation. Part of the answer is that you have to state the case better; if

90% of drugs now come from natural sources, you have to tell the pharmaceutical companies this. Do those who work in the pharmaceutical industry think it will get any more leads from organic chemistry?

Farnsworth: The ability to isolate compounds depends on the ability to screen. The screening capacity now greatly exceeds the ability of the major pharmaceutical companies to provide synthetic compounds of their own; therefore they are looking for anything to keep up with their screening capacities. So the companies are interested in plants, but mainly from a numbers viewpoint. I don't know of any company apart from Shaman Pharmaceuticals that really takes the time to think things through and select plants that are used by people for certain ailments. Most companies will just screen anything that comes through the pipeline.

Cragg: We at the NCI are now opening our repository of extracts to carefully selected organizations who abide by the same policies as we do concerning compensation and collaboration. We have had about 30 enquiries and so far we've selected 10 organizations; so there's a lot of interest in this.

McChesney: In discussions with the pharmaceutical industry, specifically with reference to the development of plant-based substances as drugs or as pharmaceuticals for marketing either directly or as a basis for such a material to be marketed, we have encountered certain problems. A very great problem is the perception of the industry that, at best, they can get a little of the substance, resolve the structure, and then prepare the compound chemically. They think they can never get enough of the substance directly from plants to do the biology and to take the compound through the development process. If you analyse the quantities of plant material that are required, and you take as an example the effort that was exerted to develop a system to supply the quantities of taxol that are now required to meet market demand, you see some of the reasons for the perception that the industry has. If we have real commitment to plant-based natural products as a new source, a return to their previous prominence in the discovery and development of new pharmaceuticals, it is important that we address the issue of how to obtain large quantities of the pure bioactive material for the development process. It is not enough to give the industry 25 mg of a natural product and say 'here's a new drug'. They need kilograms of that material to evaluate for all of the issues involved in the production and development, and to have assurance that should it be successful they can meet market need, which may be hundreds or even thousands of kilograms per year.

Farnsworth: I agree with you except on one point. If you isolate 25 mg of pure active compound, you can do the *in vitro* assays and possibly conduct a limited study in animals. This will indicate whether you should go after partial or complete synthesis of the active compound. In most cases, synthesis is where your kilogram quantities will come from; I don't think they will come from the plant. Most companies feel that the skill in chemistry is such that all these problems of supply can be solved eventually—the taxol case showed it. You can get enough taxol from the needles of *Taxus* species and there are steps in place to prepare it by partial synthesis and possibly by tissue culture. So science is improving and moving ahead fast enough to solve these problems of supply, if you have a promising compound.

McChesney: But you don't get to the point of having a winner unless the industry overcomes the perception of supply as being an issue. As yet, there is no truly complex agent such as taxol, or even the steroid hormones that are used very extensively, that is prepared by total synthesis. Those materials are prepared by modification of plant-derived materials. We use this argument to try to dissuade the industry from the perception that supply will be limiting. It is limiting only in that we have not paid attention to the issue, in my judgement, but the industry doesn't see it yet entirely that way.

Peter: One important fact is the time required to get enough material for quick follow-up investigations, even though these do not yet require kilogram quantities. In most cases it is already quite difficult and time consuming to get gram amounts. There is no quick and easy solution to this problem. Therefore, the primary objective of most larger companies is to gain information on structure–activity relationships from random screening of plant products. The most interesting aspect of secondary metabolites from plants is their unusual and often very complex structures, which would not be synthesized in the laboratory unless important information on some interesting biological activity were already well documented. So our main interest is to use plant metabolites and extracts as a complement to the chemical libraries in the random screens. The information on new structure–activity relationships of complex structures can be put to use immediately: the medicinal chemists can do a substructure search to find out which structural elements are important, then perform exploratory modifications. In a next step they try to develop structurally related products that are more easily accessible. Only if this approach is not promising will the long way to develop the natural product itself be pursued.

This complex situation is nicely illustrated by the case of taxol: if the NCI had not invested a lot of time and effort in the isolation and pre-clinical investigations of this compound, it would still not be commercially available. Because managers in pharmaceutical companies are more and more compelled to look for shorter-term results, it is increasingly difficult to get approval to continue a project which carries a relatively high risk and requires more than 20 years to complete.

Cragg: I agree with Dr Peter; taxol would have died without the NCI effort over a couple of decades. This is why we regard it as our mission to pick up these difficult leads, where industry, with its obvious and understandable profit motive, would not devote the resources to something which is an apparent high-risk compound. It is our mission to take those leads and see if anything can be developed, and of course we have hit something very good with taxol.

The supply issue is very important. We feel somewhat burnt by the issue of taxol development. Suddenly, it was shown to be effective in the treatment of ovarian cancer and we couldn't meet the demand. Fortunately, the problem was fairly easily solved, because of the other *Taxus* species available and the renewable resources in terms of needles. Our policy now is that the moment we enter a drug into the first phase of pre-clinical development, we address the supply issue. For example, with michellamine B in Cameroon we are looking into the feasibility of cultivation, even though we don't know at this stage whether the drug is going to be efficacious or not.

Schwartsmann: During drug development, supply is not the only major problem. For anticancer drugs, formulation is usually a limiting step. There are many compounds that people would wish to develop, but which simply cannot be formulated for human use. In the last five years, I would say that the main blocking factors in the development of anticancer drugs were supply and formulation. We are trying to circumvent these.

I would like to add a general comment regarding anticancer drug development and folk medicine. It is very hard to identify interesting drugs for cancer purely on the basis of folkloric information. I simply don't trust the way people describe diseases in the local communities. In that setting, cancer tends to be described mainly in terms of signs and symptoms. Furthermore, the spectrum of diseases recognized by indigenous cultures is not the same as we normally see in developed countries. I don't think they have the same type of illnesses and they probably die earlier of infections and other problems. Burkitt's lymphoma is easy to detect in Africa because it's frequent and very well characterized. However, common tumours in adults are quite hard to identify by folk medicine. The main role for natural products research relates to the huge number of novel structures that can be provided for testing. Novelty is where Nature can help in drug discovery. Scientists are trying to develop more sophisticated systems for screening, using human cell lines and more complex panels of cell lines, rather than relying on a few mouse leukaemia cell lines as people did before. This is a more promising approach. The NCI now uses 60 human cell lines. Once you have selected relevant molecular targets, you can do a lot *in vitro* by screening with a molecular approach, using very specific targets, such as protein kinase C inhibitors, oncogene products, etc. Novel chemical structures can be screened in these new testing systems.

Bohlin: We are talking about potential natural products for drug development; it's also most important to consider bioactive molecules of natural origin that can be used as pharmacological probes, e.g. forskolin and tetrodotoxin, and as models for total synthesis. This will give us a greater chance of success.

My other comment is about the scientific disciplines connected to this discussion. Pharmacognosy is rarely mentioned—why not? Pharmacognosy is an interdisciplinary field, containing some botany, some chemistry and pharmacology. We are now talking about ethnobotany or ethnomedicine; why not talk about pharmacognosy?

Farnsworth: We have seven full professors of pharmacognosy at our college and one associate professor, but I have better things to do with my energy than to argue words, I argue deeds. If a duck quacks and waddles it's a duck, you don't worry about what it's called.

Iwu: I would like to comment on the contribution of the ethnobiological or ethnobotanical teams. The deficiencies in the information they bring back have actually been retarding the development of drugs from plants. In the USA, the interest in pharmacognosy declined over two decades. Having picked up an interest in natural products, they had to create a new discipline, whereas in Europe and in Africa pharmacognosy was still treated as a multidisciplinary field. If we want to enrich pharmacognosy with ethnobotanical or ethnobiological enquiry, it will be a question of adding ethnography to the training of pharmacognosists. This would be much more useful for drug development because the student will know that it is not enough to record that a plant is used by the Kayapó to treat a particular disease. He will know that it is important to differentiate an aqueous preparation from an exudate, because from that point of view the whole chemistry changes. This kind of information is missing because the multidisciplinary nature of the discipline has been completely lost. Most of the reports we get are from individuals, either an anthropologist or a medical person working alone. In most cases there is no real teamwork and this affects the kind of information that is brought back.

Craveiro: I would like to comment on the confusion in the literature concerning pharmacognosy data. About 3–4 years ago, we were studying a plant from the Euphorbiaceae family, called *Croton sonderianus*. This was brought to our laboratory with the information that it is a very effective contraceptive. It is used by a population about 300 km from Fortaleza in Ti-an-gua county. After we started these studies, we decided to find out why this plant is used by that local population and not by another population. Then we learned that the long sticks that this plant yields are kept by the wife next to the bed and when the husband comes at night during her fertile period, she just hits the husband three times with the stick and that is a very effective contraceptive!

Balée: I would like to bring up a couple of considerations concerning the characterization of ethnomedical knowledge in indigenous context, which have implications for the search for new drugs. Within the sample of plants known to have been used medicinally by certain cultures, we need to identify the most likely candidates for screening. One criterion would be, is the use of that plant falsifiable? Does application of the plant in the indigenous context actually have some kind of therapeutic effect on the human body or is the use of the plant, although called by the same word that is used for medicine in the Western sense, actually for luck in hunting, luck in love, luck in increasing one's wealth and so forth—as a sort of catalyst that can't be subjected to falsification?

A second consideration is that this has to be a comparative undertaking. The most likely candidates would be plants that are used in the same way for therapeutic effects on the human body, by different groups widely separated in space, where the diffusion of this knowledge or borrowing in the past most likely would not have taken place. An example relates to a genus of ants well known in Amazonia. The genus is *Pseudomyrmax*. These ants live in the petioles of a couple of genera of trees in different plant families, *Tachigali* and *Triplaris* in the Amazon. The ants sting: when one touches a plant, they come out and sting one rather ferociously. The Ka'apor who live in the extreme east of the Amazon use these ants as a febrifuge; that's from a species of *Tachigali*. The Sirionó, with whom I worked in Bolivia, 1300 miles to the west, used the same genus of ants from *Triplaris* for the exact same reason. I don't know of any groups in between those 1300 miles who use those ants as a febrifuge. This is the kind of ethnomedical knowledge in a comparative context that might be worth investigating.

Farnsworth: It's a well known fact, going back over 200 years of documented literature, that drinking a glass of hot water will reduce moderately elevated body temperature. Febrifuge just means lowering fever, the lowering of temperature; you don't know whether there's anything in that hot water, i.e. the plant material, that is exerting an effect or whether it's just the hot water.

Balée: They pass the leaves of the plant, in which the ants are located inside the petioles, over the patient. The ants come out of the leaves onto the patient's chest and back. The method is identical in both cases.

Farnsworth: My point is that you don't see that kind of detail in publications.

Elvin-Lewis: Among the Indians we work with in the Amazon, the real use for ants is to treat pulpitis. This is when you have an inflamed root canal, really a sore tooth. Our dentists treat pulpitis with formic acid in the same way: ants contain formic acid and that's the real use for ants, at least in the context of the Jívaro.

Lewis: About five years ago we made the following observation. There was a group of Jívaro who had walked about two days south of their typical territory to the edge of the northern region of another group of Indians, the Candoshi, whom they do not like. In fact there is constant conflict between these two groups. One of the Jívaro men was bitten by a fer-de-lance and he was in very bad shape. A few hours later, a group of Candoshi men arrived who were not typical of the tribe, but rather they were bicultural, having been raised in both Jívaroian and Candoshi cultures. The group said: we have an ethnomedicinal remedy against this type of snake bite. I don't know whether the remedy was efficacious or not, but the Jívaro allowed them to try it, because of course they spoke Jívaro even though they were from bicultural villages. Within a day or so the inflammation was down, the pain was reduced and he limped back to the village where we were. This was intercommunication, for as a species, we are communicators, a fact often dismissed in ethnobotanical and ethnomedicinal literature. Even though 1300 miles seems rather far, we don't know what time element might be involved in this distant transfer of information. The Indians that we worked with certainly travel a lot and even among enemies or those speaking a distinct language they are able to intercommunicate.

The Jívaro were really excited about this new snake bite remedy and within a few weeks they had adopted it as theirs. When we returned on later visits, this was no longer a Candoshi snake bite remedy, but it had become at least in part a Jívaro remedy.

Balée: I agree that people communicate uses of plants quite readily, especially in areas that are adjoining. But in the case of the ant species, we don't find this use to my knowledge in the area in between these two areas which are 1300 miles apart. Also, these two languages are distantly related, mutually unintelligible languages of the same language family, Tupí-Guaraní. I think there's a strong possibility that this use of the ant species has been retained through time, which is another indicator of a potential candidate for biochemical research.

Prance: Bill Balée's point about comparative data, where there is the same use in two different places, is important. Another good example is where the same genera, such as *Panax* and *Podophyllum*, are used both in Chinese medicine and North American Indian medicine.

Farnsworth: There's no doubt about that. Our database, NAPRALERT, is broken up by ethnographic regions. For example, a plant used in Pakistan does not come out of the same cultural milieu as a plant used in India; it is more closely related to use in the Middle East. WHO asked a group of anthropologists to divide the world into broad cultural groups. We have a code in the computer so you can analyse all data within those cultural groups. The only weakness is that if you have a monotypic species that only grows in one country or one geographic area, you can't do this kind of analysis. We have a species that grows in 10 countries and you find the same use over and over in each of those countries; this is good information.

Martin: It is commonly accepted that if the use of a plant has an independent origin in many different cultures, this is good evidence that it could possess biological activity. On the other hand, if the information has been borrowed and has diffused across a broad region, this can also be an indication of biological activity. Somebody just doesn't borrow something on blind faith. When you borrow something, you begin a new process of empirical verification. If cultures have borrowed information and incorporated it into their medicine, this is also an indication perhaps of biological activity.

Jain: We also come across instances of common uses for the same species or related genera. At the Pacific Science Congress, I presented a comparison of indigenous uses of plants in India and the Hawaiian islands. There are several similarities. Two days ago, I was in Rio de Janeiro and several of the species growing in India are also in the medicinal plant garden in Rio. The uses of the plants in Brazil are exactly the same as those among the indigenous people in India. In some cases, the genus was common, but a different species was used in India compared with Brazil.

All ethnobotanical information from which leads have been developed in the last 15–20 years has come from plant taxonomists. Most of the big names in plant taxonomy over the last 15 years have become ethnobotanists. Their main interest in field work at that time was floristic work. They developed an interest in the economic uses of the plants, they noted the uses of the plants from which the chemical leads had been provided. Traditional healers are not ethnobotanists. If you are studying agricultural or genetic resources, traditional healers are no help in ethnobotany.

The system of passing information from one generation to the other is a good example. But there are various ways of passing information from one generation to another in different situations in the country. We have recorded from the ancient Indian literature that during festivals there used to be groups of people who knew about uses of plants who would meet and share their information. Each individual then went back to his community and told them about curative properties of certain plants. But when personal interest came into it, and it became a situation of information being passed from father to son and so on, people became secretive.

Elvin-Lewis: There is a common weed, *Phyllanthus amarus* (*niruri*) that grows throughout the tropics. This has antiviral activity against hepatitis B and is valued in India and elsewhere in Asia for its use to treat hepatitis. When we were studying the pharmacopoeia of the Jívaro, we found that this plant, although not commonly the plant of choice for this purpose, was used to treat hepatitis. So when you have a cosmopolitan weedy plant, sometimes people find the common use.

Balée: Sometimes different peoples arrive at the same therapeutic use independently, I agree entirely with that too. There's a snake bite remedy that the Ka'apor use. The Guajá, who are related to the Ka'apor, also use this as a remedy for snake bite. The words in the two languages, although closely related for the same species, are quite different. I can't affirm that it actually has activity against snake bite, but the therapeutic procedure is stated in such a way that it could be tested for activity against snake bite in terms of the indigenous uses in these two different cultures. In this case, it is possible that these two different groups came to the use of this tree independently, because the words are different even though the languages are related.

Posey: We are talking about artifacts of data collection as well; absence of information doesn't mean absence of a certain practice. We don't know what happens in the region between the two areas that you are talking about. We do know about Ka'apor and there are recorded uses of that ant in the literature cited in Lenko & Papavero (1978). So the fact that the ant is used in between by non-Indian populations is documented.

The other factor is the elaborate trade network that united until very recently very distant tribes. You should not rule out pre-Columbian as well as Columbian trade and interchange.

Cragg: The widespread occurrence of certain plants that are used for similar treatments is very interesting from a chemical point of view. We know that chemical secondary metabolites vary considerably depending on the environment in which the plants are grown. Steve, does the content of proanthocyanidins in the source plant, *Croton*, vary considerably from one area to another?

King: The content of proanthocyanidins is remarkably homogeneous, but the amounts of minor metabolites present do vary.

Prance: An interesting question is: how do the Indians find new things in the Amazon? We have heard several examples from that region. I've always thought that the people in the forest do need very basic things for their existence: an arrow poison, a fish poison, some sort of narcotic, medicines, foods, etc. When migrations take place, sometimes the plants the people are used to are not present in the new location, but they experiment and find a new arrow poison, a new fish poison, a new narcotic, etc. It is amazing how this has happened over the time when there's been human occupation of the Amazon region. The desire to have these essential products has certainly led to many new discoveries.

Reference

Lenko K, Papavero N 1978 Insecto no Folclore. Universidade de Sao Paulo, Brazil (Conselho Estadual de Artes y Ciências Humanas)

Basic, quantitative and experimental research phases of future ethnobotany with reference to the medicinal plants of South America

Walter H. Lewis and Memory P. Elvin-Lewis

Department of Biology, Washington University, One Brookings Drive, St. Louis, MO 63130-4899, USA

Abstract. Ethnobotany of the future will encompass what we perceive as three interrelated research phases. *Basic ethnobotany* includes the compilation and organization of information about biota obtained from indigenous and other peoples, such as obtaining data about useful plants and animals, understanding how peoples manage their environments and learning about their lexicons and classifications. This is what we try to do in the best possible way, directly in the field from original sources. These results can then be organized in many ways once species determinations are completed. They may also be organized using other types of information, the most obvious being chemical, medical and linguistic. *Quantitative ethnobotany* develops methods to allow quantitative description and to evaluate and analyse primary data sets. Original field research must be sufficiently structured and consistent, for example in relation to forest habitat and composition or to oral exchanges between informant and listener, so that statistical techniques may be used to test proposed hypotheses rigorously. This aspect of ethnobotany is in its infancy, yet it can be broadly utilized to comprehend more meaningfully and usefully ethnobotanically valued plants, particularly in the exceedingly diversified environments of tropical regions where because of community isolation practitioners are still most knowledgeable. *Experimental ethnobotany* involves the use of biota in search of products for industrial, medical and other purposes. Plant ethnomedicinal findings may set the stage for targeting materials which can be meaningfully analysed for chemical activity using appropriate biodirected assays. This approach in search of new pharmaceuticals is woefully underutilized today to the detriment of human health and a number of new strategies should be considered for future advancements in drug discovery. These aspects of ethnobotany will be evaluated largely in relation to current and future research in South America.

1994 Ethnobotany and the search for new drugs. Wiley, Chichester (Ciba Foundation Symposium 185) p 60–76

There is great scope for new drug discoveries based on traditional medicinal plant use throughout the world. In a recent review, Lewis (1992) outlined by medical category several hundred plants currently used in modern medicine and pharmacy, illustrating recent selections of natural products and their incorporation into modern pharmacopoeias. He also showed how the culturally intact South American Jívaro use plants now as they have for perhaps thousands of years and as cultures did worldwide for survival on a daily basis. However, 'serious dangers exist for the survival of such peoples and their cultures, and the ecosystems which nurture them and provide Western and traditional medicines with novel plant products for human well-being everywhere. In this race against ecosystem destruction, researchers in many disciplines must rally to provide the impetus to save global diversity while, at the same time, accelerating studies of ethnomedicine in consort with biomedical and chemical teams for developing new natural products and drugs needed by humans into the next century'. As about three-quarters of the biologically active plant-derived compounds presently in use globally have been discovered through follow-up research on folk and ethnomedicinal uses (Farnsworth et al 1985), it is imperative that ethnobotanically directed research to find active pharmaceuticals be instituted on a broad scale, including the development of new strategies.

Ethnobotany in the three phases outlined below is fundamental to the success of programmes associated with the search for new pharmaceuticals.

Basic ethnobotany

Basic ethnobotany includes the compilation and organization of information about biota obtained from indigenous and other peoples in many parts of the world. This is what a handful of researchers do and do well, by obtaining ethnobotanical information from intact indigenous peoples with a medical system which utilizes the rich and varied taxa of their environment. This primary source is the best, but today it is scarce and growing rarer. Information obtained from studying populations consisting of less intact and more acculturated peoples often results in fewer species with uses less understood. Depending on the extent of acculturation, the partial or remnant system may also incorporate methods from other sources of traditional medicine as well as from Western medicine. The extent to which new pharmaceuticals might be developed using this information clearly depends on the amount of original information retained and how its utilization impacts on the group's overall health. Because validation of the system as it has evolved can still take place, including observations on specific plant use, preparation, treatment modalities and judgement of efficacy, we consider these disclosures as original sources, and thus primary ethnobotanical data as from intact peoples. Secondary information is that obtained from the literature, herbarium collection labels and from people who have learned or heard of plants and their uses, but who lack personal experience with them.

The usefulness of this dichotomy is illustrated by the specific bioreactivity we obtained in testing plants with anti-dental plaque, antiviral, antimalarial, antiinflammatory and healing activities. Generally, plants valued the most can, in specific screens, be shown to possess functioning principles related to their ethnodental or ethnomedical use. This correlation was first recognized by studying plants used to maintain dental health, since then it has been extended to other maladies. For example, when disease descriptors are used to select for testing plants with specific or related antiviral activity, highly significant correlations can be achieved, particularly from primary data (Lewis & Elvin-Lewis 1994). However, when general antiinfective agents or other medicinal plants are included in the assay, the chance of obtaining activity decreases, as does significance. In our experience, the better one understands the disease systems and the aetiological agents or physiological events which cause them, the more definitive are the assays and the more significant the results.

These results then translate into highly focused and targeted information of enormous potential relevance in natural products research. 'As Balick (1990), Cox (1990), Farnsworth (1990), Lewis (1992), Malone (1983) and Plotkin (1988) argued, ethnobotany provides the mechanism for rapid assessments of the pharmaceutical potential of species, an obviously significant procedure when there is insufficient time and funds to test the majority of species.' (Lewis & Elvin-Lewis 1994).

Once the species of voucher specimens have been determined, or if possible their prior determinations confirmed, compilation and organization by taxa, vernacular names, use categories and other groupings provide enormously valuable data for direct utilization or comparison with different data sets. For instance, vernacular names of the Bignoniaceae, used by the Jívaro and other peoples of the upper Amazon basin in Peru and Ecuador to identify widely used medicinal plants in that family, provide valuable ethnolinguistic information about indigenous classification, as well as meaningful indications of those most prized for use. Only one species with a name was not used in some way. Named plants, therefore, target useful plants in the lexicon of an indigenous people. Nineteen species belonging to 17 genera were identified among our collections; only one species collected once by the Mayna Jívaro proved nameless; 14 were both named and used by the Jívaro, chiefly for medicinal purposes. Their names for these plants were almost exclusively monomials and generally applied to a botanically recognized genus with one or more species. Thus, *kuwinap* is their unique name for *Jacaranda copaia* (Aublet) D. Don and *J. glabra* (D.DC.) Bur. & K. Sch. and *tái(i)* is the name for *Arrabidaea chica* (H. & B.) Verl. (Lewis et al 1988). The two *Jacaranda* species are not distinguished by name, even though they are recognized as distinct; the same part of the plant, the leaves, is used in a similar or identical way to treat infection with leishmania, cuts/wounds and skin ailments like mites and scabies. The use of the same name for different species and even genera, provided that the plant part used and the medicinal use are sufficiently alike, suggests they might share a basic chemical core structure but possess different allied compounds and analogues. A battery of similar compounds might exist in Nature, which would need only isolation and characterization, a much simpler procedure than the usual trial and error modification of compounds.

In the future, more than a handful of ethnobotanists are needed to ensure that primary ethnobotanical data will be obtained before these resources are irretrievably lost by acculturation, gradual assimilation and even genocide. A part of this effort must be the training of more students in all aspects of ethnobotany.

Quantitative ethnobotany

Quantitative ethnobotany goes beyond the compilation and organization of source information by developing methods for quantitative description and to evaluate, analyse and compare relative primary data sets (Prance 1991). Original field research must be sufficiently structured, for example, in relation to forest habitat and composition or to oral exchanges between informant and listener, that statistical techniques may be used to test proposed hypotheses (Phillips & Gentry 1993a,b).

In 1975 a survey applying epidemiological techniques to ethnobotanical data was conducted to determine whether Ghanaians employed any criteria in their selection of plants used for dental hygiene and if any of these choices impacted on their dental health. Secondary sources had already identified over 100 species used in this way, but nothing was known regarding what factors dictated their selection or if availability was the predominant criterion. The survey was initiated among the known 11 tribes and linguistic groups in southern Ghana, facilitated by Ghanaian collaborators who used a comprehensive, standardized questionnaire of 118 entries. Such detail was needed to determine the sociological, ethnobotanical and dental health parameters required to calculate frequencies of uses among the 887 informants, who were well distributed by age and gender. This allowed preferences to be identified and ordered by tribe, age, etc. Analysis using a Statistical Analysis System computer program (Adu-Tutu et al 1979, Elvin-Lewis et al 1980) showed that several selective forces influenced choice and that the most popular species conferred the best dental health. Several species of five genera predominated and three species of *Garcinia* accounted for 77.6% of use by the total population. Subsequent studies on antimicrobial potential showed that the most popular species possessed a broad spectrum of activity against many oral odontoperiopathogens. Some of the plants contained astringent and healing tannins with added antiplaque activities or significant amounts of fluoride which affected the cariogenic process (Elvin-Lewis 1983).

Prance et al (1987) summarized their research among four tribes in Brazil, Bolivia and Venezuela using an ethnoecological forest inventory over a 1 ha parcel of forest. All trees 10 cm or greater in diameter at breast height were marked and specimens were collected and determined. These specimens were presented to indigenous informants for information on use. Prance et al then calculated the percentage of tree species on each hectare that was useful to each group, which ranged from 48.6% (*Panare* of Venezuela) to 78.7% (*Chácobo* of Bolivia). By dividing trees into use categories and designating the cultural importance of each species, they devised a use value for each species and family. The palms proved the most useful family for all indigenous tribes; clearly, this family deserves special consideration in terms of conservation.

The recent research of Phillips & Gentry (1993a,b) illustrated how valuable quantitative methods of collecting and processing ethnobotanical data are in providing a sound basis for testing hypotheses. For example, from data on the use of plants by non-indigenous people in Peru, they showed that informant age significantly predicts informant knowledge of edible wild plants (acquired early in life), construction and commercial uses of the forest (30–50 years) and medicinal plants (largely confined to older people). Hence, knowledge of medicinal plants is clearly the most vulnerable to acculturation, 'and should be a main focus of attempts to both record indigenous knowledge, and to encourage knowledge and pride in traditional cultural values' (Phillips & Gentry 1993b). A myriad of possibilities needs exploring using the same technique to calculate use values based on relative importance of data sets by regression analysis (Moerman 1991). Phillips & Gentry suggested testing: the hypothesis that terra firma forests are especially useful (Prance et al 1987); utilitarian/adaptationist versus folk plant taxonomies (Berlin 1990); and the likelihood that knowledge of particular environments, use classes and life forms is unequally divided between men and women.

We plan to examine these and other questions by establishing a research site in the Condorcanqui region of northern Peru on the low eastern slopes of the Andes. By using a novel, three-tiered, nested transect model, which includes large 1–2 km^2 transects examined by Landsat TM satellite receptors to determine canopy tree physiognomy, crown architecture and leaf properties, 1 ha transects and 0.1 ha transects (Fig. 1), the floral structure and composition of major forest communities will be determined. Ground reconnaissance and marking of individual trees and lianas in the large and 1 ha transects will allow use value determinations based on information from informants who live in nearby indigenous Huambisa, Aguaruna and *mestizo* communities. Identification of microhabitat preferences and occurrences of medicinally important plants will also provide valuable data for determining the potential dangers associated with species exploitation whenever demand accelerates. No such quantitative inventory is available for any tree species being used in preparing pharmaceuticals. Now, however, the opportunity exists to provide these data, so that as new species are added to our pharmacopoeia, their selective utilization and conservation can be monitored without threat of extirpation from

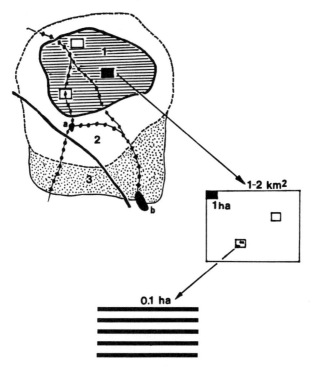

FIG. 1.   Diagram of a research site at Condorcanqui, Peru, consisting of three forest types (1 = primary; 2 = buffer or extractive; 3 = disturbed or partially cultivated) and three nested transects (1–2 km², 1 ha and 0.1 ha). Two communities are present (a and b), from which trails emanate (dots indicate reference points about 200 m apart), both near a water way (thick line).

the ecosystem in any area. Quantitative ethnoecology of this type is a necessary component of future research wherever native useful plants are exploited.

Experimental ethnobotany

The so-called experimental phase of ethnobotany searches biota for industrial, medical and other products. For example, plant ethnomedicinal findings may target materials that can be analysed for chemical activity using appropriate bioassays. This approach in the search for new therapeutic agents is greatly underutilized today.

Ethnomedicinally targeted results

In their antiviral screening programme, Van den Berghe et al (1985) reported that selection of candidates for screening on the basis of traditional medicinal information compared to several other methods 'gave a five times higher percentage of active leads', even though in some cases the same active compounds were isolated from botanically unrelated active plants. In a preliminary test using plants submitted to the US National Cancer Institute screens for activity against human immunodeficiency virus (HIV), Balick (1990) found that random plant collections yielded 6% activity, whereas those based on 'powerful plants' used by a herbal healer yielded 25% activity. However, after dereplication to remove certain components of the crude plant extract such as tannins and polysaccharides, the percentage of plants showing anti-HIV activity falls dramatically and the overall percentage is virtually identical using either random or ethno-directed collection methodology (Balick 1994, this volume). Gordon Cragg and his colleagues at the US National Cancer Institute have also shown with a much larger data set that general ethnobotanical collection does not appear to be advantageous in developing leads for HIV treatment (Cragg et al 1994, this volume).

A few examples of individually targeted plants used medicinally by the Jívaro for which the specific utility has been verified scientifically are worth examining. The use of the holly, *Ilex quayusa* Loes., as an early morning stimulant is an excellent example of empirical selection within a rich and highly varied vegetation of the only species we know to have high concentrations of caffeine, plus small amounts of theobromine and theophylline (Table 1). This trilogy of xanthine alkaloids provides the stimulation desired in the morning beverage, along with a suppressed appetite which is also important when the only full daily meal is still 12 hours away. Emesis is associated with the drinking of this leaf decoction, apparently a learned phenomenon not due to the presence of emetic compounds (Lewis et al 1991).

TABLE 1   Concentrations of methylxanthine alkaloids in dried leaf decoctions of *Ilex guayusa* from Ecuador and Peru using Jívaro extraction procedures and high pressure liquid chromotography

Collection number	Caffeine %	Theobromine %	Theophylline %
11687	1.73 ± 0.04	0.02 ± 0.004	Trace[a]
14084	1.92 ± 0.03	Trace[a]	Trace
12264	2.07 ± 0.03	0.02 ± 0.003	Trace
11492	3.11 ± 0.05	0.12 ± 0	Trace
13967	3.33 ± 0.01	0.02 + 0.003	Trace
12263	3.48 + 0.06	0.04 ± 0.003	0
12265	3.95 ± 0.06	0.02 ± 0.003	Trace
13132	7.57 ± 0.12	Trace	Trace

Modified from Lewis et al (1991).
[a]Trace = < 0.005%.

Equally impressive is the selection by Jívaro women of *Cyperus articulatus* L. or *C. prolixus* H.B.K. tops to make crude infusions drunk as oxytocic agents to aid in parturition or postpartum contractions and to reduce bleeding after childbirth. Sedge tops are inert as oxytocic agents, but on close examination we found parasitic fungal sclerotia in the apical sedge infructescences used (Lewis & Elvin-Lewis 1990). They proved to be *Balansia cyperi* Edgerton, a species classified in the Clavicipitaceae. This family contains only one other well-known genus, *Claviceps* or ergot of temperate regions. The women, therefore, select the only close relative of ergot found in their tropical ecosystem from perhaps thousands of fungal species found in the Amazon basin, and use it for the same purposes as did midwives in European folk medicine centuries ago. As we noted (Lewis & Elvin-Lewis 1990): 'Circumstantial evidence strongly suggests the presence of biodynamic principles similar to the ergot alkaloids in *B. cyperi*. Should this prove true, then the selection by Jívaro women of a plant parasite to aid in obstetrics is another sophisticated example of human ingenuity using empirical methodology'. Within a year of this publication, Plowman et al (1990) proved the existence of alkaloids in *B. cyperi* sclerotia similar to those of ergot peptide alkaloids.

The final example comprises two plants used in Peru to treat arthritis. Following local procedures, material of one plant was prepared by soaking about a handful of freshly scraped inner root bark in 4l of locally prepared sugar cane rum for four weeks. Woody root portions were then discarded and the tincture was ready for use. Tincture extracts are being examined at Monsanto/Searle Co. in a series of screens for antiinflammatory and antirheumatoid-arthritic activities, which combine enzymic and receptor-based assays with appropriate secondary and tertiary screens. Compared with over 2000 plant extracts previously tested over three years, mostly obtained from random collections and only very rarely from primary ethnobotanical sources, the two Peruvian plants contained the highest activities by far in a majority of the screens. Could this be a major breakthrough in an otherwise static antiarthritic programme? Could this research lead to the discovery of a new class of antiinflammatory agents, thereby offering relief to millions of arthritic sufferers who cannot utilize the non-steroidal therapeutics now available?

Future strategies for consideration

Ubiquitous compounds. The adage 'throwing the baby out with the bathwater' could be readily applied today to the pharmaceutical industry and other drug discovery programmes when considering bioactive ubiquitous compounds. Decades of research have focused on the discovery of unique structures and how their utility can be improved by molecular modification. This policy has indeed resulted in the development of new compounds: the development of taxol as an anticancer agent is a current example. This approach continues to be attractive, primarily because such molecules, when patented, are readily distinguishable from the milieu of phytochemicals that might be more vulnerable to patent circumvention.

Are we in an era when searching for uniqueness has taken precedence over considerations of availability, suitability and low toxicity? At this juncture, is it not reasonable to ask ourselves if our goals are truly altruistic or more self-serving to preserve the *status quo*? Ask the aspirin industry what profits there are in ubiquity. It is important to remember that it was the general utilization of salicin-containing temperate plants, like willows and poplars, by American and Eurasian peoples to treat fevers, inflammation and pain that led to the synthesis 140 years ago of a number of salicin derivates, including acetylsalicylic acid. Yet it has taken over a century for its status as the preferred analgesic and antiinflammatory agent to be challenged by acetaminophen and ibuprofen compounds.

We suggest that another ubiquitous compound might prove to be the 'aspirin' of the 21st century. For example, diverse flavones are found widely among vascular plants. These low molecular weight compounds possess a broad spectrum of pharmacological activity, but their actions have yet to be well delineated. Those flavonoids with antiviral activity are known to possess both antiinfective and antireplicative properties, through being able either to bind to virus capsid proteins or to interfere with viral RNA and protein synthesis, respectively. A few flavones have also been found to inhibit an HIV-1 protease and reverse transcriptase. In addition, the toxicity of flavonoids is typically low in humans; it is therefore possible that pharmacologically significant concentrations could be achieved in tissues and influence the outcome of viral infections.

Broad versus targeted screens: the toxicity factor.   With the introduction of *in vitro* assays using robotics, the screening of a wide range of plants for bioactivity has accelerated. When an activity is identified, it is immediately categorized as to potency, with the 'most active' frequently being selected for isolation and characterization. Usually, such studies lack prior ethnomedicinal data, so that when drug development reaches the stage of clinical or toxicity trials, compounds often fall short of expected therapeutic value because the margin between efficacy and toxicity is too narrow. The literature is replete with such examples. A better understanding of the plants themselves and how they are used ethnomedicinally would serve as an appropriate guide. Unless the species is rare, sufficient information probably exists among related taxa to anticipate the types of pharmacological activities and categories and compounds it might contain. This does not mean that unique and non-toxic compounds might exist in a highly toxic plant, or that one compound, like prostratin, might prove not to be tumorigenic when other phorbol esters are known to be so (Gustafson et al 1992). Such arguments continue to justify those who promote broad screening activities;

however, it is reasonable to ask whether or not the cost, time and effort associated with this approach are warranted at the exclusion of strategies that, through ethnomedicinally focused research, might bring more potentially therapeutic compounds to the fore.

In other cases, the biological activity of a substance may not be compatible with the delivery system and predictable adverse reactions can result. Such is the research regarding certain ubiquitous plant proteins, such as the single-chain ribosomal-inactivation proteins. One compound, trichosanthin, inhibits HIV replication in animal models of both acute and chronic infection. However, in limited clinical trials the risk of developing allergic reactions from injection of this protein was noted and, although it was claimed that symptoms of pruritis and a flu-like syndrome could be reversed through the use of non-steroidal antiinflammatory drugs and/or antihistamines, its value in long-term therapy may be curtailed as a consequence of these reactions.

Potential interaction of more than one compound.   Medicinal plants frequently contain more than one bioreactive component which can act additively or synergistically to effect the full healing potential of the remedy. Therefore, a plant valued as an antiviral agent may yield a weak or moderately antiviral substance whose activity is enhanced by the presence of immune modulators and possibly other healing or physiologically active substances. The same might be true for certain mixtures of medicinal plants. These features, valued in holistic therapies, are a challenge to drug discovery strategies involving the isolation of antiviral agents from plants identified from ethnomedicine. This is particularly relevant when significant clinical antiviral activity does not correlate well with the potency of the antiviral compounds present in the remedy. Because an important goal of antiviral drug discovery is to identify treatments low in toxicity, bioactive compounds derived from these efficacious plants should not be discarded out-of-hand without first comparing their toxicities and how they interact.

Enhancing drug discovery from ethnomedicinal sources.   One of the major difficulties in utilizing ethnomedicinal data from a wide variety of sources is the descriptors applied to disease syndromes. Part of the problem is the training of the interviewer and his/her understanding of both Western and traditional medical systems. This is more complicated when many syndromes are known to have multiple aetiologies, which, if used in screening, could easily identify a variety of very different compounds. An example of a broad antiinfective category is that of diarrhoea/dysentery, because it is impossible to know in generic terms whether or not the cause is of bacterial, viral, protozoan or parasitic origin, or if the treatment is truly antiinfective or merely palliative. Both the Samoan *Homalanthus acuminatus* (Muell.Arg.) Pax and the Malaysian *Ancistrocladus tectorius* (Lour.) Merr. were valued as antidiarrhoeal compounds and each has yielded an anti-HIV compound—prostratin and michellamine, respectively. However, it is not known whether these are active against enteroviruses, enterobacteria or enteric parasites. Without this information, the value of screening antidiarrhoeal agents for antiviral activity remains to be determined. Isolating antiviral compounds from remedies valued for respiratory infections is similarly complex, as many plants are valued primarily as antitussive agents, expectorants or rubefacients, and are unlikely to possess systemic antiviral activity. It often requires a description of the symptoms of a disease to differentiate it from those similar generically, for example measles from chicken-pox. Accurate data are important, if plant extracts used to treat such diseases are to be appropriately directed to specific antiviral screens, particularly since measles is an RNA virus and the varicella virus of chicken-pox is a DNA virus.

Traditional medical practitioners often do not recognize heart disease and circulatory problems as such. For instance, the 'frightening syndrome' among the Jívaro, characterized by shortness of breath, may represent a symptom of heart disease or one of exaggerated anxiety. Without the presence of a trained physician as part of the evaluating team, valuable information may be misinterpreted or recorded inappropriately.

The margin of efficacy over toxicity is a major problem associated with finding appropriate anticancer remedies from plants and other sources. Among a number of candidate antitumour agents isolated from plants used in ethnomedicine, the North American Indian remedy employing *Podophyllum peltatum* L. (May apple) is significant. Derivatives, like etoposide, less toxic than the native podophyllotoxin used to treat venereal warts, are valuable in treating testicular and small-cell lung cancers. Less specifically, studies at the US National Cancer Institute have shown that most potent anticancer compounds are frequently isolated from plants which are active against animals, such as those used as arrow and fish poisons or as vermifuges; as expected, these are also not well tolerated among large animals, including humans. If other types of cancer cures are to be derived from plant sources, then perhaps closer observations regarding remedy uses by traditional practitioners are in order. As differences between the malignant and normal cell are becoming better understood, it is likely that modes of treatment, including dosages which are lower than now prescribed, might be evolved should synergistic strategies with less toxic compounds be found. This therapeutic consideration is important when immunosuppression is part of the disease and where the patient would benefit from immunostimulatory activities as the cancer load decreases.

References

Adu-Tutu M, Asanti-Appiah K, Lieberman D, Hall JB, Elvin-Lewis M 1979 Chewing stick usage in southern Ghana. Econ Bot 33:320–328

Balick MJ 1990 Ethnobotany and the identification of therapeutic agents from the rainforest. In: Bioactive compounds from plants. Wiley, Chichester (Ciba Found Symp 184) p 22–39

Balick MJ 1994 Ethnobotany, drug development and biodiversity conservation—exploring the linkages. In: Ethnobotany and the search for new drugs. Wiley, Chichester (Ciba Found Symp 185) p 4–24

Berlin B 1990 The chicken and the egg-head revisited: further evidence for the intellectualist bases of ethnobiological classification. In: Posey DA, Overal WL (eds) Ethnobiology: implications and applications. Museum Paraense Emílio Goeldi, Belém, p 19–33

Cox PA 1990 Ethnopharmacology and the search for new drugs. In: Bioactive compounds from plants. Wiley, Chichester (Ciba Found Symp 154) p 40–55

Cragg GM, Boyd MR, Cardellina II JH, Newman DJ, Snader KM, McCloud TG 1994 Ethnobotany and drug discovery: the experience of the US National Cancer Institute. In: Ethnobotany and the search for new drugs. Wiley, Chichester (Ciba Found Symp 185) p 178–196

Elvin-Lewis M 1983 The antibiotic and anticariogenic potential of chewing sticks. In: Romanucci-Ross L, Moerman D, Tancredi LR (eds) The anthropology of medicine. Praeger, New York, p 201–220

Elvin-Lewis M, Hall JB, Adu-Tutu M, Afful Y, Asante-Appiah K, Lieberman D 1980 The dental health of chewing stick users in southern Ghana: preliminary findings. J Prev Dent 6:151–159

Farnsworth NR 1990 The role of ethnopharmacology in drug development. In: Bioactive compounds from plants. Wiley, Chichester (Ciba Found Symp 154) p 2–21

Farnsworth NR, Akerele O, Bingel AS, Soejarto DD, Guo ZG 1985 Medicinal plants in therapy. Bull WHO 63:965–981

Gustafson KR, Cardellina JH II, Manfredi KP, Beutler JA, McMahon JB, Boyd MR 1992 AIDS-antiviral natural products research. In: Chu CK, Cutler HG (eds) Natural products as antiviral agents. Plenum, New York, p 57–67

Lewis WH 1992 Plants used medicinally by indigenous peoples. In: Nigg HN, Seigler D (eds) Phytochemical resources for medicine and agriculture. Plenum, New York, p 33–74

Lewis WH, Elvin-Lewis MP 1990 Obstetrical use of the parasitic fungus *Balansia cyperi* by Amazonian Jívaro women. Econ Bot 44:131–133

Lewis WH, Elvin-Lewis MP 1994 Medicinal plants as sources of new therapeutics. Ann MO Bot Gard, in press

Lewis WH, Elvin-Lewis M, Gnerre MC, Fast DW 1988 Role of systematics when studying medical ethnobotany of the tropical Peruvian Jívaro. In: Hedberg I (ed) Systematic botany—a key science for tropical research and documentation. Akademial Kaido, Publishing House of the Hungarian Academy of Sciences, Budapest, p 181–196 (Symb Bot Ups vol 28)

Lewis WH, Kennelly EJ, Bass GN, Wedner HJ, Elvin-Lewis MP, Fast DW 1991 Ritualistic use of the holly *Ilex quavusa* by Amazonian Jívaro Indians. J Ethnopharmacol 33:25–30

Malone MH 1983 The pharmacological evaluation of natural products—general and specific approaches to screening ethnopharmaceuticals. J Ethnopharmacol 8:127–147

Moerman DE 1991 The medicinal flora of native North America: an analysis. J Ethnopharmacol 32:1–42

Phillips O, Gentry AH 1993a The useful plants of Tambopata, Peru. I. Statistical hypothesis tests with a new quantitative technique. Econ Bot 47:15–32

Phillips O, Gentry AH 1993b The useful plants of Tambopata, Peru. II. Additional hypothesis testing in quantitative ethnobotany. Econ Bot 47:33–43

Plotkin MJ 1988 Conservation, ethnobotany, and the search for new jungle medicines: pharmacognosy comes of age . . . again. Pharmacotherapy 8:257–262

Plowman TC, Leuchtmann A, Blaney C, Clay K 1990 (issued 1991) Significance of the fungus *Balansia cyperi* infecting medicinal species of *Cyperus* (Cyperaceae) from Amazonia. Econ Bot 44:452–462

Prance GT 1991 What is ethnobotany today? J Ethnopharmacol 32:209–216

Prance GT, Balée W, Boom BM, Carneiro RL 1987 Quantitative ethnobotany and the case for conservation in Amazonia. Conserv Biol 1:296–310

Van den Berghe DA, Vlietinck AJ, Van Hoof L 1985 Present status and prospects of plant products as antiviral agents. In: Vlietinck AJ, Dommisse RA (eds) Advances in medicinal plant research. Wissenschaftliche Verlagsgesellschaft, Stuttgart, p 47–99

DISCUSSION

Prance: I would like to remind people of the danger of referring too much to the vernacular names because they often represent groups of species, not one individual species. For example, in Amazonia *macucu* refers to members of both Chrysobalanaceae and Huminiaceae. But you are putting that fact to a profitable use, so that it can be used in your search for new cures.

Lewis: You have to put the vernacular name together with the taxonomy. If, for example, Bignoniaceae are referred to by a common name, this is a real clue that shared activity exists. Similar names based on similar utilizations are very common among the Jívaro. However, shared names cannot be the sole criterion for finding closely allied species or chemicals. *Ilex guayusa*, which contains high concentrations of caffeine, is called *wayus*, but the same name is also used for a species of Piperaceae. The leaf shapes are similar but little else is; the uses and chemistry also differ—there is no caffeine in the Piperaceae.

Jain: Local names for plants are extremely important in ethnobotany, because they are the only means of communication with the local people. They are quite often not only indicative of the structure or the properties of the plant, they also give information about the movements of plants and movement of users of plants from one region to another.

The only danger is in using the local name to determine the botanical name. When the same local name is used for different plants, it can be difficult to decipher which plant is meant. Some field workers have heard mention of a familiar local name and taken it for granted that it is the same plant as they learned before. This is where the danger lies. The other way around, where there are plants for which there are several common local names, there is no problem.

There is another danger in relying on local names for plants and that concerns the accuracy of these names. We all know that multidisciplinary teams are ideal, but they are not always possible to organize or support. Therefore botanists, particularly field botanists and taxonomists, took it on themselves to record ethnobotanical data during their normal taxonomic or botanical work. They are not medical men; they are not linguists. Then in the ethnobotanical lists we have funny local names for plants. When an informer doesn't know the name of a plant, he says in his own language, 'I don't know'. This becomes the 'local' name for the plant.

These brief indications of uses of plants given by taxonomists and field botanists on ethnobotany lead to ethnobotanical information. No ethnobotanist or taxonomist has ever been acknowledged when this ethnobotanical information has led to the discovery of a drug. The grievance of laboratory scientists is that the field botanists do not supply enough information for exploitation. But the point of a field botanist is that his information is used and not acknowledged.

Dagne: Usually, people have names for plants if they have some use for them. Sometimes, vernacular names do go to a species level; there are many instances where peasants have very precise names for various members of a genus, identifying the species. But this is the case only if they have uses for the plants.

Balée: What is the extent to which these monomial names are always, to use Brent Berlin's terminology, used for 'underdifferentiating' biological species? Is there sometimes more than one word for a single biological species—not counting domesticated species?

Lewis: For domesticated species there are often trinomial and quadranomial names. I would say that about 15–20% of undomesticated, but used, species are binomial, most of the rest are 80–85% monomials. Brent Berlin found that among the Aguaruna Jívaro monomials were also in the majority.

Martin: I like your taxonomy of primary, quantitative and experimental phases of ethnobotany because it emphasizes important aspects of the different phases of research. Would you consider adding a fourth phase that would complete the cycle, namely applied advocacy ethnobotany? This phase implies going beyond just obtaining data and returning the results to communities for the benefit of local development, conservation and other practical ends.

Lewis: No doubt; this should occur at every phase. We, for instance, will only do our ethnomedicinal and ecological studies in relation to and in collaboration with the Huambisa Aguaruna Jívaro, whose knowledge is indispensable. How we compensate them and how we return information about our research progress at each stage, I am not certain. These aspects must be considered by us all. If something comes from our work, there must be compensation in some way. It is difficult with people who have no structure we recognize into which we could incorporate funds, but there are various ways in which one can compensate. One is to help to treat tuberculosis, which our civilization gave the Jívaro. They do not appear to have found a plant efficacious against this new disease. They may yet find one, but in the meanwhile too many young people are dying from tuberculosis; we could help if funds were available to treat them with antibiotics.

We are entering an important phase, where we should be thinking about compensation, not 10–20 years down the pipeline when there may be major rewards from the development of new compounds, but right from the start of projects. Initially, as we seek the help of indigenous and other peoples, we have to provide some assistance at the community level, whatever that might be.

Then, should we need kilograms of refined compounds, requiring plant material, there might be considerable compensation. This will be possible in the programme we have suggested, because we shall know when exploitation reaches a point at which it could be dangerous. If indigenous and *mestizo* peoples are involved in the initial phase of quantitative ethnobotany, including the ecology, then they will appreciate that the material is not infinite. Among the indigenous Jívaro, there already is a perception of when plant material can be exhausted.

Farnsworth: Part of your rationale is that if you need large amounts of a plant material, you will put the natives to work cultivating it. The canopy of rain forest is so dense that you would have to chop down other plants in order to cultivate the one that you want. Where will this cultivation take place?

Lewis: There are holes in primary forest—wind blows, trees that fall and so forth. But we want to leave the primary area as pristine as possible. The manipulation that will be necessary to develop a nursery of a primary forest-loving plant will take place in the buffer zone (Fig. 1). The nursery will have to be in a shade-dominated area or the seedlings will not be successful. There are gaps which can be used, but we shall generally use those in the buffer zone, which (by and large) will be primary forest.

Barton: I don't notice any discussion about the epidemiology or the efficacy of materials used. Also, do you have any sense of whether the typical screening process used by pharmaceutical firms is likely in some significant way to miss compounds that may be important? Is there any systematic bias?

Elvin-Lewis: We have found whether you do this for chewing sticks, or for the medicinal plants of the Jívaro, the most important thing is determine the preference of one plant remedy over another. By ordering according to choice, we have found that the most favoured species are usually the most efficacious, at least in the context of the people who use them. So you should screen the most favoured plants first. We have applied this hypothesis of 'the most favoured is the most efficacious' to a number of screens for antiinflammatory activity, at the US National Institute of Allergy and Infectious Diseases, with industry and elsewhere. We've been fortunate in finding that bioactivity is almost always confirmed. If you do broad screening of medicinal plants, the results are not statistically significant—there is no logical reason why they should be. By selecting the most favoured plants for a specific disease and linking this to a complementary screen, you are likely to get significant 'hits'.

You can also get verificiation of a plant's uniqueness or generality worldwide by searching as many bibliographical sources and databases as possible, to see if the plant is unique to a group of people, is general to a region or has cosmopolitan use.

Jain: We are trying to look at this preferential ordering of plants in India. If several plants are used to treat the same disease, we are now trying to find out which plants the indigenous people prefer. And if the same plant is used for several diseases, for which disease is it considered to be best.

Schwartsmann: Walter Lewis described the way in which the *mestizos* in Peru prepare tinctures for use in the treatment of arthritis. Gordon, when you receive a plant that, for example, needs a special method of preparation, how is this information treated when you prepare extracts?

Cragg: We value that sort of information on how the local people use the plant and the method of preparation. We will follow that method of preparation exactly with part of a plant sample, then we will use our standard extraction protocol on the remainder to see whether there's any difference.

Walter, you mentioned that the plant source is the bark of the root, is that a potential supply problem? It sounds ominous to me.

Lewis: Yes, the need for roots, and the concomitant loss of woody plants in these instances, suggests that we might have to go into a quick synthetic programme. The root bark is the most difficult part of a plant to obtain, yet it is widely used in traditional medicine. One extraction involved 4l of *aguariente* (sugar cane rum) and a handful of inner bark per litre. The whole conservation problem associated with the demands that such a discovery could engender is horrendous. This is another reason we are convinced that we must develop an ecological medicinal plant study in consort with drug development. It is also an example where an immediate nursery programme ought to be initiated. The plant grows very rapidly; in about 5–10 years (I am partly guessing for we know little about its husbandry), we could have mature fruiting and flowering material.

Cragg: Is the active component confined to the root bark?

Lewis: I do not know. All I know is what the *mestizo* population uses.

King: Your combination of medical specialists and botanists greatly facilitates understanding of what's going on. Having an epidemiological profile in advance of field work gives the physician or the medical component of a team an idea of whether the diseases are self-limiting, whether they resolve without treatment by themselves, or whether there is a placebo effect. Even a trained physician benefits by having an idea of what diseases generally occur and the natural history of those diseases. This greatly enhances the estimates of efficacy. You really can't give a good judgement on efficacy without the laboratory equipment at times, but you can draw much better conclusions about use and disease with just the epidemiological profile. I strongly advocate that as a tool; we do this for the combination of physicians and ethnobotanists who work in the field for our company.

Elvin-Lewis: In order to do this accurately, you need to know in very precise terms the normal course of self-healing for the type of disease under study. The more objective rather than subjective criteria you use to determine clinical efficacy, the better. Ideally, these types of studies are best conducted in collaboration with physicians, botanists and epidemiologists. Logistically, this is not always feasible, unless the disease is well defined and commonly present.

Lewis: We realize that even though together we have a degree of expertise, we still do not understand certain phenomena. For instance, traditional medical practitioners only infrequently recognize heart diseases and circulatory problems. Or how do we comprehend what is call the 'frightening syndrome' among the Jívaro? It is usually associated with shortness of breath, but does it represent a heart disease or is it anxiety? These are very different interpretations. The addition of a medical expert who could examine such examples first-hand could be very important.

Another difficulty is with cancer, but there may be ways to recognize certain cancers. For instance, the traditional practitioner has very little chance of identifying leukaemia. But the immune suppression associated with leukaemia might lead to candidiasis or thrush, which would be recognized by the traditional practitioner. So rather than just seeing thrush, we should also be thinking of leukaemia or perhaps of a lymphoma. In consort with a physician who is medically trained in tropical diseases, we may find indirect ways of identifying diseases otherwise difficult to assess.

Ethnopharmacological search for antiviral compounds: treatment of gastrointestinal disorders by Kayapó medical specialists

Elaine Elisabetsky and *Darrell A. Posey

*Laboratório de Etnofarmacologia, Departmento de Farmacologia, Universidade Federal do Rio Grande do Sul, Porto Alegre (RS), *Program of Ethnobiology, Museu Paranese Emílio Goeldi (CNPq), Belém (PA), Brazil and *Visiting Scholar, Institute for Social & Cultural Anthropology, St. Antony's College, University of Oxford, UK*

Abstract. The *Mēbengokrê* (Kayapó) of Brazil have a highly developed medical and pharmacological tradition based on diverse specialization in knowledge and practice. Shamans (*wayangas*) are prepared to treat all kinds of diseases including those related to spirits; curers (*mākuté pidjà mari*) can deal only with diseases not related to spirits. Both utilize plant- and animal-based remedies, among other practices. Elaborate disease categories (*kanê*) include those known as *hak-kanê* (bird disease) and *tep-kanê* (fish disease). The complexity of the two categories defies easy description but both use gastrointestinal disorders as basic indicative symptoms. Given that an important percentage of gastrointestinal disorders are caused by viruses and that new antiviral drugs are sorely needed, plants used by the Kayapó for *hak* and *tep* diseases are presented and discussed as potential leads in the search for antiviral compounds.

1994 Ethnobotany and the search for new drugs. Wiley, Chichester (Ciba Foundation Symposium 185) p 77–94

Known antiviral compounds still present significant drawbacks, such as a narrow spectrum of activity, limited therapeutic usefulness and variable degrees of toxicity (Van den Berghe et al 1978). On the other hand, the prevalence of virally related diseases is of growing concern; therefore, the development of new and better antiviral compounds is desirable. It has been shown that viruses respond differently to plant extracts and suggested that natural products are preferable to synthetic compounds as sources of new antiviral agents (Van den Berghe et al 1978, Vlietinck & Van den Berghe 1991).

The study of flora in general, and medicinal plants in particular, has been considered a fruitful approach in the search for new drugs (Svendsen & Scheffer

1982, Samuelson 1989, Farnsworth 1990). Plant collections for drug discovery can follow different approaches, including the collection of plants at random, collections guided by chemotaxonomy and collections based on ethnopharmacological data. Because medical systems as products of particular cultures are enormously varied in terms of health practices and beliefs, detailed ethnography is needed to select plants that may be sources of cross-culturally effective drugs. It is through the correlation of traditional therapeutic practices with Western biomedical concepts that species can be selected and scrutinized for particular pharmacological activities (Elisabetsky & Setzer 1985). Selection of species that are claimed by humans to have a given clinical activity may constitute a valuable short-cut for drug discovery. Ethnopharmacologically based strategies have been applied to several therapeutic areas, such as cancer (Duke 1986, Cordell et al 1991), immunomodulators (Labadie et al 1989), allergy (Elisabetsky & Gely 1987, Wagner 1989), contraceptives (Xiao & Wang 1991), analgesics (Elisabetsky & Castilhos 1990), antimalarial agents (Phillipson & Wright 1991, Brandẽo et al 1992), antidiarrhoeal/antimicrobial compounds (Caceres et al 1990, Heinrich et al 1992a,b) and antiviral agents (Vlietinck & Van den Berghe 1991).

Gastrointestinal disorders are frequently associated with viral diseases: rotavirus is responsible for at least 50% of infections that lead to acute diarrhoea (Krej 1988). We offer in this paper an analysis of Kayapó treatment of gastrointestinal disorders, which might be useful for selection of plant species as potential sources of antiviral compounds.

Intellectual, cultural and scientific property

The authors of this paper embrace the principles of the Covenant on Intellectual, Cultural and Scientific Property developed by the Global Coalition for Biological and Cultural Diversity (see p 227). The data were obtained with full consent of the Kayapó people. The paper is published in the spirit of joint partnership with the Kayapó to advance knowledge for the benefit of all humanity. Any information used from it for commercial or other ends should be properly cited and acknowledged: any commercial benefits that should accrue directly or indirectly should be shared with the Kayapó people.

Methodology

Research for this paper was carried out in Gorotire, the largest (over 600 persons at the time of this study) of the 15 villages that compose the Kayapó (*Mẽbengokrê*) nation, one of the major Jê-speaking groups of Brazil. The village is located at the south-eastern edge of the Amazon Basin on the banks of the Rio Fresco (Pará State). D. A. Posey has conducted ethnobiological research in Gorotire since 1977. This paper is based on a collaborative effort with E. Elisabetsky, who made two trips to the village (one month each) in November,

1983, and April, 1984. Plant collections during these trips were made by Dr Anthony Anderson; these are currently deposited in the Goeldi Museum Herbarium. Additional visits to the village by E. Elisabetsky in 1985 and 1986 provided complementary data and concepts.

Our principal informants were Kwyra-kà, Beptopoop (two shamans or *wayanga*) and José Uté and Tereza (noted medicinal plant knowers or *mẽkute pidjà mari*). Basic ethnobotanical information was collected by E. Elisabetsky and D. A. Posey in the field, with subsequent ethnopharmacological interviews in the project house. Additional sessions were held to discuss the more general concepts of Kayapó diseases, their symptoms and cures. Both concepts and plants were discussed with these four key informants, as well as with members of the village in general.

Between 1987 and 1989 contacts with Kayapó Indians, both in Belém and at the Funai Hospital in Icoaraci, were frequent. Time spent with Paiakan and his family, at the time living in Belém, was used for clarifying concepts on the Kayapó life in general and their medical system and practices in particular.

Kayapó medical system

It is beyond the purpose of this paper to describe fully the intricate medical system and practices of the Kayapó (Bamburger 1967, Elisabetsky & Posey 1988, 1991). Points relevant to understanding medical uses of species dealt with in this paper are discussed.

Disease epidemics often preceded by months or years what was considered 'first contact' with indigenous groups. Trade routes and extensive travelling brought remote indigenous peoples into contact with groups already infected with *kuben kanê* (non-Indian diseases) (Posey 1987). Traders or raiding parties then spread the diseases and/or brought them back to their villages. 'Permanent contact' with the Kayapó was not established until 1936. Missionaries and FUNAI (the Brazilian Indian Agency) routinely used Western medical treatment to secure contact with Indian groups. According to Horace Banner[1], the inhabitants of Gorotire had little choice but to establish peaceful relations with the White Man, because they were weak from *kuben kanê* and had been reduced to only 250 people; within a year that number had fallen to only 85. Such a demographic collapse obviously had profound effects on all aspects of Kayapó society, especially their health and medical beliefs. Medicines brought into the region by the Unevangelized Fields Missions were attributed with saving entire Kayapó villages (such as Kuben-krã-kein and Kokraimoro) in the 1950s. Such

---

[1]Horace Banner was one of the first Protestant missionaries to live with the Kayapó. He left a number of journals and manuscripts, which D. A. Posey was able to study in 1981, with the kind permission of his widow, Eva Banner, under a grant from the American Philosophical Society.

'miracles' convinced the Kayapó that *kuben* medicine was powerful, although they never abandoned their traditional cures.

The Kayapó are unanimous in insisting that prior to the arrival of *kuben kanê* Indians died only from old age or accidents, not from diseases. With the arrival of the White Man, however, deadly new diseases appeared along with a general weakening (*mẽtykdjà*) of the Indians that permitted their own diseases to become more powerful and deadly. Today, there is generally a clear distinction between what is a non-Indian disease (*kuben kanê*) treated with non-Kayapó medicine and what is an Indian disease (*mẽbengokrê kanê*) treated with traditional medicine.

Sociological considerations

Medical specialization must be viewed within the sociological (emic) context of local concepts of life, death, illness and curing (Fabrega 1975). Curers amongst the Kayapó can be classified into two basic categories: *mẽkuté pidjà mari* (medicinal plant knowers) and *wayanga* (shaman). The former deal only with non-spiritual diseases, while the latter also manipulate spirits in their curing methods; both cure with a variety of plants and plant concoctions. Our survey of the village in 1984 showed that 5% of the population were considered as *wayanga*, with 26% of the population being considered as practitioners in one or more disease specialties.

To understand the nature of the power of the *wayanga* to cure, one must understand how a person can become a shaman. This transformation was explained to D. A. Posey by Beptopoop in Gorotire in 1978 as follows:

Listen! Those who become sick from strong fevers lie in death's position; they lie as though they are dead. The truly great ones, the truly strong person who is a *wayanga*, shows the sick how to leave their bodies. They leave through their insides. They pass through their insides and come to be in the form of a stone. Their bodies lie as in death, but beyond they are then transformed into an armadillo. As an armadillo, they assume good, strong health and they pass through the other side, over there (pointing to the east). Then they become a bat and fly—ko,ko,ko,ko,ko... (the noise of flying).

Then they go further beyond in the form of a dove. They fly like a dove—ku,ku,ku,ku . . . (the sound of a dove's flight). They join the other *wayangas* and all go together. 'Where will we go? What is the way? Go to the east, way over there.' ku,ku,ku,ku . . .

And way over there is a spider's web . . . Some go round and round near the spider's web and they just sit permanently. The true and ancient shamans must teach them how to fly through the web. But those who have not been shown how, try to break through the web and the web grabs their wings thusly (the narrator wraps his arms around his shoulders). They just hang in the web and die. Their bodies are carried by their relatives and are buried without waiting, for the spider's web has entangled them, wrapped up their wings, and they are dead. Those who have been caused to know themselves, however, go round the spider's web. They sit on the mountain seat of the shamans and sing like the dove—tu,tu,tu,tu . . . They acquire the knowledge of the ancestors. They speak to the spirits of all the animals and of the ancestors. They know (all).

They then return (to their bodies). They return to their homes. They enter and they breathe. And the others say: He arrived! He arrived! He arrived! And the women all wail: '*ayayikakraykyerekune*'. (And the shaman says) 'Do not bury me, I am still alive. I am a *wayanga*. I am now one who can cure: I am the one who smokes the powerful pipe. I know how to go through my body and under my head. I am a *wayanga*.'

There is far too much in this story to analyse entirely (see Posey 1982a), but a few major points are relevant to this discussion. First, the *wayanga* is capable of leaving his/her body (*kà*) and being transformed into other physical forms. The energy (*karon*) can be stored temporarily in rocks, but inevitably gets transformed into armadillos, doves or bats. The spider's web represents the barrier between the visible and invisible worlds. Armadillos are persistent animals that know to burrow under the web; doves are powerful fliers and can break through the barrier; while bats are such skilful fliers that they can manoeuvre through the strands.

The most powerful shamans can transform themselves into not just one of the animals, but all of them. Once on the other side of the spider's web, after they have passed through the endless dark chasm, they enter into the spectral frequency for each animal (*mry-karon*). Some shamans learn the secrets of only one or a few animals and their energies, while others 'know all' (in the words of the myth). They have learned about all of the spectral frequencies and their respective animal energies.

Upon return to their bodies, the *wayanga* begin to 'work with' (*nhipex*) the animal energies encountered in their transformation. There are literally dozens of different specialists (Elisabetsky & Posey 1985, 1991). The basis of the 'work' is to maintain a balance between animal energies and human energies (Posey 1982b). Eating the meat of, coming in contact with, or even dreaming about animals can cause an imbalance in these energies, as can a well-elaborated list of antisocial actions. *Wayanga* use a great variety of techniques for restoring balance, but plants are the most common 'mediators' (Elisabetsky & Posey 1991). Plants have qualities that can either harm or help the balance between human and animal energies—indeed, some Indians say that all plants have curative values.

In any case, the Kayapó respect both plants and animals, because their energies are keys to the health of the Kayapó society. Permission is asked when taking the life of an animal and songs of appreciation are offered to the spirits of the dead animals. Likewise, annual rituals extol the importance of plants and instill a great sense of respect for their overall role in the socioecological balance (Posey 1982b). The Kayapó have no question about their existence and future health being dependent upon plants and animals and the forces of nature.

Kayapó diseases that include gastrointestinal disorders

Hak kanê (bird disease) and *tep kanê* (fish disease) are two major classes of diseases that include gastrointestinal symptoms as important markers. *Hak kanê* is associated with dizziness and diarrhoea. *Tep kanê* is associated with diarrhoea, yellow body and generalized pain. Gastrointestinal disorders, with or without diarrhoea, can have both spirit and non-spirit causes. Spirit-related diseases are difficult to evaluate in ethnopharmacological research, because cultural factors are too complex to be easily interpreted. All Kayapó diseases, however, include some treatment practices that lend themselves to field and laboratory evaluation.

TABLE 1   Plant species used to treat *tep kanê*

Collection no.	Species	Family	Part of plant	Route
897	*Ruellia* sp.	Acanthaceae	Whole plant	External, oral
564	*Annona coriacea* Mart.	Annonaceae	Fruit, flower, bark	Oral
618	*Annona crassifolia* Mart.	Annonaceae	Seeds	External
524	*Mandevilla tenuifolia* (Mikan) Wood.	Apocynaceae	Whole plant	External
724	*Aristolochia* sp.	Aristolochiaceae	Tuber	Oral
767	*Helosis cayenensis* Spreng.	Balanophoraceae	Whole plant	Oral
766	*Cleome guianensis* Aubl.	Capparidaceae	Leaf	External, oral
796	*Veronia herbacea* Rusby	Compositae	Tuber	External, oral
1003	*Croton aff. agraphilius* M. Arg.	Euphorbiaceae	Leaf	Oral
727	*Coutoubea ramosa* Aubl.	Gentianaceae	Fruit, stem	External, oral
655	*Arachis* sp.	Leguminosae	Whole plant	Oral
649	*Sauvagesia erecta* L.	Ochnaceae	Whole plant	Oral
788	*Passiflora alata* Dryand	Passifloraceae	Whole plant	Drops in eyes, ears, nose and mouth; external, oral
731	*Piper snethlagei* Yucker	Piperaceae	Leaf	External
695	*Polypodium phyllitidis* L.	Polypodiaceae	Whole plant	Oral
782	*Psychotria lupulina* Benth.	Rubiaceae	Whole plant	Oral
745	*Anemia oblongifolia* Sw.	Sauruceae	Leaf, whole plant	External, oral
679	*Zingiber officinale* Rosc.	Zingiberaceae	Whole plant	External, oral sniff
911	*Xyris* sp.	Xyridaceae	Whole plant	Oral

For the Kayapó, gastrointestinal disorders can have spirit-based sources transmitted by the wrong food that is eaten by a child's father or some other relative. Food taboos help differentiate age grades and lineage groups. Dietary infringements cause a spiritual/social imbalance that, in turn, causes the child to become sick without any direct contact with the relative. Sometimes food can be improperly hunted, collected, cleaned or prepared and, through eating the food, the person gets sick; a mother can contaminate her child through her milk. All circumstances are common with *hak* and *tep kanê*.

Whether a disease is spirit related or not is not always obvious to the specialist (*mēkute pidjà mari*). Frequently, the illness is considered to be spirit based when normal cures are unsuccessful or symptoms worsen after treatment. Simultaneously occurring sicknesses also complicate diagnosis and usually *wayanga* are called in for evaluation and treatment. A series of *wayanga* may be consulted before the correct specialist is found.

Kayapó pharmacy

The Kayapó are precise in regard to their traditional pharmacological technique and posology. Modes of preparation of medicines include: plants prepared with cold or warm water, plants mixed with cold water then left to boil, plant sap extracted by squeezing, plants heated over fire, plants crushed and mixed with *Genipa americana* (*genipapo*) and charcoal or with *Bixa orellana* for body painting. The principal ways to administer a medicine are: cold tea, hot tea, topical baths of specific parts or the whole body, external topical application (heated leaf or sap) over affected areas, drops in eyes, nose or ears, rubbing on the face or affected areas, wrapping bark around affected parts of the body, sleeping on top of a plant, sniffing, inhaling the smoke and stuffing in the nose.

Each treatment includes the time of day a medicine will be given, almost always between one and five times daily. Times are indicated by pointing to the sky, with the temporal points being: sunrise, mid-morning, noon, mid-afternoon and sunset. A specific number of days is prescribed, depending upon the diagnosis by the curer. Most treatments require medication for between one and five days, although some are indicated for 'use as long as needed'. Dosage is adjusted for each patient, especially infants and children. Certain curers have preferences for modes of preparation and application of their medicines.

Results and discussion

Plants used for *tep kanê* include 19 species, distributed among 18 genera and 18 families (Table 1); most treatments are internal. Plants used for *hak kanê* include 34 species, distributed among 33 genera and 21 families (Table 2); most of these treatments are also internal. Plants used for diarrhoea include seven species, distributed among six genera and five families (Table 3); all species are given orally. All the plants listed are prepared in water.

TABLE 2   Plant species used to treat *hâk kanê*

Collection no.	Species	Family	Part of plant	Route
884	*Mandevilla cf. seabra* (R.?) K. Schum	Apocynaceae	Flower, leaf	Oral
776	*M. tenuifolia* (Mikan) Wood.	Apocynaceae	Whole plant	Oral
893	*Aristolochia* sp.	Aristolochiaceae	Whole plant	Oral
631	*Barjonia* sp.	Asclepiadaceae	Whole plant	Oral
755	*Blepharodon* sp.	Asclepiadaceae	Leaf, sap	External, oral
873	*Arrabidaea cf. cinnamomea* (D.C.) Sandw.	Bignoniaceae	Whole plant	Oral
656	*Protium unifoliolatum* Engl.	Burseraceae	Whole plant	Oral, sniff
678	*Terminalia* sp.	Compositae	Whole plant	External, oral
690	*Wulffia baccata* Kunt.	Compositae	Whole plant	External
651	*Rourea induta* Planch.	Connareceae	Whole plant	External
752	*Sapium poeppigii* Hemsl.	Euphorbiaceae	Whole plant	External, sniff
760	*Sebastiania corniculata* Muell. Arg.	Euphorbiaceae	Leaf	Sniff, stuff nose
780	*Coutoubea ramosa* Aubl.	Gentianaceae	Whole plant	External, eye drops, nose drops
667	*Olyra latifolia* L.	Gramineae	Whole plant	External
634	*Cassia* sp.	Leguminosae	Whole plant	External
751	*Desmodium adscandens* DC.	Leguminosae	Whole plant	External, oral
923	*Periandra heterophylla* Bentz.	Leguminosae	Whole plant	Drops in eyes, ears, nose, mouth; oral
721	*Phaseolus* sp.	Leguminosae	Whole plant	Drops in eyes, ears, nose and mouth; sniff
689	*Spigelia anthelmia* L.	Leguminosae	Whole plant	Eye drops, nose drops
1010	*Utricularia oliverana* Steyerm.	Lentibulariaceae	Ashes of whole plant	Oral
662	*Byrsonima aerugo* Sargot.	Malpighiaceae	Whole plant	Oral
666	*Dyplopterys pauciflora* Niedenzo	Malpighiaceae	Whole plant	Oral
659	*Miconia barbigera* DC.	Melastomataceae	Whole plant	Drops in eyes, ears, nose and mouth; external, oral

continued

TABLE 2 *continued*

Collection no.	Species	Family	Part of plant	Route
774	*Cissampelos tropae-olifolia* DC.	Menispermaceae	Whole plant	External, oral
675	*Heliconia psittacorum* Sw.	Musaceae	Whole plant	External
691	*Ouratea hexasperma var. planchonii* Baill.	Ochalaceae	Whole plant	Eye drops, nose drops, oral
685	*Oxalis barrelieri* Willd. ex Zucc.	Oxalidaceae	Whole plant	External
775	*Desmoncus* sp.	Palmae	Whole plant	Eye drops
635	*Borreria* sp. G. F. W. Mey	Rubiaceae	Whole plant	Eye drops; drops in eyes, ear and mouth
684	*Faramea egensis* M. Arg.	Rubiaceae	Whole plant	Oral
949	*Geophila gracilis* DC.	Rubiaceae	Whole plant	Drops in eyes, ears, nose and mouth
623	*Palicourea quadrifolia* (Rudg.) Steyerm.	Rubiaceae	Whole plant	Eye drops, nose drops, oral
979	*Psychotria* sp.	Rubiaceae	Whole plant	Oral
754	*Spiranthera odoratissima* St. Hil.	Rutaceae	Leaf	Drops in eyes, ears, nose, mouth

Several classes of natural substances such as alkaloids (e.g. castanospermine, lycorine and papaverine), polyphenolic compounds, phenolic glycosides, tannins, lectins, protein polysaccharide complexes, sulphated polysaccharides, mixtures of sugars, proteins and inorganic elements, flavonoids, flavones and saponins are reported to posses antiviral activity (Ieven et al 1982, Van Hoof et al 1984, 1989, Vlietinck & Van den Berghe 1991). Flavones, a class of flavonoids, are of special interest because they have attractive antiviral mechanisms of action, a pronounced and broad spectrum of activity and do not show induction of resistance.

Most genera employed by the Kayapó for the treatment of gastrointestinal disorders include species that contain classes of compounds relevant to antiviral activity, or are related to species used by other peoples for viral diseases or gastrointestinal troubles. Among medicinal species used by several Amazonian Indians, for instance, *Ruellia colorata* is used by the Kofán as a vermifuge and vomitive and *Ruellia aff. malacosperma* is used for diarrhoea, measles and fever (Schultes & Rauffauf 1990). Among the Annonaceae, Heinrich et al (1992a,b) report that *Annona muricata* is used in Mexico for diarrhoea; Caceres et al (1990) report that *Annona cherimola, A. muricata* and *Annona reticulata* are used in Guatemala for diarrhoea; and Grenand et al (1987) found that *Annona ambotay* and *Annona haematantha* are used in French Guyana as febrifuges. Also in French Guyana, *Aristolochia staheli* is used as a febrifuge; *Aristolochia leprieurii* is used for diarrhoea and *Aristolochia trilobata* for hepatitis and malaria (Grenand et al 1987); Ieven et al (1979) report that *Aristolochia elegans* and *Aristolochia forckelii* have antibacterial activity against *Escherichia coli*, *Staphylococcus aureus* and *Pseudomonas aeruginosa*. *Helosis guyannensis* is used for diarrhoea and dysentery. Among the Bignoniaceae, *Arrabidea chica* is used by the Tikuna for conjunctivitis (especially in children), whereas they use *Arrabidea xanthophylla* for serious conjunctivitis (Schultes & Rauffauf 1990). Several *Protium* species are used for relieving nasal congestion associated with colds (Schultes & Rauffauf 1990). *Wulffia baccata* is used in Guyana for colds, nausea and as a febrifuge (Grenand et al 1987). *Byrsonima* species are rich in tannins and *Byrsonima ciliata* is used by the Kubeo for diarrhoea and is said to be very effective (Schultes & Rauffauf 1990). Several species of *Eugenia* (like *Eugenia uniflora*) are used in Brazil for diarrhoea; their effect is usually attributed to the high tannin content of many species of this genus. *Eugenia florida* and *Eugenia patrissi* are used by Indians for treatment of respiratory problems (Schultes & Rauffauf 1990). Heinrich et al (1992b) report antibacterial and antifungal activity for *Eugenia acapulcensis*, used orally to treat diarrhoea and dysentery by the Mixe in Mexico. Among the Gentianaceae, Grenand et al (1987) report that *Coutoubea ramosa* is used as a vermifuge and febrifuge. *Miconia barbigera* is used in Guyana to treat dysentery (Grenand et al 1987). *Cissampelos pareira* is used in Mexico for diarrhoea and dysentery (Heinrich et al 1992a) and in Guatemala for diarrhoea, dysentery, stomach pains and worms (Caceres et al 1990). Ieven et al (1979) report that *Heliconia psittacorum* has activity against *S. aureus*, *E. coli* and *P. aeruginosa*. Very little is known about the chemistry of Ochnaceae, but tannins and flavonoids were reported in *Sauvagesia* species; *Sauvagesia erecta* is used by the Siona for stomach-ache and other *Sauvagesia* species are used by the Kofán for stomach pains (Schultes & Rauffauf 1990). According to Grenand et al (1987), *S. erecta* is used in French Guyana as a febrifuge. Species of Passifloraceae are known to contain alkaloids, phenols and tannins. *Passiflora cumbalensis* and *Passiflora killipiana* are used for treating fever, whereas *Passiflora phaeocaula* is used for conjunctivitis and *Passiflora serratodigitata* for eye inflammation (Schultes & Rauffauf 1990). According to Caceres et al (1990), *Passiflora ligularis* is used for diarrhoea, dysentery, stomach pains and indigestion in Guatemala. *Passiflora edulis* is active against *P. aeruginosa* (Ieven et al 1979) and devoid of antiviral activity (Van den Berghe 1978). Grenand et al (1987) report the use of *Passiflora coccinea* for conjunctivitis and *Passiflora laurifolia* as a vermifuge. Species of Piperaceae are commonly used in many medical systems. Ethereal oils, mono- and sesquiterpenes, phenyl propanoids, polyphenols, lignans and alkaloids were reported to be present in the family and alkaloids are common in the genus

TABLE 3   Plant species used to treat diarrhoea

Collection no.	Species	Family	Part of plant
961	*Psittacanthus biternatus* Blume	Loranthaceae	Leaf
969	*Struthanthus marginatus* (Desr.) Don	Loranthaceae	Whole plant
537	*Eugenia punicaefolia* DC.	Myrtaceae	Fruit, root
555	*Palicourea cf. crocea* Schlecht.	Rubiaceae	Flower, leaf
603	*Selaginella penniformis* (Lam.) Hieron	Selaginellaceae	Sap, whole plant
617	*Qualea grandiflora* Mart.	Vochysiaceae	Leaf
540	*Q. multiflora* Mart.	Vochysiaceae	Leaf

All plant extracts are given orally; those of *S. penniformis* may also be applied externally.

Piper. *Piper arboretum* is used to treat stomach poisoning; *Piper augustus* and *Piper caudatum* are used as carminatives; *Piper futuri* is for 'sick stomach' and *Piper macerispicum* for stomach pains (Schultes & Rauffauf 1990). Faramea (Rubiaceae) are little known chemically but alkaloids were found in two Brazilian species: *Faramea anisocalyx* is used to treat food poisoning; *Faramea glandulosa* and *Faramea salicifolia* are used for fever (Schultes & Rauffauf 1990). *Palicourea* species are known to be bioactive and sometimes highly toxic; *Palicourea buntigii* is used for respiratory problems, *Palicourea corymbifera* for persistent cough and chest ailments, *Palicourea crocea* as an emetic after food poisoning from fish or meat, and *Palicourea guianensis* as a vermifuge (Schultes & Rauffauf 1990). The genus *Psychotria* is rich in bioactive alkaloids: *Psychotria brachiata* is used for problems in breathing, *Psychotria capitata* for severe colds, *Psychotria egensis* as an emetic, *Psychotria poeppigiana* for pulmonary ailments and *Psychotria rufescens* for dysentery. *Psychotria lupulina* is reported to yield strong positives in alkaloid tests (Schultes & Rauffauf 1990). From the relatively unknown Vochysiaceae, deoxyflavones were isolated; *Qualea acuminata* is used as a vermifuge (Schultes & Rauffauf 1990).

Therefore for at least 18 genera there is additional evidence for biological activity and/or related use relevant to Kayapó treatment and usage. Interestingly, for the Kayapó nearly all plants take their names from the diseases they are used to treat. Thus the plant 'families' of *tep kanê* and *hak kanê* are based not on morphological characteristics, but rather on functional similarities that cover a wide range of Western taxonomic botanical families.

Conclusion

In the search for plant-derived antiviral agents, the screening of a relatively low number of randomly selected plants has afforded a remarkably high number of active leads in comparison with screening of synthetic compounds. Comparing different approaches of plant collecting, Vlietinck & Van den Berghe (1991) showed that folk-based collections give a five times higher rate (circa 25%) of active leads, whereas random collections offer fewer leads but more novel compounds.

It is noteworthy that natural substances interfere with a range of viral targets, which can mean that they show mechanisms of action complementary to those of existing antiviral drugs. Because natural products are known to yield prototypic drugs, with innovative mechanisms of action, the end-points of screens must be carefully selected and interpreted in order to avoid false negatives. *In vivo* assays continue to be the stepping-stone between *in vitro* evaluation and human trials. Too often, the therapeutic ratio of active compounds is inadequate. This may be in part because the concentrations in target tissues are not sufficient under dosing conditions owing to species-specific characteristics of absorption, tissue distribution, metabolism and excretion (Vlietinck & Berghe 1991). As a result, election of species claimed by humans to be therapeutically useful in the treatment of viral conditions might be a very significant gain in the research and development of antiviral drugs.

The Kayapó have great faith in their medicines. They offer us an interesting list of candidates as potential sources of antiviral compounds. The diversity of genera and species included in these lists is unlikely to be obtained by following any chemotaxonomy-based strategy. These species should be considered with the seriousness deserving of the original discoverers, because their medical concepts are still the best guides to biomedical evaluation and understanding of the parameters of their diseases.

Acknowledgements

The authors wish to acknowledge assistance from Maria Aparecida Correia dos Santos, Ionara Siqueira Rodrigues, Luciane dos Santos Costa in the literature survey and Sr. Carlos Rosário, the very able botanical field assistant, whose participation in this study was invaluable. Dr Anderson is currently Program Director for the Ford Foundation in Brazil. This work was supported by CNPq.

References

Bamburger J 1967 Environment and cultural classification: a study of the northern Cayapó. PhD thesis, Harvard University, Cambridge, MA, USA
Brandão MGL, Grandi TSM, Rocha EMM, Sawyer DR, Krettli AU 1992 Survey of medicinal plants used as antimalarials in the Amazon. J Ethnopharmacol 36:175–182
Caceres A, Cano O, Samayoa B, Aguilar L 1990 Plants used in Guatemala for the treatment of gastrointestinal disorders. 1. Screening of 84 plants against enterobacteria. J Ethnopharmacol 30:55–73
Cordell CA, Beecher CWW, Pezzuto JM 1991 Can ethnopharmacology contribute to the development of new anticancer drugs? J Ethnopharmacol 32:117–133

Duke JA 1986 Folk anticancer plants containing antitumor compounds. In: Etkin NL (ed) Plants in indigenous medicine and diet—biobehavioral approaches. Redgrave, New York, p 70–90

Elisabetsky E, Castilhos ZC 1990 Plants used as analgesics by Amazonian caboclos as a basis for selecting plants for investigation. Int J Crude Drug Res 28:49–60

Elisabetsky E, Gely A 1987 Plantes médicinales utilisées en Amazonie comme fond potentiel de noveaux agents thérapeutiques dans les cas d'allergie, thrombose et inflammation. J Agric Trop Bot Appl 36:143–151

Elisabetsky E, Posey DA 1985 Pesquisa etnofarmacológica e recursos naturais no Trópico Umido: o caso dos índios Kayapó e suas implicações para a ciência médica. Anais do I Simpósio Internacional do Trópico Umido. EMBRAPA, Belém, vol 2:85–93

Elisabetsky E, Posey DA 1988 Use of anticonceptional and related plants by the Kayapó indians (Brazil). J Ethnopharmacol 26:299–316

Elisabetsky E, Posey DA 1991 Conceito de animais e seus espíritos em relação a doenças e curas entre os índios Kayapó da aldeia Gorotire, Pará. Boletim do Museu Paraense Emílio Goeldi, Série de Antropologia, Belém, Parà, vol 7:21–36

Elisabetsky E, Setzer R 1985 Caboclo concepts of disease, diagnosis and therapy: implications for ethnopharmacology and health systems in Amazonia. In: Parker EP (ed) The Amazon Caboclo: historical and contemporary perspectives. William and Mary University Press, Williamsburgh, p 243–278 (Stud Third World Soc Publ Ser 32)

Fabrega H 1975 The need for an ethnomedical science. Science 189:969–975

Farnsworth NR 1990 The role of ethnopharmacology in drug development. In: Bioactive compounds from plants. Wiley, Chichester (Ciba Found Symp 154) p 2–21

Grenand P, Moretti C, Jacquemin H 1987 Pharmacopées traditionelles en Guyane. Editions de l'ORSTON, Paris, p 379–382

Heinrich M, Rimpler H, Barrera NA 1992a Indigenous phytotherapy of gastrointestinal disorders in a lowland Mixe community (Oaxaca, Mexico): ethnopharmacologic evaluation. J Ethnopharmacol 36:63–80

Heinrich M, Kuhnt M, Wright CW, Rimpler H, Phillipson JD, Schandelmaier A, Warhurst DC 1992b Parasitological and microbiological evaluation of Mixe Indian medicinal plants (Mexico). J Ethnopharmacol 36:81–85

Holmstedt B, Bruhn JG 1983 Ethnopharmacology—a challenge. J Ethnopharmacol 8:251–256

Ieven M, Van den Berghe DA, Mertens F, Vlietinck AJ, Lammens E 1979 Screening of higher plants for biological activities. I. Antimicrobial activity. Planta Med 36:311–321

Ieven M, Vlietinck AJ, Van den Berghe DA, Totté J 1982 Plant antiviral agents. III. Isolation of alkaloids from Clivia miniata Regel (Amaryllidaceae). J Nat Prod 45:564–573

Krej GJ 1988 Diarrhea. In: Wyngaarden JB, Smith LH (eds) Cecil textbook of medicine. WB Saunders, Philadelphia, PA, p 680–687

Labadie RP, van der Nat JM, Simons JM et al 1989 An ethnopharmacological approach to the search for immunomodulators of plant origin. Planta Med 55:339–348

Phillipson JD, Wright CW 1991 Can ethnopharmacology contribute to the development of antimalarial agents? J Ethnopharmacol 32:155–166

Posey DA 1982a The journey of a Kayapó shaman. J Lat Am Indian Lit 6(3):13–19

Posey DA 1982b Time, space and the interface of divergent cultures: the Kayapó Indians face the future. Rev Bras Antropol 25:89–104

Posey DA 1987 Contact before contact: typology of post-colombian interaction with Northern Kayapó of the Amazon Basin. Boletim do Museu Paraense Emílio Goeldi, Série Antropologia, Belém, Pará, vol 3:135–154

Samuelson G 1989 Nature as sources of drugs. Acta Pharm Nord 1:111–116

Schultes RE, Rauffauf R 1990 The healing forest. Medicinal and toxic plants of the northwest Amazonia. Dioscorides press, Portland, OR

Svendsen AB, Scheffer JJC 1982 Natural products in therapy: prospects, goals and means in modern research. Pharm Weekbl (Sci Ed) 4:93–103

Van den Berghe DA, Ieven M, Mertens F, Vlietinck AJ 1978 Screening of higher plants for biological activities. II. Antiviral activity. Lloydia 41:463–471

Van Hoof L, Van den Berghe DA, Hatfield GM, Vlietinck AJ 1984 Plant antiviral agents. V. 3-Methoxyflavones as potent inhibitors of viral-induced block of cell synthesis. Planta Med 459:513–517

Van Hoof L, Totté J, Corthout J et al 1989 Plant antiviral agents. VI. Isolation of antiviral phenolic glucosides from *Populus* cultivar Beaupre by droplet counter-current chromatography. J Nat Prod 52:875–878

Vlietinck AJ, Van den Berghe DA 1991 Can ethnopharmacology contribute to the development of antiviral agents? J Ethnopharmacol 32(1-3):141–154

Wagner H 1989 Search for new plant constituents with potential antiphlogistic and antiallergenic activity. Planta Med 55:235–241

Xiao P-G, Wang N-G 1991 Can ethnopharmacology contribute to the development of anti-fertility drugs? J Ethnopharmacol 32:167–177

DISCUSSION

Farnsworth: I have been involved with the WHO for several years. I always figured that it was worth my time if I could see a 0.5 or 1% improvement in developing countries in their science or whatever, but I'm really not sure I have seen that. In developed countries, if you are a good scientist, with good ideas, you can get funded. In a developing country, if you are a good scientist with good ideas, you can't get funded or it's very difficult. To improve this situation, you have to work with your Ministry of Health. The Ministry has to go, with good documentation, to the WHO Regional Office or a similar international agency and convince them that there's a dire problem in developing countries in establishing safety first and efficacy second for a lot of these traditional preparations. The only way that could be done, in my opinion, is if centres of excellence are established—one on each continent, one in Africa, one in South America, one in South-East Asia, etc. These would provide training facilities and facilities for scientists who don't have the capability to test extracts for safety and efficacy. My bet is that within 10 years the amount of patent income generated by these activities could make them self-supporting.

This has to be organized by people in developing countries. These organizations don't listen to people from the developed countries, concerning developing countries' problems. I proposed this in a paper at the first meeting of Directors of WHO Collaborative Centres for Traditional Medicine. I had a time table and all the steps that needed to be done, suggestions for funding; when the report was published, my paper was the only one that was deleted from the series! The drive has to come from within; you have to be willing to spend the political time to convince Ministries of Health to do something about this.

In Brazil, you have Linalool, which is an excellent anticonvulsive agent. But for a compound to be of interest to the developed countries, it not only has to be better than existing anticonvulsants, it has to have a different mechanism of action or no drug company will touch it. Linalool is a compound that is suitable for making analogues and so forth, so maybe it will be developed.

I know somebody who is working with a very closely related monoterpene, that at very low doses, given orally to rats, completely prevents conception. But it's such a simple compound, just a hydroxylated monoterpene, that nobody will look at it in a developed country. I wouldn't give up, but you have to organize, you have to give political strength and will, and you have to present a plan. Carl Djerassi did this here in Brazil. He proposed that there would be a Centre of Excellence and Chemistry set up here, supported by the chemical societies of the developed world who could afford it.

Elisabetsky: There is a two-way cooperation with pharmaceutical industries. There's a lot of work being done within the industry that is good for Maurice Iwu's purpose or Alaide de Oliveira's purpose. There are known compounds that might be good sources of local drugs, for which the pharmaceutical companies have identified the activity but we don't know it yet and they have not developed it because they don't want it. We could be cooperating with these industries; we might have data to which they don't have access.

Lozoya: I agree. I think we have been using the wrong methodology in the evaluation of medicinal plants used in traditional medicine. After 20 years of discussions about the same problem, today we must finally tell the truth. The truth is that Western methodology, based on animal experimentation and avoiding direct clinical evaluation of popular herbal remedies, is wrong. Because of this approach, the Western world has no new drugs, whereas the Chinese have hundreds of them. We are losing the game. American companies and European companies are falling behind; they have been falling behind for the last 20 years. The Chinese, Koreans and Japanese have been working hard for 20 years in the direct clinical evaluation of their cultural heritage. The problem is that in Western society, scientists and companies do not believe the common people; they think they are culturally superior.

We must follow a methodology that considers the isolation of an active compound as a final step after clinical evaluation of the natural herbal remedy as used by the people for many years. In the past five years, in Mexico, we have tried this different approach. Now we are having some success in obtaining new drugs. The key to this approach is first to do good medical studies of the popular herbal practices. If biologists or ethnobotanists do not incorporate good medical research into their work in this field, the problems will remain, because in the study of a plant drug the medical information is required first. Fortunately, now everybody here talks about the importance of the ethnomedical approach in the selection of plant-derived drugs to be studied; 15 years ago this was not the case. The classical pharmaceutical approach was predominant and the use of plants was scrapped under the 'economic botany' approach. Popular knowledge was considered irrelevant. Finally, today everyone understands that people are not stupid and that information from traditional medicine is very useful because it represents an intelligent selection of plants on a cultural basis.

The important change in our strategy in Mexico is that today we start with clinical studies and not with animal studies. The Chinese did this in the 1950s, although they don't talk about it much. They decided to believe the people, to take the plants selected by traditional medical practice and introduce them directly into the hospitals and clinics to be used in the traditional way. The plants that succeeded in the clinical evaluation were later selected for basic chemical and toxicological studies in scientific research centres.

In the West, we are going backwards. We use as the first step the research centre where chemists and pharmacologists must isolate the active compounds! But the question is: which are the active compounds? To answer that, hundreds of screening studies are required *in vitro* or *in vivo* because everybody in this part of the world thinks that to give a 'herbal tea' to a patient is unethical, only because they culturally do not recognize the value of popular tradition.

Elisabetsky: That kind of analysis may be appropriate for the Chinese, but it's not appropriate for other countries.

Lozoya: Why not? Today, we have the possibility to develop completely new and different methodologies. Now is a good time to discuss our mistakes openly, because we are not succeeding in this area of research. Please remember that taxol has come to market only after 23 years of work. Before taxol, our commercial successes were vincristine and vinblastine. Between those, Western science has discovered nothing from medicinal plants.

de Oliveira: We followed that methodology in the 1970s in Brazil. First, plants were investigated in laboratory animals using the preparation that was used by the local population. Then clinical trials were carried out. Ten out of 50 plants were shown to have efficacy and low toxicity. The intention was to introduce those plants into the primary health service. But the programme was so good that it was stopped!

Lozoya: Why was it stopped?

de Oliveira: Ask the politicians, I don't know.

Schwartsmann: I would like to challenge the statement that drug development has been a failure, particularly for anticancer drugs. The survival rates of patients with many types of tumours, especially the rapidly growing ones like leukaemias, lymphomas or germ-cell tumours, have improved substantially. Twenty years ago, the two-year survival rate in children with leukaemia was less than 10%. Now we can cure over 70% of children with leukaemia.

Presently, the challenge is to take advantage of the tremendous knowledge we have gained in molecular biology and drug testing and apply it to the search for agents effective against slowly growing cancers. This is why the US National Cancer Institute (NCI) has changed its policy over the last few years and is now using a complex panel of cell lines from human solid tumours.

Regarding the use of traditional medicine as a source of new drugs, we have to make a distinction between common illnesses and complex ones such as cancer. We cannot use the same type of symptom-oriented approach for the latter and expect a high level of success. Indigenous people usually do not recognize cancer as such; they interpret diseases on the basis of signs and symptoms.

Nor do I accept that in China they are having more success with traditional medicine than we are in the West. I don't see evidence of this in the scientific literature. If you take cancer as an example, the Chinese treat tumours in a similar way as we do. They use chemotherapy, radiotherapy and some form of non-specific immunotherapy. When they are successful, they are using almost the same type of drugs as we are using.

Bohlin: Nature as a source of new lead compounds has proven in recent years to be very successful. Many important molecules with specific pharmacological activities have been identified and developed; for example, forskolin, phorbol esters, peptides and peptidomimetics.

Elisabetsky: I am much more optimistic than my friend, Xavier Lozoya. We are re-thinking our approaches now. The NCI is changing. Shaman Pharmaceuticals is developing a completely new methodology that uses ethnopharmacology for the purposes of the industry. We in the developing countries are finally giving up waiting and getting smarter at using our First World colleagues to get money for ourselves. These are new developments and I am optimistic that we are still at the beginning.

Martin: Elaine, you said that efficacy, effectiveness and bioactivity are not the same thing. What do you see as the differences among them?

Elisabetsky: They are three levels of looking at the same thing, but they are three very different levels. You can find that a certain preparation is not active in the original bioassay, which means you looked at 2–3 doses, maybe in one animal model, or you looked at a certain enzyme which happened to be available in your laboratory, and you found no activity. This does not mean that the treatment is not effective within the culture in which it is used.

Efficacy and effectiveness are different in pharmacological terms. Some pharmacological activity is associated with toxicity; bioactivity translates into toxicity as well. The balance between the two is the therapeutic value of a drug. When you talk about efficacy, you are evaluating the pharmacological effects. Effectivity refers to a much broader view, it really describes the cost–benefit to the patient.

Iwu: We have to be careful how we determine the effectiveness of these therapies. An example is the case of the old Chinese herb, *Artemisia*. If it had been screened by any of the standard methods at the time its activity was discovered, it would have failed. The classical method used at the time artemisinin was discovered consisted of pretreatment of the plant material with hexane or petroleum ether to 'defat' the material and subsequently extraction with alcohol. Since artemisinin is very hydrophobic, pre-extraction with hexane would have amounted to throwing the proverbial baby out with the bathwater. One of the most intriguing aspects of the ethnopharmacology of *Artemisia* is that in ancient Chinese medicine the plant was used as an aqueous decoction. As early as AD 340 the Chinese herbalist Ge Hong indicated that certain fevers could be treated by drinking a concoction of one handful of *qing hao* in a litre of water. By this extraction method only an insignificant amount of the most active endoperoxide will be in the aqueous extract, and yet the drug was effective as an antimalarial. Initial attempts in the 1970s to establish the antimalarial activity of *Artemisia* extracts were disappointing, largely due to the adoption of the classical extraction–screening method described earlier. In 1972, Chinese scientists, using a modified technique, isolated the major active constituent which they named *qinghaosu* (artemisinin). The structure of this compound was established in 1980 and was found to represent a hitherto unknown class of sesquiterpenes with an unusual endoperoxide moiety. The pure compound was very active against drug-resistant *Plasmodium falciparum* and was remarkably non-toxic at doses of up to 5 g per kilogram of body weight in rats.

Then modern science came along with all its arrogance and discarded 500 years of science using this drug. It was purified by phytochemical isolation to give a wonder drug. We are now realizing that with the high efficacy of this new wonder drug also comes high neurotoxicity. The Chinese were using it as an extract that had only a small amount of artemisinin mixed with a lot of flavonoids and other compounds.

It cannot be over-emphasized that the fact that two different substances have the same effect and may share a common constituent does not make them identical—similar, but not identical. The compound artemisinin may be present in both the pharmaceutical agent called 'artemisinin' and the crude drug called 'artemisia extract'; both are essentially different medicines, with different activity, pharmacokinetics and toxicity profiles.

Natural product chemistry in north-eastern Brazil

A. A. Craveiro, M. I. L. Machado, J. W. Alencar and F. J. A. Matos

Laboratory of Natural Products/Department of Organic and Inorganic Chemistry, Universidade Federal do Ceará, Caixa Postal 6822, Campus do Pici, 60458-978, Fortaleza, Ceará, Brazil

Abstract. The north-eastern region of Brazil comprises about one third of the country's territory. It is a semi-arid region with a flora rich in aromatic, toxic and medicinal plants. Screening of aromatic plants led to the investigation of about 2000 samples of essential oils from plants from the region and from abroad. Studies done by the Universidade Federal do Ceará and other research groups in the region discovered several new substances with distinct pharmacological activities. Recent examples are: schultezin, hydroxy-bisabolol, *trans*-annonene, (−)-hardwickic acid, *trans*-cascarillone, nor-cucurbitacins, oleanolic saponin and chalcone dimer. A social programme called the Living Pharmacy was created to teach poor people how to cultivate and use medicinal plants correctly. A project to develop interaction between the university and industry also arose from these studies.

1994 Ethnobotany and the search for new drugs. Wiley, Chichester (Ciba Foundation Symposium 185) p 95–105

The north-eastern region of Brazil comprises 1 542 000 km^2 or 30% of Brazilian territory, located between 1° and 17° latitude North and 31° and 47° longitude West. One third of the country's population lives in this region. The region is considered poor and under-developed even by Brazilian standards. The climate is hot and dry with only two seasons, the rainy season locally called 'winter' and the dry season called 'summer'. Occasionally, the rainy season does not occur for one to four years and severe drought occurs with terrible social and ecological consequences. The major part of this region is semi-arid and is locally named *caatinga*; in it are found many types of native species from several families, including the Leguminosae, Anacardiaceae, Euphorbiaceae, Gramineae, Malvaceae and Verbenaceae. Aromatic plants are common among these families. The local folk medicine is very rich; everywhere, it is easy to find primitive herbalists, called by the people *raizeiros*, selling the medicinal plants of the region and teaching people how to use them. Unfortunately, traditional knowledge is being replaced year after year, giving way to popular misconceived information. A programme for screening the local flora, botanically, chemically and pharmacologically, especially for essential oils of the aromatic flora, was started by the Universidade Federal do Ceará in 1971, supported by the Brazilian Federal Government and the Banco do Nordeste, a regional development bank. Other research groups in the states of Pernambuco, Rio Grande do Norte, Paraiba, Bahia, Sergipe, Alagoas, Piaui and Maranhão have also studied the phytochemistry of the regional plants. Therefore, the work that will be described here is only a fraction of the total effort being done in the region; the paper will be restricted to the achievements of our group's principal research programmes in the last few years.

Studies on essential oils

This programme deals with determination of the volatile chemical constituents of aromatic plants of the north-east Brazilian flora as well as of exotic plants cultivated in the region for different purposes. Since 1976, more than 2000 samples of essential oils, mainly extracted from local plants, and about 200 samples received from outside researchers have been analysed. The results of 83 analyses selected from 575 were condensed in a book (Craveiro et al 1981a). Besides the analytical data, the book included the scientific and popular names of the plants, a list of all places and collection dates, a chromatogram of each oil with the name of the identified constituents and other pertinent information about extraction methods, and statistical data concerning the Brazilian essential oil market.

The analyses were performed in a combined gas chromatography and mass spectrophotometry system coupled with a computer programmed for automatic identification of the oil's chemical constituents. Computer programs based on real and simulated Kovats chromatographic retention index were developed to optimize the identification work. The following are examples showing the potential importance of these oils in the fields of pharmacy, cosmetics and other industries.

Canelinha oil: anethole type

Anise-like oil obtained with 1–2% yield from leaves of a chemical variety (i.e. a variant of a species whose chemical content differs according to the environment) of *Croton zehntneri* Pax et Hoff. (Euphorbiaceae) growing at 300 feet on a sedimentary mountain in Tianguá County in Ceará State (Craveiro et al 1989a).

Pepper-alecrim oil

This oil is very similar to Mexican oregano oil or thymus oil, because of the presence of thymol in concentrations of 30–80%. It can be obtained from the leaves of *Lippia sidoides* Cham. (Verbenaceae) in yields ranging from 2–4%. This oil possesses very effective antiseptic properties and shows strong activity against bacteria of the vaginal tract. This species of *Lippia* grows in the *caatinga* near the city of Mossoró in the state of Rio Grande do Norte. It is now being cultivated from stalks of plants grown in the Medicinal, Aromatic and Toxic Plant Garden of the Universidade Federal do Ceará (Lemos et al 1990a, Mendonça et al 1989, Craveiro et al 1981b).

Candle-wood oil

The plant that produces this oil was identified as *Vanillosmopsis arborea* Baker (Compositae), a shrub common at 200 feet on the hillsides in the Araripe sedimentary mountains near Crato, a city in the south of Ceará State. The essential oil contains 69% of α-bisabolol and is obtained in a 0.6% yield from the wood. There are new developments to promote its use as a substitute for chamomile oil in the cosmetic industry (Craveiro et al 1989b).

Canelinha oil: estragole type

This tarragon-like oil contains 90% *p*-allylanisole (estragole). Obtained in 4% yield from the leaves of another variety of *Croton zehntneri* growing near the coastal region in the county of Granja, Ceará (M. I. L. Machado, L. H. Barreto and A. A. Craveiro, unpublished results 1989). This oil has a strong appeal to industry as a source of *p*-allylanisole, because of the high yield from the plant.

Spearmint oil

This oil is obtained in low yield from leaves of *Mentha spicata* L., an exotic aromatic herb which is cultivated in the Medicinal, Aromatic and Toxic Plant garden of the Universidade Federal do Ceará. The major chemical constituent of this oil is piperitenone oxide (21%), the same compound as found in *Mentha X villosa* Huds., which is supposed to be the active principle of this medicinal plant. The plant is a very active herb used against the intestinal parasites *Amoeba hystolitica* and *Giardia llamblia* (Magalhães et al 1992, Santana et al 1989).

Studies on medicinal and toxic plants

Chemical research in this area is being done by researchers of the Laboratory of Natural Products and the Department of Organic and Inorganic Chemistry, using modern analytical instruments. In several cases, the work forms part of MSc and PhD theses of students enrolled in the Graduate School of Organic Chemistry at the Universidade Federal do Ceará. Work in this field has led to publication of a book with selected information about the chemical constituents of Brazilian medicinal plants (Sousa et al 1991). In recent years, the major achievements have been to describe the following compounds.

Schultezin 1

This is a new natural alkaloid isolated from *Schultezia guianensis* L. (Nóbrega & Craveiro 1988), a small herb which is considered toxic to cattle. It has interesting pharmacological activity in the heart muscle, which is being studied by researchers in the Physiology and Pharmacology Department of the Universidade Federal do Ceará.

Hydroxy-bisabolol 2

This is obtained from the wood of the trunk of the aromatic plant *Vanillosmopsis arborea* (Asteraceae). This is a medium-sized tree that grows widely in the mountains in the south of Ceará State. It produces an essential oil rich in α-bisabolol (Matos et al 1988), a substance with reported antiinflammatory activity (Harborne & Williams 1977).

Trans-annonene 3, (−)-hardwickic acid 4 and trans-cascarillone 5

Pharmacological screening of the hexane extract of the roots of *Croton sonderianus* Muel. Arg. revealed antimicrobial activity when tested *in vitro*

	R
3	CH$_3$
4	COOH

5

(McChesney & Silveira 1990). Several ent-clerodane terpenes were isolated from this extract. The plant is a very common Euphorbiaceae shrub native to north-eastern Brazil.

Nor-cucurbitacin glucosides 6,7

The roots of *Wilbrandia* sp. (Cucurbitaceae) are used in folk medicine as anti-rheumatic agents. A mixture of two novel cucurbitacins 6,7 isolated from the roots showed antiinflammatory activity in experimental studies (Matos et al 1991).

	R_1	R_2	
6	H	Ac	Δ^{23}
7	H	H	23,24–dihydro

Oleanolic saponin 8

The ethanolic extract of *Sapindus saponaria* L. (Sapindaceae) was shown by silica gel chromatography to contain a new triterpenoid saponin 8. The acetylated saponins show antimicrobial activity against *Pseudomonas aeruginosa*, *Bacillus subtilis* and *Cryptococcus neoformans* (Lemos et al 1990b).

R
8 Ac

Chalcone dimer 9

New chalcone dimers were isolated from the alcoholic extract of *Myracroduon urundeuva* Fr. All. (= *Astronium urundeuva* Engl.) (Anacardiaceae). This plant

9

is known locally as *aroeira*; it is used traditionally in folk medicine for its antiinflammatory properties, especially post partum. Recent studies demonstrated that it also has antiulcer activity (Bandeira 1993).

The Living Pharmacy programme

This is a social medicine programme organized by the Universidade Federal do Ceará and maintained with a grant from the Royal Botanic Gardens of Kew, UK. The institutions are linked through the theme 'local plants for local people' created by researchers at Kew. The programme was developed on the basis of the knowledge accumulated during the last 20 years in the areas of ethnobotany, pharmacognosy, natural products chemistry and natural products pharmacology. The goal is to provide the local communities with medicinal plants selected from the regional flora because of their proven therapeutic efficacy and low toxicity and to teach the people how to use them. The name Living Pharmacy means a place where people can get and prepare their own remedies from living plants cultivated by themselves under scientific guidance. Some examples of the 30 medicinal plants existing in the living pharmacies are: *Lippia sidoides* Cham., which is used as a local antiseptic preparation; *Justicia pectoralis* Jacq. var. stenophylla Leon., used as a remedy against asthma and bronchitis; *Mentha X villosa* IIuds., used for the treatment of amoebiasis and giardiasis; *Psidium guajava* L., used to treat diarrhoea, especially in children; *Malpighia glabra* L., used because of its high vitamin C content, and *Eclipta alba* Haack, which has protective effects for liver cells and increases non-specific immunity (Matos 1991).

Technological development park—PADETEC

As an extension of these works, the Universidade Federal do Ceará has created a centre for transferring the knowledge generated in the university to industry. A technological development park called PADETEC has been built on the university campus; it consists of both a research centre for product development and a centre where industries can be created on the basis of new ideas and products.

Research in the area of natural products and fine chemicals that has industrial appeal for product development can be transformed into new enterprises or products using PADETEC's facilities. It is hoped that this centre will contribute to the many benefits already derived by the people of north-eastern Brazil from their local flora.

Acknowledgement

The authors are indebted to Professor Afranio Gomes Fernandes from the Biology Department of the Universidade Federal do Ceará for his collaboration.

References

Bandeira MAM 1993 Contribuiçào ao conhecimento quimico de plantas do nordeste (*Myracrodruon urundeuva* Fr. All. = *Astronium urundeuva* Eng.) Aroeira do Sertao. MSc thesis, Universidade Federal do Ceará, Fortaleza, Ceará, Brazil

Craveiro AA, Fernandes AG, Andrade CHS, Matos FJA, Alencar JW, Machado MIL 1981a Óleos essenciais de plantas do nordeste. Edições Universidade Federal do Ceará, Fortaleza, Ceará, Brazil

Craveiro AA, Alencar JW, Matos FJA, Andrade CHS, Machado MIL 1981b Essential oils from Brazilian Verbenaceae. Genus *Lippia*. J Nat Prod 44:598–601

Craveiro AA, Matos FJA, Alencar JW, Machado MIL, Teixeira LL 1989a Óleos essenciais na produção industrial. Quim Ind 19:60–64

Craveiro AA, Alencar JW, Matos FJA, Sousa MP, Machado MIL 1989b Volatile constituents of leaves, bark and wood from *Vanillosmopsis arborea* Baker. J Essent Oil Res 1:293–294

Harborne JB, Williams CA 1977 Vernonieae—chemical review. In: Heywood VH, Harborne JB, Turner BL (eds) The biology and chemistry of Compositae. Academic Press, London, vol 1:523–537

Lemos TLG, Matos FJA, Alencar JW, Craveiro AA 1990a Antimicrobial activity of essential oils of Brazilian plants. Phytother Res 4:82–84

Lemos TLG, Mendes AL, Sousa MP, Braz Filho R 1990b New saponin from *Spindus saponaria*. Fitoterapia 63:515–517

McChesney JD, Silveira ER 1990 Ent-clerodanes of *Croton sonderianus*. Fitoterapia 61:172–175

Magalhaes RA, Machado MIL, Craveiro AA, Alencar JW, Matos FJA 1992 Oxido de piperitenona obtido do óleo essencial de *Mentha spicata* Huds. Abstracts from the 15th Annual Reunião da Sociedade Brasileira Quimica. Caxambú, Minas Gerais

Matos FJA 1991 Farmacias vivas. Ediçoes Universidade Federal do Ceará, Fortaleza, Ceará, Brazil

Matos MEO, Sousa MP, Matos FJA, Craveiro AA 1988 Sesquiterpenes from *Vanillosmopsis arborea*. J Nat Prod 51:780–782

Matos MEO, Machado MIL, Craveiro AA, Matos FJA, Braz Filho R 1991 Nor-cucurbitacin glucosides from *Wilbrandia* species. Phytochemistry 30:1020–1023

Mendonça VLM, Fonteles MC, Aguiar LMBA, Craveiro AA 1989 Toxicidade e alergenicidade do óleo de *Lippia sidoides* cham para utilização em cosméticos. Cosmet & Toiletries 1:26

Nóbrega M, Craveiro AA 1988 New alkaloid from *Schultesia guianensis*. J Nat Prod 51:962–965

Santana CF, Almeida ER, Santos ER, Souza IA 1989 Ação da *Mentha crispa* (Giamebil) em pacientes humanos portadores de protozoozes intestinais. Monograph. Instituto Oswaldo Gonçalves de Lima, Departamento de Antibióticos, Universidade Federal de Pernambuco, Recife, Pernambuco, Brazil

Sousa MP, Matos MEO, Matos FJA, Machado MIL, Craveiro AA 1991 Constituintes químicos ativos de plantas medicinais Brasileiras. Edições Universidade Federal do Ceará, Fortaleza, Ceará, Brazil

DISCUSSION

Cragg: The National Cancer Institute is very happy to test any novel compounds which any group might produce, such as your group Professor Craveiro. We will test them on a confidential basis and return the results to you. If the results were significant, we would then collaborate in the development of that particular compound towards clinical trials. We would certainly welcome your participation in the screening programme.

Elvin-Lewis: Professor Craveiro, you have one plant in which you consider the most bioreactive compound to be thymol. Listerine, which is a mouthwash in the United States, has thymol as one of its main substances—thymol and eucalyptol, both essential oils. Both are known not to have substantivity, which means that they don't adhere well to any surface that might be infected with any odontopercopathogen. Therefore, if you are going to use that type of compound, you have to have a constant or repetitive delivery system, such as mouth washing two or three times a day.

Craveiro: We have also in Brazil one mouth gargle that contains only thymol, called Gargol. This is not as potent as the antiseptic that is made from *Lippia sidoides*. I cannot say for sure whether there is any similar effect among the compounds present in the oil, but Professor Matos has observed in his field studies where it is used by the people that the activity is not due only to thymol.

Matos: I would like to tell you a little more about the Living Pharmacy programme here in Fortaleza. The programme selects plants of the region, both wild plants and cultivated plants, that have proven therapeutic efficacy and low toxicity. We grow these plants in the Medicinal Plant Garden of the Federal University of Ceará. These plants, together with information on their activity, are offered to local communities, governmental or non-governmental. Some of them were not used by local people before, but after experimental, chemical and pharmacological assay and a bibliographic review, they were introduced into popular use in several communities. Each community has established its own Farmácia Viva (Living Pharmacy). Some of the plants are also selected for study in the chemical and pharmacological programmes that Professor Craveiro has described and for complementary taxonomical and agronomical studies.

King: Programmes like this can be logically applied to many urban populations that require improved health care. There is a moral, ethical and human imperative to bridge those worlds. This is a beautiful example of how health care and research can be integrated.

Barton: Professor Craveiro, you mentioned the start-up companies that are commercializing these products. What products are they hoping to develop? Are they actually picking up on your research or are they doing other things?

Bohlin: We are talking about two different things here. My opinion is that drug development should be done by drug companies. Scientific research at a university should provide a broad training and produce excellent PhD students—that is our product. However, an important task in our research area is to develop new methodology to improve selection of plants, resulting in new lead compounds with unique structures and activities for drug development.

Craveiro: I cannot say anything about product development outside the university, although I recognize that the product of the university is knowledge, training, etc. But if you have a product that can be developed by the university and use that to start a new business, I cannot see anything wrong with that. This is happening in the US and in the UK. We have established this science park so that if industry is willing to invest money, we can transfer the technology from the university to the industry. For people who are interested but have no money to start a business, we provide financial assistance. The scheme is funded by the Brazilian government and by local banks, but we have to pay back the loan. We have seven small companies, five in the area of natural products, one in electronics and one in analytical services. All these interact and help each other—one gives a product to another one to formulate and bring to market.

Iwu: I find this very exciting. This stresses the point that in Third World countries we cannot afford esoteric science or science for its own sake. We have to combine the role of the scientist, the advocate and the incubator of small-scale industry, otherwise our research takes place in a vacuum. I'm happy that this is an integrated approach, whereby there is real benefit from what is done in the classroom to the people in the field. This is the only way we can justify the huge amount of money being spent on maintaining universities in Third World countries.

Craveiro: Professor Iwu is one good example of this interaction. We are advising one US company concerning essential oils. We were invited to select a director for research and development; we are fortunate that Professor Iwu accepted our invitation. The goal of this institution is to start a cooperation directly with our university through PADETEC, which is the technological development park.

Martin: Is there a link between these newly developed companies, the Living Pharmacy and other applied projects? Is there a return of a percentage of the profits or an agreement to provide funds for these initiatives?

Craveiro: No. The Living Pharmacy project has no income yet. It's a social programme financed by the government. Professor Matos has many difficulties running this programme. He now has help from Kew Gardens in London, for which we are very grateful. What is happening now is that some people in the community who are taught how to grow and use the plants then make small preparations which they sell among themselves.

When the companies become fully commercial, part of the money stays within the university. The university holds some of the shares, participates in the sales of these products, and sometimes participates in the technology. This money is used to start other companies and also to help basic research. Some of the analytical facilities in our university are being financed by income generated in the technological development park.

Elisabetsky: You say that the university derives royalties from these companies. But if Brazil doesn't accept that natural products can be patented, how can they accept royalties?

Craveiro: The royalties come from the electronic company and the one for analytical services.

Elisabetsky: This is a catch 22 situation in Brazil. I've tried, through the Universidade Federal do Rio Grande do Sul to make an agreement with a pharmaceutical company. The company was willing to make a joint cooperation agreement, we would study the extracts together and share royalties. The response from the university was that I could only make a service agreement. I could service the company by doing some experiments in my laboratory, but I could not set up a joint cooperation agreement.

Iwu: I have studied the pharmaceutical patents system in Brazil. The issue is not only patents. There are actual drugs being produced and sold and money is being made. There are also extraction companies that are involved in natural products. You don't need a patent to produce and sell. It is cheaper to make these products here than in the United States or in Europe, so these can be sites of production. The more vexing issue is the lack of equity in most North/South trade.

McChesney: Often the most valuable thing that a university has is not the patents it owns, because those are public knowledge and therefore can be got around with a little clever science and thinking, it is the technology that the university may own. Utilization of the technology to make a specific product may be more effective than relying on patent protection. You protect more effectively by technology and trade secret kinds of operations.

The activities that Dr Craveiro described, and the concern that Elaine Elisabetsky voiced, are not unique to Brazil or to other Third World countries. We had exactly the same problem in the state of Mississippi until two years ago. Technically, the conflict of interest law in the State of Mississippi meant that I could only teach classes and do research on grants; technically, the university could not accept contracts, because that would constitute a conflict of interest. With the increase in pressures that universities are being placed under, in terms of resources, but more importantly in the social expectation of return to their communities in the form of economic and social development, there is an increasing willingness to re-think those conflict of interest issues. We actually had to re-write the law to allow the university to take up equity positions in technologies developed by the university, so those technologies would get into the market place. This is something one might wish to explore, whatever particular circumstance you face in your own environment.

Amazonian ethnobotany and the search for new drugs

Richard Evans Schultes

The Botanical Museum, Harvard University, 26 Oxford Street, Cambridge, MA 02138, USA

Abstract. Tropical rain forests offer enormous prospects for the discovery of new drugs for use in Western medicine. The Amazon supports 80 000 species of higher plants and a diverse Indian population. Focusing attention on those plants used as medicines by indigenous peoples is the most efficient way of identifying the plants that contain bioactive compounds. There is an urgent need for more ethnobotanists and ethnopharmacologists to be trained to document as much information as possible before it and the plants are lost through destruction of the rain forest and acculturation of the indigenous peoples. Ethnobotanical studies have identified plants documented by early travellers; these include *Paullinia yoco* and *Ilex guayusa* which are used as stimulants and have been shown to be rich in caffeine. Studies of the hallucinogen prepared from *Banisteriopsis caapi* have shown that the native people know which plants to add to the mixture to lengthen and intensify the intoxication produced by the β-carboline alkaloids in the plant. Three major snuffs are used in the Amazonia; the plants from which they are derived have been identified. One of the snuffs also has antifungal and curare-like activities; chemical analysis on the active principles has not been done. Several plants are considered as prime candidates for scientific study as sources of useful chemicals for medicine or industry. These include some used to prepare teas or other infusions for treatment of various symptoms of senile dementia.

1994 Ethnobotany and the search for new drugs. Wiley, Chichester (Ciba Foundation Symposium 185) p 106–115

There are numerous definitions of ethnobotany. The usual and simplest definitions suggest that it is the study of the uses of plants in primitive societies in both modern and ancient times. Given the many rapidly expanding subdivisions of this interdisciplinary field, a more inclusive definition seems to be necessary. It would encompass the study of the uses, technological manipulation, classification, indigenous nomenclature, agricultural systems, magico-religious concepts, conservation techniques and general sociological importance of the flora in primitive or pre-literate societies (Schultes 1992).

One of the most active aspects of ethnobotanical research is the search for potentially valuable medicinal material with chemical constituents from which new curative agents may be created for modern pharmacopeias and for the benefit of all mankind. This aspect is often termed ethnopharmacology.

Tropical rain forests, with their exuberant vegetation, which is usually extremely rich in species but incompletely studied, offer the greatest possibilities for the discovery of promising new drugs. The Amazon is an exellent example: 2 722 000 square miles, containing the largest drainage area in the world and one-sixth of the fresh water of the globe, support 80 000 species of higher plants or approximately 15% of the world's flora and an extraordinarily diverse Indian population speaking some 500 languages. The flora increases in complexity and richness towards the western parts near the eastern Andean slopes.

If chemists expect to obtain by random collecting sufficient quantities of 80 000 species and all their ecotypes from this area of such difficult travel and transportation, the task may never be completed. It would be more productive to concentrate on those plants that the indigenous peoples have found, over millennia of experimentation, to be bioactive and which they have used in their medicine as ameliorators, stimulants or curative or psychoactive agents.

An Indian will naturally not seek out a plant, prepare it and imbibe or otherwise employ it to lessen pain or 'cure' ills, unless it produces some physiological effect. If a plant has any physiological or psychic effect, it must have at least one bioactive chemical constituent. We should learn what the constituents are. They may never be useful in our medicine; they may—as in the case of tubocurarine or rotenone—be put to a wholly different use; or, rarely, they may be employed—as quinine has been—for the same purpose as amongst the natives. Occasionally, the chemical structure may be manipulated by the synthetic chemist to produce new semi-synthetic molecules, some of which might be of interest.

That the Amazonian flora represents an extraordinarily rich and hitherto neglected chemical laboratory cannot be doubted. A few years ago, I surveyed five annual volumes of several phytochemical journals and discovered that, in that period, some 250 new alkaloids from Amazonian species had been reported. Since alkaloids represent only one type of the numerous secondary organic compounds in plants, the chemical wealth that lies awaiting in this vegetal laboratory is at once obvious.

Urgency for ethnobotanical conservation

There is an urgency to train many more field ethnobotanists. This precious knowledge seems in many regions to be doomed to extinction with the rapidly increasing acculturation and Westernization resulting from construction of roads, airstrips and dams, with consequent tribal displacement, missionary pressure, warfare, tourism, industrial penetration, local greed on the part of settlers and even well-intentioned governmental efforts to 'civilize' the natives. Loss of this knowledge—and even the physical annihilation of whole tribes—hinders not only the search for new drug plants but also our efforts to conserve the environment and the flora. Realization of this impending loss has recently given a sense of urgency to the need for ethnobotanical conservation.

Examples of ethnobotanical contributions to the search for new drugs

During my 47 years of field work and that of my students amongst the Indians of the north-west Amazonia, especially in Colombia, some 1600 species employed by the natives as medicines, hallucinogens or poisons have been collected and identified (Schultes & Raffauf 1990). Although only a few have ever been chemically analysed, a number of extremely interesting and unexpected results have come from the botanical and chemical studies of these bioactive plants.

Yoco is a wild forest liana from the bark of which is prepared a stimulating drink of great importance to numerous tribes of the Putumayo of Colombia and adjacent Ecuador (Schultes 1942). References to its domestic and medicinal uses in reports of travellers and missionaries going back to the 17th century (Patino 1967) mention its importance in tribal life and include much ethnobotanical information (Schultes 1979).

The bark of *yoco* was analysed by French scientists (Rouhier & Perrot 1926) and found to have a high concentration of caffeine. The plant was not identified botanically, however, until 1942, when it was described as a new species of the Sapindaceae: *Paullinia yoco*. It is the only caffeine-rich species the bark of which is the part used for its bioactive effects (Schultes 1942).

Another caffeine-rich plant the identity and properties of which ethnobotanical studies have clarified is the cultivated *Ilex guayusa*. This is used by numerous tribes in Ecuador as a morning stimulant and as a vomitive before magico-religious ceremonies (Schultes 1972b). It was described botanically in 1901 from sterile material and was believed never to flower. There was some uncertainty that it was a species, until ethnobotanical investigation found fertile trees in Provincia Pastaza, Ecuador, in 1979 (Shemluck 1979).

The most important hallucinogen in the western and south-western Amazonia of Bolivia, Brazil, Colombia, Ecuador, Peru and Venezuela is a drink prepared basically from the bark of a malpighiaceous liana (Naranjo 1953, 1986, Schultes & Raffauf 1992). It was identified in 1852 by the British plant explorer and ethnobotanist, Richard Spruce (1978, 1908). It is known variously as *ayahuasca*, *caapi*, *natema*, *pinde* and *yajé*. Spruce, unlike most botanists of the last century, wrote a detailed ethnobotanical report on this psychoactive plant, including his experiences in participating in an Indian ceremony and partaking of the intoxicating brew. He collected and described the liana as *Banisteria caapi* (now called *Banisteriopsis caapi*). Furthermore, unlike botanists of his time, he even collected material for chemical analysis. This was lost in transit but eventually arrived at the Royal Botanic Gardens, Kew. The analysis was not done until 1969 (Schultes et al 1969).

Ethnobotanical studies of this hallucinogen have led to the discovery of significant aboriginal knowledge of the appropriate plant additives to use as admixtures with the original drink from *Banisteriopsis*. Many different plants—mostly themselves toxic—are employed by various tribes as additives, but two are pre-eminent and widely used: leaves of the rubiaceous *Psychotria viridis* and of the malpighiaceous *Diplopterys cabrerana* (formerly known as *Banisteriopsis rusbyana*) (Instituto Indigenista Interamericano 1986, Rivier & Lindgren 1972, Schultes & Hofmann 1980). These additives lengthen and intensify the intoxication produced by the *B. caapi* plant. The active constituents of *B. caapi* are β-carboline alkaloids (Deulofeu 1967), harmine, harmaline and tetrahydroharmine. The alkaloids in the leaves of *Psychotria* and *Diplopterys* are tryptamines, which are inactive when taken orally unless taken with a monoamine oxidase inhibitor. The β-carbolines in *B. caapi* are in effect inhibitors of this enzyme. The aboriginal discovery of these two most important additives from a flora of 80 000 species indicates the depth of Indian understanding of the properties of plants in their ambient vegetation. How much more can ethnobotanical research learn of value, if it be intensified in such regions as the Amazonia?

Recent ethnobotanical research has clarified earlier confusion in the anthropological literature concerning the snuffs of the Amazonia (Cooper 1949). We know now that three major snuffs are involved: tobacco, *yopo* (*Anadenanthera peregrina*) and the recently identified *nakwana*, *epena* or *yakee* from several species of the myristicaceous genus *Virola*.

Tobacco is widely employed in the western Amazonia in the form of snuff or *ambíl* (a concentrated liquid placed in the mouth over the gums); it is rarely smoked except in certain ceremonies. *Yopo* is historically the most important snuff; it is used mainly in the northernmost parts of Brazilian Amazonia and the Orinoquia of Colombia and Venezuela. It was formerly employed in the West Indies where it was called *cohoba* (von Reis Altschul 1972, Safford 1916).

A snuff long confused with *yopo* is widely prepared in the Orinoquia of Venezuela and Colombia, the Vaupés of Colombia and the north-western part of the Brazilian Amazon. The source of this hallucinogenic snuff is the red, resin-like exudate of the inner bark of *Virola calophylla, Virola calophylloidea* and especially *Virola theiodora*; other species of this genus may be employed locally. The source of this snuff was definitively identified from ethnobotanical research in 1954 (Schultes 1954). The hallucinogenic activity is due to high concentrations of several tryptamines, in some preparations as high as 11% (Schultes & Holmstedt 1968).

Several discoveries resulted from my ethnobotanical research into the source of the *Virola* snuffs. The primitive Waika Indians in Brazil, who employ the snuff with unusual frequency, value the resinous exudate on darts as a weak kind of curare for hunting small animals. The other interesting use amongst these people is the application of the 'resin' over 10 to 15 days in the treatment of fungal infections of the skin with apparent success, whether as a cure or suppressant is still not known. Studies have not yet been carried out on the active principles for the curare-like or the antifungal effects.

The interesting point here is the frequency with which ethnobotanical research on one aspect of the use of a plant leads to other unexpected utilitarian aspects that may present new opportunities for phytochemical or other investigations.

Suggested Amazonian plants worthy of study for new drugs

Of the 1600 species reported as medical or toxic plants in the north-west Amazonia, several appear to be worthy of scientific study as sources of useful chemicals either in medicine or industry. Those of top priority, in my opinion, are the following.

The Indians of the Colombian Amazonia have a reputation for caring as much as possible for the aged and infirm. In my ethnobotanical field work, I found 23 species of plant employed in one way or other in treating senile dementia. Of these 23, the following seem to be outstanding and worthy of technical investigation in view of the direct statements of native informants (Schultes 1993).

Mandevilla steyermarkii (Apocynaceae) is a strict endemic: the roots of this beautiful plant are decocted for administration 'to the sick and aged'. *Tabernaemontana heterophylla* (Apocynaceae) is used by Tukano Indian medicine men; they prepare a tea of the leaves for 'old folk who are slow and forgetful'. The fruit of *Gnetum nodiflorum* (Gnetaceae) is boiled by the Kubeo to make a tea, which is prescribed for three or four days for elderly men 'who cannot walk straight and totter and fall'. *Vismia tomentosa* (Guttiferae) is also made into a tea by the Yukunas and Makunas. They believe that the elderly 'who suffer difficulty in understanding instructions and have physical degeneration and difficulty in talking' are helped by a week-long administration of a tea of the leaves. *Piper schultesii* (Piperaceae) is valued by the Karijonas, who gather leaves from the Sierra de Chiribiquete and dry them for use. The plant material is allowed to stand in water for a day and may be put into *chicha* (a fermented drink of maize) for the elderly who 'sit without talking all day, staring into space'.

The Tikunas of the Río Loretoyacu boil the leaves of *Tournefortia cuspidata* (Boraginaceae) to prepare a tea given twice daily over two weeks to the elderly who 'shake all over'. Several species of *Souroubea* (Marcgraviaceae)—*S. crassipetala*, *S. guianensis* var. *corallina* and *S. guianensis* var. *cylindrica*—are frequently employed by Indians in the Colombian Vaupés to make a leaf tea to calm extremely nervous elderly people and to hasten sleep when taken warm in the evening (Schultes & Raffauf 1990).

The leaves of *Caryocar microcarpum* (Caryocaraceae) appear to be toxic to the leaf-cutting ant, killing the ants when they attempt to cut the leaves. Since these insects cause great crop losses in the tropics, the responsible chemical constituent could be a boon to agriculture.

Conclusion

The intensification of ethnobotanical field research is urgently needed because of acculturation in most areas where people of primitive societies are living and because of the increasing devastation and destruction of rain forests in the tropics. From both points of view, an incredible chemical laboratory still practically untouched by scientific investigation for new therapeutic agents is awaiting urgent attention.

I end with a plea that all scientists involved make every effort possible to encourage the training of more ethnobotanists or ethnopharmacologists willing to carry on field work with people who still have the knowledge of plants and their properties which they have inherited from generations unnumbered.

References

Altschul von Reis S 1972 The genus *Anadenanthera* in Amerindian cultures. Botanical Museum, Harvard University, Cambridge, MA

Cooper JM 1949 Stimulants and narcotics. In: Steward JH (ed) Handbook of South American Indians. US Government Printing Office, Washington, DC, vol 5, p 525–558

Deulofeu V 1967 Chemical compounds isolated from *Banisteriopsis* and related species. In: Efron D, Holmstedt B, Kline NS (eds) Ethnopharmacologic search for psychoactive drugs. US Government Printing Office, Washington, DC (Publ Health Serv Publ 1645) p 393–402

Instituto Indigenista Interamericano (ed) 1986 América indigenista. Instituto Indigenista Interamericano, México, DF, vol 46

McKenna DJ, Towers GHN, Abbott F 1984 Monoamine oxidase inhibitors in South American hallucinogenic plants. Tryptamine and beta-carboline constituents of ayahuasca. J Ethnopharmacol 10:195–223

Naranjo P 1953 Ayahuasca: ethnomedicina y metología. Ediciones Labri Mundi, Quito

Naranjo P 1986 El ayahuasca en la arqueología ecuatoriana. In: Instituto Indigenista Interamericano (ed) América Indigena. Instituto Indigenista Interamericano, México, DF, vol 46:117–127

Ott J 1993 Pharmacotheon. Natural Products Company, Kennewick, WA, p 223–231

Patino VM 1967 Plantas cultivadas y animales domesticados en Amèrica equinoccial, vol 3: Fibras, medicinas, miscelaneas. Imprenta Departmental, Cali, Colombia, p 259–262, p 244–253

Rivier L, Lindgren J-E 1972 Ayahuasca, the South American hallucinogenic drink: an ethnobotanical and chemical investigation. Econ Bot 26:101–129

Rouhier A, Perrot 1926 Le 'yocco', nouvelle drogue simple à caffeine. Bull Sci Pharmacol 33:537–539

Safford WE 1916 Identity of cohoba, the narcotic snuff of ancient Haiti. J Wash Acad Sci 6:548–562

Schultes RE 1942 Plantae colombianae. Yoco: a stimulant of southern Colombia. Botanical Museum, Harvard University, Cambridge, MA

Schultes RE 1954 A new narcotic snuff from the northwest Amazon. Bot Mus Leafl Harv Univ 16:241–260

Schultes RE 1972a De plantis toxicariis e Mundo Novo tropicale. XI. The ethnotoxicological significance of additives to New World hallucinogens. Plant Sci Bull 18:34–40

Schultes RE 1972b Ilex guayusa from 500 AD to the present. Etnol Stud 32:15–138

Schultes RE 1979 Discovery of an ancient guayusa plantation in Colombia. Bot Mus Leafl Harv Univ 27:143–153

Schultes RE 1992 Ethnobotany and technology in the northwest Amazon: a partnership. In: Plotkin M, Famolare L (eds) Sustainable harvest and marketing of rain forest products. Island Press, Washington, DC, p 7–13

Schultes RE 1993 Plants in treating senile dementia in the northwest Amazon. J Ethnopharmacol 38:129–135

Schultes RE, Hofmann A 1980 The botany and chemistry of hallucinogens, 2nd edn. CC Thomas Publishers, Springfield, IL

Schultes RE, Holmstedt B 1968 De plantis toxicariis e Mundo Novo. II. The vegetal ingredients of the myristicaceous snuffs of the northwest Amazon. Rhodora 70:113–160

Schultes RE, Raffauf RF 1990 The healing forest. Medicinal and toxic plants of the northwest Amazonia. Dioscorides Press, Portland, OR

Schultes RE, Raffauf RF 1992 Vine of the soul. Medicine men, their plants and rituals in the Colombian Amazonia. Snyergetic Press, Oracle, AZ

Schultes RE, Holmstedt B, Lindgren J-E 1969 De plantis toxicariis e Mundo Novo tropicale. III. Phytochemical examination of Spruce's original collection of Banisteriopsis caapi. Bot Mus Leafl Harv Univ 22:121–132

Shemluck M 1979 The flowers of Ilex guayusa. Bot Mus Leaft Harv Univ, vol 27, nos 5–6

Spruce R 1978 Richard Spruce and the potential for European settlement of the Amazon; an unpublished letter. Bot J Linn Soc 77:131–139

Spruce R 1908 Notes of a botanist on the Amazon and Andes. (Wallace AR: ed) Macmillan, London, vol 2:414–425, 453–454

DISCUSSION

King: Virola is well known to many scientists. We have looked at its antifungal activity; it does suppress the growth of skin and other fungal pathogens, but it's not fungicidal. The compounds are known and have been published (Gottlieb 1979). It works and it may not be a good drug for Western markets but it's quite efficacious in local applications.

Jain: Training is something that we should think about very seriously. Compared with laboratory training in pharmacology or tissue culture or genetics, it is not expensive to train ethnobotanists. Particularly, because the need for training of ethnobotanists is greater in developing countries, where usually things are still inexpensive and training can be done at less cost. We have been conducting a training course in ethnobotany in Lucknow in India in alternate years since 1986. This lasts for a week or 10 days and there are about 30 students in each group. Of the 120 who have been trained in the last seven years, about 70% have taken up ethnobotany as a profession. The responsibility for training should not fall only on educational institutions, or government or research institutions; it should also fall on people like those here who in the future want the phytopharmaceutical industry to benefit from ethnobotanical findings.

Lozoya: There are numerous definitions of ethnobotany, but I disagree absolutely with the definition of Professor Schultes. He is ignoring 20 years of very important discussions by many of the people who are sitting around this table. In Latin America, we have been fighting for 20 years to replace language that refers to us as primitive societies. Mexican traditional medicine is not primitive. I propose that we use one of the common definitions of ethnobotany already used by American, French, German, Mexican, Brazilian, Chinese and Japanese universities, that give us a better perspective for the end of the 20th century.

Balée: Professor Schultes' definition of ethnobotany is still the Harshberger's definition, to a large extent, of the 19th century in that the focus is on 'primitive' or non-literate societies. Schultes tries to amplify the importance of the definition in other ways, and I agree with the topical material. We could more broadly define ethnobotany as the total relationship between people and plants. It need not be, for example, non-literate societies only that have ethnobotanical systems. What about Chinese ethnomedical systems, Ayurvedic ethnobotany in India or Mayan ethnobotany, all of which are based on literate traditions?

The big question is whether ethnobotany is seen as a science or not. Are non-Western therapeutic uses of medicinal plants, for example, in the materia medica of China, India and ancient Greece before there was 'real' science, to be considered as ethnobotany or not? Or does ethnobotany simply involve the study of relationships between plants and people? Ethnobotany should not be limited to one or another kind of society; it should be applicable to humankind generally, wherever plants are used and are present in the environment.

Jain: This is not the forum for a discussion on the definition of ethnobotany but certainly the word primitive should be deleted. Ethnobotany is now considered almost unanimously all over the world as the direct relationship between human beings and plant resources. Many advanced human societies in the world have very poor direct relationships with plants, relationships or uses which do not fall into the category of economic botany or medical botany. Therefore a direct relationship between plants and human societies is ethnobotany.

Iwu: We should state clearly that some words are insensitive and should not be used. We should not describe a people's culture without any respect. Professor Schultes has made a wonderful contribution to this discipline, but I take strong exception to the continued use of the term primitive in describing non-European cultures or ethnobotanical work.

Peter: Professor Schultes applied the term 'primitive' to communities that are still rooted in nature. It is not limited to any specific region and could apply to rural communities in many parts of the world. However, since it can easily be (mis)interpreted as a demeaning cultural qualification, it should not be used any more in the context of ethnobotany.

Cragg: Professor Schultes has a tremendous respect for the people with whom he has worked for decades. I don't think he would have meant this in a demeaning sense at all; it is an unfortunate choice of words.

Martin: Ethnobotany is not just about the people we are studying, it's also about the plants and the ecosystems where we work. Professor Schultes concentrates on tropical humid forest, primary forest at that. Ethnobotany also focuses on temperate forests, arid areas and on secondary vegetation. In fact, the majority of useful plants are found in secondary vegetation. Professor Craveiro's presentation showed that medicinal plants come, not just from tropical humid vegetation, but also from arid areas (Craveiro et al 1994, this volume). We can expand the definition of ethnobotany to include the people other than indigenous people and plants beyond those found in primary tropical forest.

Iwu: I agree. The problem with Professor Schultes' definition is not just the choice of words, it is the whole line of thinking that has governed ethnobotany up till now. Walter Lewis mentioned dividing the forest into three zones: the core region, the buffer zone and an outer, more interactive zone (Lewis & Elvin-Lewis 1994, this volume). It is difficult to get any foreign conservation group, such as The World Wide Fund for Nature or the British Overseas Development Agency to agree to explore areas where the human interaction takes place. Everybody is interested in so-called travel book wilderness—'where no man has ever gone before'. In our cultural system, those areas are already protected, because we don't go there. So the area where there is very high intensity of human use, where ecological conservation is needed, is not studied.

The type of ethnobotany described by Professor Schultes was very useful work but for a different era. It is an era that has gone and the old school needs to re-educate itself. This brings us to the issue of training. How suitable are the programmes offered in Western institutions to meet the demands of ethnobotany? We have just conducted a training programme for ethnobotanists. This took 10 weeks and was for graduates who had already worked in the field. They need to learn about our value systems, about our religion, about our cosmology. These are people who grew up in Nigeria and they still need 10 weeks of intensive training to be able to interview and talk with the local people. Imagine for how much longer a foreign ethnobotanist needs to be educated before being able to enter my world.

I am happy that the New York Botanical Garden is doing a random screen of the USA's own flora. Nobody has ever done any serious study of the flora in their own backyard, because it doesn't fit into what we call ethnobotany. We have to clarify this concept because it gives the impression in Third World countries that ethnobotany is a colonial exercise—that people are coming to get the remnants of what is left in our culture and they are putting nothing back. Some people don't want the lifestyle of the Indians they are studying to be better because that means change.

Posey: I totally agree, even to the point one can say that the concept of ethnobotany is an oxymoron. Traditional peoples do not separate plants from ecosystems; that is all integrated. We ought to be finding some basis for the re-integration of science, not the particularization of it. We have to start seeing all of this in a much larger ecological context—plants in the context of the environment.

McChesney: Plants are utilized in very broad ways. This symposium has focused on pharmaceuticals—new drugs. The outcome of research on plants depends primarily on how we ask the question relative to the biological activity. Will that biological activity address a disease state or will it address an agricultural application, for example? There is increasing concern over the environment, in the developed world as well as in the developing world, for example over the persistence of our current pesticides in the environment and their potential toxicity. We need new approaches to maintain the productivity of agriculture. Plant products and natural products hold real promise in that area and those issues ought to be addressed as well as issues directly affecting health.

Martin: Professor Schultes mentioned the urgency for ethnobotanical conservation. He has played a really critical role since the 1930s in bringing to our attention the importance of local knowledge and the fact that it is a tradition which should continue into the future.

Conservation is a difficult term to use for holding on to peoples' knowledge. I would prefer to use a term which stresses the continuing development of ethnobotanical knowledge. When we think about conservation of forests, we know there's a division between *ex situ* and *in situ* conservation. If you take things out of the forest and preserve them in botanical gardens, or practise other forms of *ex situ* conservation, you take a very small sample of genetic diversity. The tendency now is towards *in situ* conservation, so that the ecosystem can continue evolving and continue to develop with new speciation. Let's extend that metaphor to the knowledge about plants. If we just extract the knowledge and put it in books, which is a kind of *ex situ* conservation of that knowledge, it won't grow and develop. We should be aiming for *in situ* development of this knowledge, so that ethnobotany can be a living tradition, not just an academic tradition.

References

Craveiro AA, Machado MIL, Alencar JW, Matos FJA 1994 Natural product chemistry in north-eastern Brazil. In: Ethnobotany and the search for new drugs. Wiley, Chichester (Ciba Found Symp 185) p 95–105
Gottlieb OR 1979 Chemical studies on medicinal Myristicaceae from Amazonia. J Ethnopharmacol 1:309–323
Lewis WH, Elvin-Lewis MP 1994 Basic, quantitative and experimental research phases of future ethnobotany with reference to the medicinal plants of South America. In: Ethnobotany and the search for new drugs. Wiley, Chichester (Ciba Found Symp 185) p 60–76

African medicinal plants in the search for new drugs based on ethnobotanical leads

Maurice M. Iwu

Phytotherapy Research Laboratory, Bioresources Development and Conservation Programme, University of Nigeria, Nsukka, Nigeria; and Visiting Research Associate, Walter Reed Army Institute of Research, Washington, DC 20307-5100, USA

Abstract. In the African world view the natural environment is a living entity whose components are intrinsically bound to mankind. Dietary plants, spices and common herbs dominate the materia medica, in contrast with modern orthodox medicine which uses many regulated poisons. Drug development based on ethnobotanical leads has followed two paths: the classical approach of identification of single plant species with biologically active compounds and the characterization and standardization of traditional recipes for reformulation as medicines. The first approach has led to the recognition of many African plants as medicines and the isolation of several biologically active molecules; examples range from the well known physostigmine (from *Physostigma venonosum*) used for the treatment of glaucoma to the recently identified antiviral agents from *Ancistrocladus abbreviatus*. The second approach which aims at optimization of mixed remedies as formulated dosage forms is perhaps more relevant to the needs of the poor rural populations but has remained largely ignored. Drug development programmes based on ethnobotanical leads must provide for just and fair compensation for individual informants and local communities.

1994 Ethnobotany and the search for new drugs. Wiley, Chichester (Ciba Foundation Symposium 185) p 116–129

Culturally, Africa would appear to be extremely heterogenous with well over 2000 distinct tribes and as many languages and dialects; however, certain common threads run through most traditional medical systems in the continent. An example is the near total reliance on plants for sources of ingredients for the formulation of remedies. Animal parts, insects and certain clays are used in a rather limited sense. Another common feature is the strong influence of religion in the diagnosis of diseases and prescription of remedies.

The continent, with its long history of human civilization and centuries old record of the use of plants as medicine, is a rich source of leads for the development of new therapeutic agents. Indeed, many modern pharmaceuticals and everyday herbs owe their origin to Africa. Many scientific groups are currently exploring African flora for new compounds with pharmacological activities. Such efforts have led to the isolation of several biologically active molecules that are in various stages of development as pharmaceuticals (Iwu 1993). There is therefore little doubt that a systematic ethnobotanical study of African plants is a viable approach for the development of new therapeutic agents; the unsettled questions, however, concern the methods to be followed for such studies. What criteria should be adopted to protect the cultural integrity of the host communities from being completely destroyed by the investigators? What type of drugs should be developed? Should the health needs of developed countries be the sole criterion for determining the usefulness of the medicinal plants? In a fundamental sense is it ethical to isolate the cure from the system that produced and nurtured it? Even the basic question of what constitutes a medicine has to be addressed: should we be forced to accept the reductionist Cartesian model of therapeutics as the basic framework for ethnobotanical studies?

This paper will focus on Africa and will be largely limited to our experience in the development of drugs based on ethnobotanical leads. I shall begin with a brief discussion of the fundamentals of African traditional medicine and world view, followed by an overview of drug development programmes based on the utilization of African plants. I shall conclude by giving insights into some of our current projects in ethnobotany and biodiversity prospecting, including the Salvage Ethnography Project, the KIBORD (Kates Institute of Bio-organic Research and Development) initiative and the Bioresources Development and Conservation Programme, as well as our joint projects with Shaman Pharmaceuticals Inc. and the Walter Reed Army Institute of Research. These projects differ significantly in their objectives, scope and methodology but unfortunately time will not allow a detailed discussion.

Several factors have limited the search for new drugs from African plants. Three of these have seriously undermined otherwise well conceived projects. The first is the inadequate appreciation of the relationship between indigenous African communities and the environment. There is a strong belief in the *sacrality* of the Earth, according to which, not only is the Earth considered sacred but precise rules and rituals are prescribed for the proper use of its bounties. It is therefore very difficult to separate the purely physical properties of plants from their spiritual attributes. The second limiting factor has been the near total devastation of waves of colonial rule and the enduring disruptive effect of the more aggressive and dominant European culture. For example, most of traditional medicine consists of mixtures of various herbs, whereas European drugs are mainly isolated compounds obtained from single plants. When ethnobotanical surveys are conducted in Africa, it is usually not to record the general relationship between the local communities and plants but to discover whether any of the plants contain chemicals for development as drugs for European medicine.

A third limiting factor, which is perhaps global, is the fact that the early investigations of African remedies were conducted by anthropologists who were more concerned with exotic tales of 'primitive' tribes with bizarre habits than undertaking the often dry chore of recording countless remedies in jaw-breaking languages. Even today, although the multidisciplinary nature of ethnobotany is espoused by nearly all those involved in the subject, most ethnobotanical studies are conducted not by multidisciplinary teams with qualified medical practitioners, botanists and ethnographers but by individual botanists with limited medical knowledge or worse still by pharmacists and chemists with very poor training in plant taxonomy. A study of African medicinal plants must begin with the broad consideration of the role of plants in the practice of traditional medicine. A proper ethnobotanical investigation aimed at drug development should take into account the medical system in which the plants are used.

Traditional African medicine belongs to what have been classified as *personalistic* systems in which supernatural causes ascribed to angry deities, ghosts, ancestors and witches predominate, in contrast with the *naturalistic* systems where illness is explained in impersonal, systemic terms (Bannerman et al 1983). In the African system of medicine, healing is concerned with the utilization of human energy, the environment and the cosmic balance of natural forces as tools in healing. In the African world, the natural environment is a living entity, whose components—the land, sea, atmosphere, and the faunas and floras—are intrinsically bound to humans. Plants therefore play a participatory role in healing. A healer's power is determined not by the number of efficacious herbs he knows but by the magnitude of his understanding of the natural laws and his ability to utilize them for the benefit of his patient and the whole community. Treatment therefore is not limited to the sterile use of different leaves, roots, fruits, barks, grasses and various objects like minerals, dead insects, bones, feathers, shells, eggs, powders and the smoke from different burning objects for the cure and prevention of diseases. If a sick person is given a leaf infusion to drink, he or she drinks it believing not only in the organic properties of the plant but also in the magical or spiritual force imbibed by Nature in all living things and in the role of his or her ancestors, spirits and gods in the healing processes (Iwu 1990). The African healer, therefore, could play one or more of the following roles: medicineman, diviner, adjudicator, protector against natural and supernatural forces, and enhancer of success.

Another major characteristic of the African materia medica is the dominant use of edible plants as medicines, in contrast with the modern orthodox medicine in which drugs are essentially poisons that, if taken in regulated doses, may be useful in treating diseases. Everyday culinary plants when processed in a prescribed manner, often different from their nutritional use, provide the traditional healer with most of his remedies.

Medicinal plants and drug development in Africa

Medicinal plants are used in five main ways in Africa: 1) as ingredients for the preparation of traditional remedies; 2) as herbs in medicinal soups and teas; 3) for the preparation of pharmaceutical galenicals; 4) as phytomedicines prepared in standardized forms but retaining essential features of their traditional use; and 5) as sources of biologically active compounds for the development of pharmaceutical dosage forms. In the first type, the plants are usually collected fresh and when needed, except for those that require prior drying or are available only in certain seasons or distant locations, in which case they are collected and stored until needed. This method accounts for the bulk of medicinal plant use in the continent. Herbal teas and medicinal soups account for the second major type of plant use. Only a few standardized phytomedicines are manufactured locally in Africa. Pilot projects to manufacture local plant medicines as standardized drugs have been initiated in Rwanda, Botswana, Egypt, Mali, Nigeria, Kenya and many other countries. Local investigation of medicinal plants for the isolation of pure compounds has been limited to phytochemical analysis of plants rather than a systematic and purposeful programme of drug development.

The Organization of African Unity's Science & Technology Research Council has pioneered a series of projects aimed at the standardization of plants used in traditional African medicine and the evaluation of plants as sources of biologically active compounds. The study has resulted in the publication of an African pharmacopoeia, a two volume compendium of plant medicine from the continent and methods for their standardization (Organization of African Unity 1986). In 1978, the World Health Organization initiated a medicinal plants evaluation programme. Many African plants were included in the more than 20 000 plant species identified in that study. The Nigerian Society of Pharmacognosy reviewed the study of medicinal plants in Nigeria and selected 12 plants for further development (Sofowora 1986).

One of the most active natural products networks, NAPRECA (Natural Products Research Networks for Eastern and Central Africa), has been coordinating studies on medicinal plants of Eastern and Central Africa. NAPRECA, which held its 5th Symposium in September 1993, publishes a newsletter on various aspects of medicinal plant study for its members.

Many of the transnational pharmaceutical companies have in-house drug development programmes that include several plants collected from Africa. Over a dozen European, Japanese and US companies have obtained their raw source materials from Africa. In nearly all cases, these companies treat the host countries where the plants are collected as global warehouses to be exploited at will and abandoned when the supplies are depleted.

Salvage Ethnography Project

The Institute of African Studies, the Faculty of Arts and the Department of Pharmacognosy, all of the University of Nigeria, Nsukka, in 1982 initiated a

'Salvage Ethnography Project' which was aimed at providing documentation on the *Nka-na-Nzere* of the Igbo people of south-eastern Nigeria. *Nka-na-Nzere* does not have an equivalent phrase in English; a rough translation would be 'the art and norms of the Igbo people'.

The project also allowed us the opportunity to develop a framework for the interdisciplinary collaboration that was essential for the objectives we had set ourselves. Under the general direction of the late Dr Donatus Nwoga, a professor of English and folk Igbo literature, the project collected information on proverbs, music, oral history, ethnobotany, indigenous biotechnology, ethnomedicine, literature, foods, customs, visual arts and other aspects of life in Igbo land.

Among the important lessons learned from that project was that meaningful community participation is essential for the success of such projects. Organizational innovations were formulated to integrate the efforts of the various disciplines involved in the programme and also to circumvent the traditional bureaucracy and crippling compartmentalization of the administrative structures in academic institutions.

Biotechnology Development Agency

Following the success of the Salvage Ethnography Project and our experience in the evaluation of traditional medicines, in 1989 a group of scientists, non-governmental agencies and the private sector established a consortium to develop a programme of resources management based on the application of modern biological techniques. The approach was to apply modern methodology to the study of traditional biological resources. It was clear to us that a fundamental factor in the threat to African biodiversity is the declining economic value of the environmental resources.

In broad terms, the main objectives of the cooperative programme were to develop methods for sustainable utilization of tropical plants and, more specifically, to collect, collate and codify available information on the uses of African plants, with special reference to indigenous food crops, medicinal and aromatic plants, and industrial crops.

A major aspect of this programme is the KIBORD project, a private sector initiative which, in collaboration with the Department of Pharmacognosy, University of Nigeria, Nsukka, has been investigating tropical African plants as possible raw materials for the cosmetics, pharmaceutical and food flavour industries.

Bioresources Development and Conservation Programme

The Bioresources Development and Conservation Programme was formed in 1991 as the conservation wing of the Biotechnology Development Agency.

It has since become an independent international agency. Its focus is presently on the south-eastern rain forest region of Nigeria, Western Cameroon and the Republic of Guinea. The eastern region of Nigeria presents varied ecological zones; by maintaining biodiversity plots in several areas of the region, we hope to have access to diverse plant species for future drug development work. The programme was designed right from the beginning to address the real concerns of the rural dwellers, whose plight was often linked to previous 'top-down' experiments designed and implemented with minimal input from those whose lives were directly affected. The programme adopts a 'bottom-up' approach in its efforts to empower the poor and powerless rural dwellers to enable them to derive maximum benefits from their environmental resources and their labour.

Three major projects have been formulated under this programme. The first is the compilation of ethnobotanical information from our study area and an inventory of species in the Oban-Boshi-Okwangwo forest complex and the Korup region of Cameroon. The second project is to assess the economic value of the species in the forest complex. The third is the establishment of long-term nature plots to study forest dynamics.

It is perhaps too early to assess the long-term impact of these projects but within the short period of operation of the Bioresources Programme we have observed some tangible results. We do not yet have a finished product but we have begun a process with a clear vision as to its probable outcome. A major difference from similar efforts in other parts of the continent is that the programme was home grown, initiated and managed in its entirety by indigenous staff. It was therefore possible to integrate the decision-making process into the community. Science and technology are viewed as useful tools to be adapted to the cultural framework of productive activities, not as modern alternatives to the contributions of members of the community.

Preparatory to the above projects, we embarked on several activities. First, we hosted an international conference in collaboration with the RainForest Alliance, early in 1993, to address the related issues of biodiversity conservation and the industrial utilization of medicinal plants, at Enugu, Nigeria. The conference attracted participants from several countries and from diverse disciplines, providing local scientists the rare opportunity of discussions with experts in various fields. We started a training programme for ethnobiologists and field taxonomists. The first graduates of this programme will form the core group in our concentric model of building a network of ethnobiologists, field taxonomists and ethnographers. With the help of Shaman Pharmaceuticals Inc., we sponsored one of our ecologists to participate in the first ever training workshop on biodiversity monitoring organized by the Smithsonian Institution, US. We have so far conducted 10 sampling studies and our main ethnobotanical inventory of the region will commence in January, 1994.

Collaboration with Shaman Pharmaceuticals Inc.

The arrangement provides for joint efforts in all aspects of drug development. A joint team of Shaman staff and Nigerian scientists is engaged in a field ethnobiomedical survey. Selected plants are collected directly from the local communities and payment and compensation are effected in three modes. Firstly, a small cash payment is made directly to the informant/collector. Secondly, the community is assisted in its development projects. Thirdly, the medical member(s) of the team consult with the local healers and help them in treating some acute, life-threatening conditions. The international composition of the team helps us in our campaign to popularize the use of plant drugs. What could be more convincing than to have a Western-trained 'European' physician coming back to the village to correct the mistakes of earlier white missionaries.

It is also in our agreement that if and when a drug is developed from any of the leads provided by us, the royalty will be distributed among the informant, the community and the cooperative. The role of the scientists in this arrangement is essentially that of facilitating the contact between Shaman and the healers, not as middlemen or brokers. Because we are involved in all aspects of the drug development process, we are in a position to continue the development of some of the plant drugs as intact phytomedicines if they are found to be active but do not meet Shaman's criteria for future development. Perhaps the greatest impact of this collaboration is on our staff development and capacity building programme. We have been assisted by the company in our training of conservation staff at the Smithsonian Institution and they sponsored two of the lecturers at our first training course for ethnobiologists and field taxonomists.

Tropical Diseases Chemotherapy Project with US Army Medical Command

Life in the tropics is not really the idyllic haven many armchair pundits would like us to believe. Various types of parasitic diseases plague all the countries in the tropics. Unfortunately, because most of the people living in these countries are poor and unable to afford costly prescription drugs, the diseases that affect them are of little interest to pharmaceutical companies. Therefore, while malaria remains the number one killer disease in the world, no new drugs are being developed to treat it. Coincidentally, the US Army is interested in developing new antiparasitic drugs as part of its strategic programme to protect US troops. We have been collaborating with the Walter Reed Army Institute of Research in the development of new drugs for the treatment of malaria, leishmaniasis and trypanosomiasis. We have over a dozen candidate compounds in various stages of development. This project has led to the identification of indole alkaloids of *Picralima nitida* as a new type of chemical in the treatment of chloroquine-resistant malaria and possibly as the first broad-spectrum antiprotozoan agent for the treatment of leishmaniasis and trypanosomiasis.

There is no financial compensation for either the Nigerian scientists or the informants but we are adequately rewarded by the fact that we are developing drugs for the treatment of diseases that affect us. The intellectual rights are retained by us and members of our team have unrestricted access to modern facilities at the Walter Reed Institute.

Biological prospecting

I believe that this paper will not be complete if I do not comment on the current debate about the ethics of biodiversity prospecting. Drug-discovery programmes based on natural products offer one of the most feasible approaches to increase the net worth of forests while standing. It has been argued that while it is important to demonstrate the economic value of biological resources to a country's social and economic development, biological resources are in a sense beyond value because they provide the biotic raw materials that underpin every major type of economic endeavour at its most fundamental level (Oldfield 1984).

Development of drugs based on ethnobotanical leads has followed two paths: the classical approach of identification of single plant species containing biologically active compounds and the characterization and standardization of traditional recipes for reformulation as medicines. The first approach has led to the recognition of many African plants as medicines and the isolation of several biologically active molecules. The second approach, which aims at optimization of mixed remedies as formulated dosage forms, is perhaps more relevant to the needs of the rural populations but has remained largely ignored. Table 1 shows a list of plants that have been considerably investigated and may be used locally in primary health care. The thrust has been to promote plants that provide feasible returns on investment, while little attention has been paid to plants that contribute to the socioeconomic well being of the rural communities.

Most biodiversity prospecting programmes have followed the first model, in which raw biological materials are collected from developing tropical countries with the promise that if new pharmaceutical agents are discovered from the materials so collected, the pharmaceutical company will share the benefits with the donor country. In this scheme, it is expected that the donor country agency or contact will use the proceeds from the venture to support conservation programmes and foster economic development in areas where the product was harvested. Most of the companies participating in the development of plant-based drugs have taken steps to ensure adequate compensation for their partners in developing countries and some have mechanisms for protecting intellectual property rights of the indigenous peoples who provide them with information on plant use. Walter Reid et al (1993) have recently reviewed the various arrangements in place for biodiversity prospecting.

TABLE 1   Nigerian medicinal plants with potential applications in primary health care

Plant	Constituent(s)	Activity/indications
Aframomum melegueta	Essential oil, shagoal, gingerol	Antimicrobial, rubefacient
Ageratum conyzoides	Ageratochromone	Wound healing
Azadirachta indica	Nortriterpenoids	Antimalarial, antipyretic, seed insecticidal
Balanites aegyptica	Steroidal glycosides, furanocoumarines	Laxative, anti-inflammatory, molluscicidal
Bridelia ferruginea	Coumestans, flavonoids	Antifungal, mouth infections
Butyrospermum paradoxum	Fatty acids	Emmolient, anti-inflammatory
Cajanus cajan	Amino glycosides, phenylalanine	Management of sickle-cell anaemia
Carica papaya	Proteolytic enzymes (volatile oils in leaves)	For fevers, antidiabetic
Cassia spp.	Anthraquinone, glycosides	Laxative
Cola nitida	Caffeine, aromatic acids	Tonic
Cymbopogon citratus	Volatile oils	Diuretic, tonic
Dorstenia multiradiata	Leucoanthocyanidins	Antifungal, antiviral
Dracaena mannii	Saponins	Local antifungal, anti-protozoan
Eucalyptus globulus	Essential oil	Local antiseptic, colds, rubefacient
Garcinia kola	Biflavonoids	Antihepatotoxic, antiviral, adaptogen, plaque inhibitor
Morinda lucida	Anthraquinones	Antimalarial, jaundice
Ocimum gratissimum	Terpenes, xanthones	Antiseptic, coughs, fevers
Picralima nitida	Indole alkaloids	Antimalarial, broad-spectrum antiprotozoan
Piper guineense	Lignans, alkaloids	Antimicrobial, insecticidal, tonic, antiinflammatory
Psidium guajava	Essential oils, vitamins	Carminative
Sabiaceae calycina	Alkaloids, flavonoids	Wound dressing, laxative
Schwenkia guineensis	Steroidal glycosides	Oral hygiene
Sclerocarya birrea	Catechins, flavonoids, amino acids	Antidiabetic, tonic
Tamarindus indica	Ascorbic acid, citrates	Laxative, nausea
Tetrapleura tetraptera	Saponins, coumarins	Antiinfective, tonic
Uvaria chamae	Chalcones, terpenes	Antimicrobial
Vernonia amygdalina	Sesquiterpenes, saponins	Tonic, antidiabetic
Xylopia aethiopica	Diterpenes	Tonic, carminative, antiviral
Zanthoxylum xanthoxyloides	Aromatic acids	Management of sickle-cell anaemia
Zingiber officinale	Terpenes	Antihypertensive, anti-histamine

Apart from the serious issues relating to intellectual property rights raised by Shelton Davis (1993), biodiversity prospecting, as presently conducted by many organizations, may be harmful to the long-term interests of indigenous communities in more fundamental ways: for example, the so-called compensation, if not properly handled, could perturb the cultural value system of the community.

Another unsettled issue in biodiversity prospecting based on ethnobotanical leads in Africa is that of compensation of informants and communities that provide both the ethnobotanical information and the genetic materials used for drug development. The pharmaceutical companies are willing to pay only extremely low prices for plant samples. There is also the issue of genetic resource piracy that has been promoted and encouraged by many US, Japanese and European agencies which give supply contracts exclusively to their national institutions. These institutions undertake plant collection expeditions through the services of either herbaria or universities while paying minimal fees for the plants. Because drug development arrangements with transnational pharmaceutical companies may be inevitable for a variety of reasons, an indispensable component of the agreement should be to make a provision for developing local capacity and strengthening the scientific base of the indigenous materia medica.

Acknowledgement

The author received financial support from the UNDP/World Bank/WHO Special Programme for Research and Training in Tropical Diseases (TDR). The opinions and assertions contained herein are the private views of the author and are not to be construed as official or as reflecting the views of the Department of the Army or the Department of Defense.

References

Bannerman R 1983 Traditional medicine and health care coverage. World Health Organization, Geneva

Davis S 1993 Pathways to economic development through intellectual property rights. Paper delivered at the 1st International Conference on Cultural and Intellectual Property Rights of Indigenous Peoples, in Wakatane, New Zealand, 12–18 June 1993.

Iwu MM 1990 Symbols and selectivity in traditional African medicine. School of Postgraduate Studies, University of Nigeria, Nsukka, Nigeria

Iwu MM 1993 Handbook of medicinal plants. CRC Press, Boca Raton, FL

Oldfield M 1984 The value of conserving genetic resources. US Department of Interior, National Park, Washington, DC

Organization of African Unity, Scientific Technical and Research Commission (OAU/STRC) 1986 African pharmacopoeia. Volumes 1 and 2. Lagos, Nigeria

Reid WV, Laird SA, Meyer CA et al 1993 Biodiversity prospecting: using genetic resources for sustainable development. World Resources Institute, INBio, Rainforest Alliance, ACTS

Sofowora AE 1986 The state of medicinal plants research in Nigeria. University of Ife
    Press, Ile-Ife, Nigeria

DISCUSSION

Balée: You mentioned the identification of indole alkaloids from *Picralima
nitida* and their use as antiprotozoan agents. What is the plant used for by the
indigenous people?

Iwu: It is used all over west Africa for sleeping sickness, for malaria. In
northern Nigeria, it is used to treat cutaneous lesions.

Lewis: Maurice, have you made some arrangement with Shaman
Pharmaceuticals regarding compensation for the intellectual property of healers
in this long-term process, if something comes out of it?

Iwu: We have adopted a sample form from Shaman Pharmaceuticals. This
records the name of the person who supplied the medicine and their village.
If there is a profit, 50% goes to the person, 50% goes to the community. In
our system, the individual doesn't really have many rights, but we respect the
Western system in this.

Lewis: Is your institution in Nigeria involved in any way?

Iwu: I teach at the University of Nigeria, Nsukka. Most of our support comes
from grants. At the moment we have not made any money as royalties from
drug development. Since 1990, we have had a lot of material support in terms
of chemicals and so on from international agencies and corporations. Also, one
of our post-doctoral students is going to work at Shaman Pharmaceuticals;
another is presently at the Walter Reed Institute in Washington.

Lewis: Is there any direct payment if a product is produced?

Iwu: If there is a profitable product, the individual healer, his community
and our organization will benefit, but not the university, because public
institutions have no clearly defined guidelines for receipt of direct payment from
a corporate body.

King: The people in the different countries choose their own specific
arrangements; it is not Shaman Pharmaceuticals' choice to say what form
compensation will take. But whenever we discover something and commercialize
it, wherever we commercialize or discover it, whether in Ghana, Nigeria or
Peru, the profit is returned to the whole group, so that even if the successful
product doesn't come from Nigeria, they will still be beneficiaries and
vice versa.

Iwu: Shaman wanted to make a cash payment to one of the villages we have
been working with. In Shaman terms this was small, but for us it was significant.
I went to the village and told them about this contribution. They said that the
best use of the money was to support their ongoing communal project.

McChesney: In terms of your focus on tropical diseases, are you using
ethnobotanical leads primarily or exclusively?

Iwu: Exclusively. The Walter Reed Army Institute is still doing the studies on the medicinal chemistry and pharmacokinetics of the isolates. We already have the products formulated in Nigeria as a standardized extract.

We are exploring the possibility of securing loans from banks to help establish local phytomedical enterprises. The agricultural loans are more attractive because the interest rates are low. The strategy is to combine the pharmaceutical development with preparing crude drugs for the local populations. It is a continuing exercise and at the moment we are testing only the efficacy and toxicity of the extracts. This is enough to produce some useful results for the people who gave us the drugs in the first place. We are still following the classical drug development protocol and hopefully some day we will be able to isolate the active constituents as pharmaceutical agents.

McChesney: You are using standardized plant preparations?

Iwu: Yes, not pure isolates.

Cox: You believe the African people are avoiding plants of high toxicity. Secondly, you say there is overlap between plants used in diet and those used in medicine within traditional African plant use. Have you looked at diet as a factor in disease causation, for example in malaria?

Iwu: You have to appreciate that in Africa there is a completely different concept about what causes or constitutes a disease and what being healthy means. In Africa, somebody is sick only because they have deviated from the norm. They are not thought to be sick because of an external physical agent; except where there is a naturalistic causation, the spritual causation is the dominant explanation.

We don't normally use medicinal plants that are poisonous, unless the healer belongs to a special category. Those who have been initiated into the cult are taught about the poisonous plants. The shameful thing is that some ethnobotanical investigators insist on being told about these powerful drugs; these people have not been initiated and they are not supposed to know this. Western medical doctors are respected because they have been initiated according to the rites of their own culture. Such people may be told about the poisonous drugs.

The Calabar bean (*Physostigma venenosum*) was popularly used in West Africa as an ordeal poison, to test whether people are guilty or not. The only people who know how to use the drug are sworn to secrecy. To break the barrier, you have to send someone whom the people will accept as worthy to learn this secret. A new use has now been found for it in the treatment of glaucoma and recently for Alzheimer's disease. When the NIH (through Indena) tried to collect this plant, there was not enough, because in the 1940s the British had banned this drug.

Cox: I am interested in the role of diet in disease, because of the low incidence of antiparasitic plants we observed in our survey (Cox, this volume). Then I noted that, for example, the betel nut has very strong antihelminthic action.

I'm wondering if some dietary elements might have a therapeutic value. Does anyone know of any drugs that have come from plants used in traditional diets, but not as medicines. Secondly, I know that the National Cancer Institute has talked about diet as a possible avenue of cancer prophylaxis; are there investigations of diet as having some therapeutic value after the onset of disease?

Cragg: I think diet is very important. The NCI is placing more and more emphasis on prevention; obviously, nutrition and diet are critical factors in the whole concept of cancer and disease prevention. Various classes of compounds have been developed, for instance the carotenoids and the retinoids in preventive medicine.

Farnsworth: The US Congress appropriated $25 million for this kind of programme in 1990. This started because of claims that bran prevents colon cancer and lowers cholesterol levels. We received some contract money to study flax seed, which was speculated to have antioestrogenic-type activity. The idea was that eating bread baked with flax meal might prevent breast cancer in women. This project is going on. In the last two years, I've been invited to at least 10 major food companies for 2–3 day workshops, to discuss whether or not they should begin a programme in this area of functional foods or 'designer foods' or 'nutraceuticals'. There is a lot of interest in this. Maybe ethnobotanists should be looking not only at medicinal plants, but also at foods that are different from those usually eaten in the West that may contribute to nutrition and disease prevention.

Lozoya: Professor Iwu, according to our experience in Mexico, one of the big problems in the development of, and the use and promotion of, these herbal remedies is that in Mexico the majority of the plants are not formally cultivated in agricultural industries. Are you working on the problem of introducing the most important of these uncultivated plants into agriculture? How are you facing the production of large quantities of herbal remedies to be used in your country?

Iwu: This is a very important question. Often when you declare that a plant is useful or has a medicinal value, you are more or less signing the death warrant of the plant, unless adequate steps are taken to guide against over-harvesting. We are lucky to have a large pool of highly trained and enlightened forestry staff in Nigeria. We have not had problems like those with *Prunus* and *Pygium africana* in Cameroon and Madagascar.

Proper project evaluation is imperative. In our project on *Physostigma*, we studied its propagation, fruiting and cultivation. Interestingly, the plant thrives only in a deep forest setting. It does well in traditional agriculture where trees are left to provide shade. The seeds are very hard and, according to folklore, the nut has to be eaten by the African porcupine (*Hystriz cristata*) before it splits open to allow seed germination. So availability of the plant depends on the porcupine population in the forest. We are presently collecting the fruits from the wild through local farmers. The project has provided a good example of marrying ecological needs to economic considerations. We have started trial cultivation of this plant. The project is a difficult one. Since most agroforestry plants need many years to mature, they are not attractive to investors. Cash-flow analysis or the problem of cost effectiveness is a bane to all such projects and will remain so because *Homo sapiens* has become transformed into *Homo economicus* by the successes of the capitalist system. The whole economic order has to be changed. The concept of what constitutes a viable project has to be looked at.

McChesney: A vast majority of the world's population depends upon the use of medicinal plants for their primary health care. We are seeing a selective depletion of these plants. As we discover new Western pharmaceuticals from the same sources, and perhaps even more importantly as Western societies become either for economic reasons or for philosophical reasons more interested in traditional medical remedies, there is an increasing commercial market in those materials. That commercial market may cause the loss of the diversity of those particular species more rapidly than for any other species. We must take care that those species are not singled out for exploitation without appropriate strategies for their maintenance or production. Perhaps more importantly, we need a general strategy to maintain the diversity of medicinal plants that are presently recognized and will be important in the future.

Reference

Cox PA 1994 The ethnobotanical approach to drug discovery: strengths and limitations. In: Ethnobotany and the search for new drugs. Wiley, Chichester (Ciba Found Symp 185) p 25–41

Two decades of Mexican ethnobotany and research in plant drugs

Xavier Lozoya

Research Unit in Pharmacology of Natural Products, National Medical Center, Mexican Institute of Social Security, Avenida Cuahutemoc 330, Mexico City 06725, Mexico

Abstract. A renewed interest in the systematic study of indigenous medicines and associated medicinal plants arose in the 1970s. In Mexico the government established a national pharmaceutical industry to make use of the valuable colonial heritage of traditional practices combined with European medical concepts and resources. In 1975 the Mexican Institute for the Study of Medical Plants was created to integrate botanical, chemical and pharmacological studies on the Mexican flora. It compiled a database on ethnobotanical information relating to Mexican medicinal plants from the medical literature of the 16th to 19th centuries. A second database contained information on medicinal plants in current use. A medicinal herbarium was established. Taxonomical studies led to classification of the 11 000 voucher specimens in the herbarium and cross-referencing of the information with other databanks. A core group of 1000 plants used in traditional medicine throughout Mexico for almost 400 years was identified. Most of these are used to treat common diseases or basic health problems, usually given orally as decoctions or infusions. 95% of the plants used traditionally are from wild species. Information was collected from almost 3000 small Indian communities over four years on three aspects of traditional medicine—the healer, the disease categories recognized and the therapeutic resources in use. Plants with reported medicinal activity were selected for laboratory screening according to the frequency and commonality of their use, geographical distribution and seasonal availability. Screening involves a collaboration between chemists and pharmacologists: plant extracts are sequentially assayed and fractionated until the pure compound is isolated. Several active compounds are usually obtained from the same extract, frequently from the aqueous fractions. Ethnomedical information influences which plants are selected for screening and the type of assay used.

1994 Ethnobotany and the search for new drugs. Wiley, Chichester (Ciba Foundation Symposium 185) p 130–152

A renewed interest in the systematic study of indigenous medicines and associated medicinal plants arose in the 1970s. This was a worldwide phenomenon encouraged by the action of the World Health Organization (WHO), which was stimulated by the impact of the 'Chinese experience' on the Western medical world at that time (WHO 1978). Suddenly, Western medicine 'discovered' that the People's Republic of China, with a human population of millions, had been solving its primary medical necessities through the use of traditional, but at that time not known to Western science, procedures and therapies, with a high degree of success. Their heterodox programmes of primary health care that combined Western and traditional resources (herbal remedies, acupuncture, 'barefoot doctors', etc) appeared to be a pragmatic solution to historical problems of sanitation, nutrition and medical assistance (WHO 1984). Very soon, this extraordinary influence reached other developing countries in Africa, Latin America and, of course, the rest of Asia. The term traditional medicine came to mean any of the indigenous medicines and medical practices of 'Third World' countries (according to the terminology of those days), which until then were considered by Western scientists as primitive, underdeveloped or obsolete procedures. By contrast, after promotion by the WHO, traditional medicines appeared to be suitable for study and development, particularly because of the applicability of their human and therapeutic resources within the framework of WHO's goal: 'Health for all by the year 2000' (Akarele 1984). During the 1970s and the early 1980s, the political conditions in several countries in Africa and the aggressive economical plans of Asian countries helped create an impressive network of WHO Collaborative Centers (WHO 1987) for the promotion and development of traditional medicines in these areas. The Centers were mainly interested in the scientific validation of plants used in local medicines. Several established direct links with new companies in Europe and Asia interested in the development of herbal remedies.

In the Americas, this renaissance in the study of traditional medicines was not particularly well supported by health institutions and governments. This was due mainly to the lack of interest shown at the time by the Pan-American Health Organization in the promotion of the WHO Traditional Medicine Programme and also to the influence in the area of US health institutions and pharmaceutical industries. The former considered traditional medicine as a primitive custom of Latin American countries and native North American minorities. The pharmaceutical companies were still rooted in the 'economic botany model' of the 1950s with respect to financing plant drug research. In other words, North American science, which deeply influences medical and academic Latin American institutions, disdained the information about the use of medicinal plants by indigenous cultures (Lozoya 1980, Farnsworth 1984). This explains, 20 years later, why North American companies came late to the search for new plant drugs, when Asiatic and European companies were already marketing new herbal remedies.

Meanwhile, scientists from some universities in the USA, particularly from faculties of biology and anthropology, resuscitated some of the once very popular ideas about the importance of the study of the use of plants among the Indian groups of Mexico, Ecuador, Brazil and other countries in Latin America (Kroeber 1939, Schultes 1941, Jones 1941, Castetter 1944, Hill 1945).

They soon capitalized on the new interest in the study of traditional medicines. They created academic structures for participation in research in this field in some Latin American countries with the collaboration of local scientists. The concept of the 'ethnosciences' was promoted, including ethnobotany, ethnopharmacology and ethnomedicine (Hernández 1970, Barrera et al 1976, Shultes 1976, Lozoya 1976, Viesca 1976, Ford 1978, Smith 1979, Gomez Pompa 1982, Toledo 1982, Bárcenas et al 1982). The avalanche of medico-anthropological and ethnobotanical studies on medicinal plants of Latin America performed during the first decade (1973–1983) occurred, in the majority of cases, without support from the international pharmaceutical industry. This situation was readily accepted by Mexican scientists because the country was trying to avoid long-term technological dependence on foreign companies. They persuaded the government of the necessity to build a strong national pharmaceutical industry.

Mexico has a vigorous heritage of pre-Hispanic Indian culture and appears as a nation with strong traditional medicine. During the several centuries of colonial life, the combination of traditional medical practices with European medical concepts and resources produced a complex cultural wealth. Nevertheless, medicine, biology, botany, chemistry, pharmacology and other related disciplines that were developed in Mexican universities were exclusively Western oriented and ideologically dependent on foreign centres of the pharmaceutical industry. Until 1975, the traditional medicine of Mexico was considered only a theme of certain anthropologist's erudition. Some psychoactive mushrooms and the use of hallucinogenic plants in ritual ceremonies gave us international recognition as an exotic society: some North American and European researchers used to confuse aspects of Mexican traditional medicine with the hippie culture of the 1960s in their own countries.

The universities and large specialized centres of Mexican official medicine were, in the 1970s, still sceptical about influences from any other part of the world, with the exception of North America. In those years, for example, acupuncture was considered an 'act of faith' and Mexican physicians would leave the conference room if someone tried to demonstrate this technique. Only several years later, when US research centres and universities started to publish the first studies on acupuncture, did the Mexican institutions show some interest in this topic. Ethnobotany was no exception. This discipline grew up during the same period under the influence of European and North American academic groups that promoted the study of the indigenous flora of developing countries as a manifestation of the 'naturalism' and interest in the use of herbal remedies that arose in their postindustrialized societies. This 'green', 'ecological' social movement born in Europe has encouraged the appearance of herbal remedies in the Western world (usually called 'soft' herbal remedies) derived mainly from the study of traditional medicines of Asia and Africa, but also from the European flora.

As experience in the study of traditional medicines accumulated, a problem arose: not all traditional medicines showed the same level of complexity and development. Chinese, Japanese and Ayurvedic medicines have systematic and integral theoretical foundations, closely related to resources and ideas used over many centuries. In contrast, for Latin American and several African medicines, the picture was incomplete because the original indigenous theoretical framework was lost or deeply modified under the influence of the European colonial past. Books, schools and practitioners of indigenous medicine were destroyed or imprisoned during centuries of persecution. The Christian religion (especially the Catholic church) implanted discriminatory doctrines against indigenous cultures and colonial governments imposed other regulations which stopped the natural development of ancestral medicinal Indian practices. Because of this, the popular empirical use of plants has only partially survived and is now the most concrete aspect of traditional medicine available.

In general terms, that was the framework of reference when the Mexican Institute for the Study of Medicinal Plants (IMEPLAM) was created in 1975. The aims of this team, supported by governmental funds, were to integrate botanical, chemical and pharmacological studies on the Mexican flora and try to save the popular knowledge on the use of plants. Six years later, in 1981, this group was incorporated into the official health sector of the government's research system under its present name of Biomedical Research Center in Traditional Medicine and Natural Products funded by the Mexican Institute of Social Security (IMSS). During this time, in Mexico interest in ethnobotany has increased notably. Several universities have been involved in the promotion and development of ethnobotanical studies, particularly directed at the recovery of information about medicinal plants. Groups of pharmacologists and phytochemists have proliferated in several Mexican universities and the study of medicinal plants has become a formal activity in national science. What are the results and conclusions drawn from almost 20 years of ethnobotanical and plant drug studies in Mexico? I shall try to answer this question in the present paper as a member of one research team that intends to cover all the steps related to the complex problem of developing a plant drug in the prevailing technological and social situation in Mexico.

A first step: the ethnobotanical information

Before the creation of IMEPLAM, contemporary information about the medicinal plants of Mexico was scarce and disorganized. It resulted from some sporadic studies of pioneer Mexican chemists and physiologists in the 1940s and 1950s, which were influenced by the discovery of diosgenin in the *Dioscorea mexicana* rhizome. This finding allowed the massive production, by Syntex (US), of steroidal hormones. In those years, investigations on medicinal plants were published mainly as chemical reports on the structural characteristics of compounds isolated from several species. Few were about pharmacological observations of plant extracts administered to laboratory animals. As I said before, no national or international pharmaceutical company was particularly interested in these Mexican studies. Because chemical synthesis was the fashionable field in the study of natural products at that time, plants in general were considered as alternative raw materials for the synthesis of new molecules. In the 1960s, the interest was focused mainly on the search for new compounds for the treatment of cancer. The commonly expressed opinion in Mexican academic circles was that to obtain such compounds from plants was an expensive and technologically difficult process, possible only in rich, developed countries. The US companies kept the technology in their own laboratories and looked at the Mexican researchers only as providers of raw plant materials; they made no investment in pharmacological or chemical research in Mexico.

When, in 1975, the Mexican government intervened politically and economically in the exploitation by foreign companies of the *Dioscorea* rhizome (*barbasco*) that was taking place in Mexico, to prevent the extinction of this natural resource, a big political problem arose. The Mexican authorities intended to establish a governmental pharmaceutical hormone industry, but the reaction of the foreign pharmaceutical companies to this decision was to abandon the utilization of the Mexican raw material. Evidently, at that time, foreign investment in the promotion and development of research on medicinal plants was not expected.

In the middle of this conflict, the government decided to promote the national study of medicinal plants in Mexico to avoid repetition of the '*barbasco* drama' that reflected the total dependence of Mexico on foreign pharmaceutical industries for the supply of drugs. In 1975, IMEPLAM was created to design and organize a methodological approach to the study of medicinal plants.

The first step was to reorganize local information about these resources. Those were the years of the novelty of computerized databanks, so it was decided to create one on ethnobotanical information about Mexican medicinal plants. IMEPLAM started compiling a historical bibliography about medicinal plants. Several ancient sources of information about the use of medicinal plants were present in literature from colonial times, but they were dispersed and there was no system of classification.

The first databank was created with information from books selected from the medical literature of the 16th to 19th centuries. Many of the plants reported in this literature had been classified botanically. We discovered that a great majority of them were still used by the present Mexican population (Díaz 1976, 1977).

A second databank was created to collate information from surveys and field work about the medicinal plants in use. This second programme lasted seven years and included the creation of the first medicinal herbarium of Mexico. The IMSS herbarium now preserves 11 000 vouchers of medicinal plants collected from around the country. Taxonomical studies allowed the classification of this collection and cross-referencing of the information by botanical name with other databanks. This simple task was extremely important: Mexican medicinal plants are known by almost 10 000 synonyms in 52 Indian languages and Spanish. At the end of these studies we reported the existence of a 'basic group' of 1000 plants used in Mexican traditional medicine throughout almost all the country for almost 400 years (for review, see Aguilar & Martínez-Alfaro 1993). We also showed that the majority of them are used for the treatment of common diseases or basic health problems, such as respiratory, skin and digestive infectious disorders, hypertension, pain or diabetes, or to induce sleep or labour. This confirmed the argument that traditional herbal medicine is closely related to a Primary Model of Health Attention (Lozoya et al 1987). The majority of the plants are used as decoctions or infusions for oral administration (Lozoya & Velázquez 1988). There is no experience of cultivating medicinal plants in the country; 95% of the plants used in traditional medicine in Mexico are from wild species.

Other catalogues of the medicinal flora of Veracruz and Yucatan were produced by other institutions (Del Amo 1979, Mendieta & Del Amo 1981), based on the same methodology. Regional medicinal herbaria were established in several states.

Together with this ethnobotanical research, information on traditional medicine was considered necessary to understand the medical system in which the plants are used. We identified three main areas of study, closely linked but evaluated separately. One was the study of the healer (traditional practitioner) as a human resource to be recognized and supported by future health policies. One was the study and systematization of the fragmentary ideological framework of traditional medicine, its categories and nosologies (classifications and descriptions of diseases). Finally, there were the therapeutic resources in use. Assisted by personnel from the Social Security network of rural programmes of medical care, we collected information about herbal traditional practices in almost 3000 small Indian communities over four years. We were fewer than 2500 persons and we established collaborations with more than 14 000 traditional practitioners. These were the first statistical data obtained by the government about practitioners of traditional medicine at a national level. They included information about the different professional categories (variants or local versions of healers, midwives, bone-setters, herbalists and others), their medical area of activity, including the population's geographical distribution, predominance of age, sex, forms of transmission of knowledge, etc. The data obtained were highly informative: in general terms, in the rural areas of Mexico, the number of people dedicated to the practice of healing with herbs was at least 4–5 times more than the number working in the official medical system. The number of plants used for medicinal purposes was huge, but there was a 'basic group' of plant remedies. Other complementary data in this field of research were considered crucial for the selection of plants to be studied in our laboratories for determination of their pharmacological properties (Lozoya 1990).

Another basic conclusion of our methodological approach was that the study of medicinal plants must be multidisciplinary (Lozoya 1984, Linares & Bye 1993, Argueta & Cano 1993). Medical doctors were essential in the ethnobotanical field studies to evaluate the information about the use of plants in traditional medicine. We observed that, commonly, biologists or botanists who lack training in medical aspects cannot interpret many of the key concepts of diseases and treatments followed by traditional practitioners using herbs. A modest and collaborative attitude of physicians, botanists and anthropologists during the field ethnobotanical work was the key to successful recording of the information.

The selection of plants to be screened in the laboratory also depended on other aspects, including distribution of the plant, its availability during the different seasons, the permanence and congruence of the therapeutic application, priority in the efficacy attributed by the practitioners and reports of toxicity. All these parameters required a detailed ethnobotanical search in which the collaboration of the traditional practitioners with the investigators was very necessary.

A second step: the screening process

The approach followed in pharmacological laboratories for the study of plant drugs was designed by the 'classic' pharmaceutical companies. This approach, which rarely considers information available from popular use, is based on the principle that it is necessary first to isolate a pure compound from the plant and then to determine its biodynamic properties. How can one isolate a pure active compound if one does not know which is active? Generally, phytochemists would isolate several compounds from plants then, according to the structural properties, ask a pharmacologist to 'check for bioactivity'. Some phytochemical screens of dozens (or hundreds) of plants are designed exclusively on the basis of technological advances in biochemistry and spectrophotometry that have facilitated the search for compounds such as alkaloids, saponins and tannins. This approach ignores or rejects the isolation of other compounds for which there is no bibliographical evidence of previously reported biodynamic properties.

Today's 'bioassay-based' methods for separating and identifying active compounds arose in the 1980s from groups like ours who decided to experiment with whole plant extracts, crude products and chromatographic fractions, assaying them in the standard biological tests used for pure compounds (for references, see Lozoya 1985). It was demonstrated that the extracts and crude fractions perform extremely well in *in vitro* assays, if the appropriate controls are used. In the beginning, we had to create our own journals to publish these types of studies that were rejected by conventional pharmacology journals. Now, there is a large bibliography on the pharmacology of crude extracts.

Continuous interaction between chemists and pharmacologists was an important feature of this work. We called this approach the 'ping-pong' method: the extract obtained by the chemist is first tested for biological activity by the pharmacologist. Only once dose-dependent effects of the extract have been demonstrated is it returned to the chemist for fractionation. After chromatography, the fractions are re-tested and the one that shows increased biological activity with respect to the original extract is returned to the chemist. This process is repeated until the pure active compounds are isolated. Many surprises arose from this approach; for example, several active compounds are usually obtained from the same extract. They frequently originate from the polar, aqueous fractions, from which it is very difficult to separate them.

The combination of the two steps—ethnomedically guided selection of candidate plants and separation of chemical constituents on the basis of a reiterated bioassay—leads to rapid isolation of the active compounds. The question now is, what must follow and why?

In the Western hemisphere, over the last two decades, very few new herbal drugs have been developed. *Vinca rosea* has long been cited as the obligatory example of a source plant identified by serendipity. Today, taxol is receiving all the attention, 15 years after its detection in a plant, having shown anticancer properties. During the same time, several authors have been demonstrating— with stoicism—that the lack of effective cooperation among researchers in the relevant biological, physical and clinical sciences has accounted, in large measure, for the lack of successful development in the USA of a significant number of new plant drugs during the latter part of the 20th century (Farnsworth et al 1985, Tyler 1986). I shall add that the lack of clinical studies of plant remedies traditionally used by the people has also contributed to this failure.

Today, in the majority of laboratories, a plant is selected using a more or less suitable bioassay in the screening process. Recently, simplification of these bioassays has accelerated the screening process but after that begins the real business—the long and expensive chemical, pharmacological, toxicological and preclinical studies of pure compounds, which are frequently initiated without a real confirmation of the clinical benefits of the suspected drug.

Here comes the hot point! There is a big difference between selecting a plant from the vast flora of this planet in the hope that it will show medicinal properties and choosing a plant *already selected* by the people and used by them successfully! There are two issues: firstly, whether we believe those peoples or not; secondly, whether medical doctors will agree to run a clinically controlled study of the popular remedy without feeling themselves to be 'healers'. In my experience, the real handicap of the Western approach to the study of medicinal plants is a *cultural handicap*, in comparison with the strategy followed by Chinese, Japanese, Indian and Korean researchers.

If we accept the advantages of the 'ethnomedical' approach, i.e. selecting for investigation plants identified from traditional, popular knowledge, we must start to apply a methodology whereby the first step is to confirm clinically the therapeutic action of the crude herbal remedy. Only after the benefits of a plant remedy (an infusion or decoction or other common form of administration in traditional medicine) have been validated scientifically, should chemical and pharmacological studies be initiated to obtain information about the active principle. The clinical information obtained using the traditional herbal remedy determines which bioassay should be used. Finally, recommendation for development of a particular plant product depends only partially on the results of the chemical and pharmacological studies; today, marketing conditions are of the utmost importance.

Practice is demonstrating that the Western 'soft herbal remedies' prepared as closely as possible to the traditional form used for centuries are preferred by the post-modern urban consumer. An African or Latin American healer will consider these products the right and intelligent way to use plants. International agencies are recognizing that the 1980s have seen a significant increase in the use of herbal remedies around the world. Many countries are now seeking assistance in identifying safe and effective herbal remedies for use in national health-care systems. In 1989, the World Health Assembly adopted a resolution to assist the health authorities of member states to ensure that the drugs used are those most appropriate to local circumstances, that they are used rationally and that the requirements for their use are assessed as accurately as possible (WHO 1991). The term 'herbal remedy' appears more frequently in the legal and medical language every year. Similarly, numerous products derived from medicinal plants appear in the stores of Western cities. Governments are facing problems in classifying and regulating herbal remedies that do not fit into the common legislative framework for pharmaceutical drugs or foods.

Finally, new scientific evidence supports a future increase in the use of such herbal remedies. The recognition and study of the synergistic effects of herbal remedies is a chapter of today's pharmacology that will open many incredulous eyes. The observed immunoregulatory effects produced by plant products or the modification of prostaglandin synthesis are examples of the change in the scientific paradigm that will soon explain many of the uses of crude plant extracts by the 'primitive healer'. A new pharmacology of natural products is being developed all around the world. The challenge for scientists working in the Americas is to accept a new approach to the evaluation of medicinal plants. We should recognize that Asian researchers have an advantage in the commercial introduction of herbal remedies directly into the clinics because their physicians do not feel ashamed of traditional medical cultures and do not consider them inferior. The Cartesian–Aristotelian way of thinking is no longer relevant because the model is trapped in contradictions that have now become evident. The science of the 21st century will be influenced more by Asiatic philosophy than by European. Tradition plus high technology is the formula and, in this sense, the indigenous cultures of Latin America are closer to the new paradigm than are Western archetypes. The future is ours.

References

Aguilar A, Martínez-Alfaro M 1993 Los herbarios medicinales de México. In: Lozoya X (ed) La investigación científica de la herbolaria medicinal Mexicana. Secretaria de Salud, México, p 89–102

Akarele O 1984 Programa OMS de medicina tradicional: progresos y perspectivas. Crón OMS 38:83–88

Argueta A, Cano L 1993 El atlas de las plantas de la medicina tradicional Mexicana. In: La investigación científica de la herbolaria medicinal Mexicana. Secretaria de Salud, México, p 103–115

Bárcenas A, Barrera A, Caballero J, Durán L (eds) 1982 Simposio de etnobotanica 1978. Memórias. Instituto Nacional de Antropologia e Historia, México

Barrera Marin A, Barrera Vázquez A, Lopez Franco RM 1976 Nomenclatura etnobotánica Maya. Coleccion Cientifica 36. Etnologia. Instituto Nacional de Antropologia e Historia, México

Castetter E 1944 The domain of ethnobiology. Am Nat 78:158–170

Del Amo S 1979 Plantas medicinales del Estado de Veracruz. INIREB (Instituto de Investigation de Recurços Bioticos), Xalapa, Veracruz

Díaz JL 1976 Indice y sinonímia de las plantas medicinales de México. IMEPLAM (Mexican Institute for the Study of Medicinal Plants), México

Díaz JL 1977 Usos de las plantas medicinales de México. IMEPLAM (Mexican Institute for the Study of Medicinal Plants), México

Farnsworth NR 1984 How can the well be dry when it is filled with water? Econ Bot 38:4–13

Farnsworth NR, Akerele O, Bingel AS, Soejarto DD, Guo ZG 1985 Medicinal plants in therapy. Bull WHO 63:965–981

Ford RI 1978 Ethnobotany: historical diversity and synthesis. In: Ford RI (ed) The nature and status of ethnobotany. Museum of Anthropology, University of Michigan, Ann Arbor, MI (Anthropol Pap 67) p 33–40

Gomez Pompa A 1982 La etnobotánica en México. Biotica 7:151–161

Hernandez XE 1970 Exploracion etnobotánica y su metodologia. Escuela Nacional de Agricultura, Secretaria Agricultura (SAG), Chapingo, México

Hill AF 1945 Ethnobotany in Latin America. In: Verdoom F (ed) Plants and plant science in Latin America. Waltham, MA, p 12–25

Jones VH 1941 The nature and status of ethnobotany. Cron Bot 6:219–221

Kroeber AL 1939 Cultural and natural areas of native North America. University of California Press, Berkeley, CA (Univ Calif Publ Am Archeol Ethnol vol 38)

Linares E, Bye R 1993 Los jardines botánicos y las plantas medicinales. In: Lozoya X (ed) La investigación científica de la herbolaria medicinal Mexicana. Secretaria de Salud, México, p 75–88

Lozoya X (ed) 1976 Estado actual del conocimiento en plantas medicinales. IMEPLAM (Mexican Institute for the study of Medicinal Plants), México

Lozoya X 1980 Plantas medicinales de México. ¿Perderemos otra vez la batalla? Rev Med Tradit (Méx) 3:69–71

Lozoya X 1984 Medicina tradicional y herbolaria. Materiales para su estudio. IMSS (Mexican Institute of Social Security), México

Lozoya X 1985 A decade of studies on traditional medicine and ethnobotany in Mexico. In: Ethnobotanik. Friedrich Vieweg & Sohn, Wiesbaden, p 417–420

Lozoya X 1990 An overview of the system of traditional medicine currently practised in Mexico. In: Farnsworth NR (ed) Economic and medicinal plant research. Academic Press, New York, vol 4:71–94

Lozoya X, Velázquez G 1988 Medicina tradicional en México: la experiencia del programa IMSS-COPLAMAR. IMSS (Mexican Institute of Social Security), México

Lozoya X, Aguilar A, Camacho J 1987 Encuesta sobre el uso actual de plantas en la medicina tradicional mexicana. Rev Med IMSS (Mex) 25:283–291

Mendieta RM, Del Amo S 1981 Plantas medicinales del Estado de Yucatán. INIREB, Xalapa, Veracruz

Schultes RE 1941 La etnobotánica: su alcance y sus objetivos. Caldasia 3:7–12

Schultes RE, Hofmann A 1979 Plants of the gods. McGraw-Hill, Maidenhead

Smith CE Jr 1979 The value and potential of ethnobotany. An Antropol 16:95–104

Toledo V 1982 La etnobotánica hoy. Revision del conocimiento y lucha indigena. Biotica 7:141–150

Tyler V E 1986 Plant drugs in the twenty-first century. Econ Bot 40(3):279–288

Viesca C (ed) 1976 Estudios sobre etnobotánica y antropologia medica. IMEPLAM (Mexican Institute for the Study of Medicinal Plants), México, vol 1

WHO 1978 Promotion and development of traditional medicine. World Health Organization, Geneva (Tech Rep 622)

WHO 1984 Primary health attendance: the Chinese experience. World Health Organization, Geneva

WHO 1987 Report of the second meeting of directors of WHO collaborating centers for traditional medicine. World Health Organization, Geneva (Trad Med Tech Rep 88.1)

WHO 1991 Guidelines for the assessment of herbal medicines. World Health Organization, Geneva (Trad Med Tech Rep 91.1)

DISCUSSION

Balick : You say there are 11 000 herbarium sheets from Mexico. Elaine Elisabetsky said that only about 8–15 published ethnobotanical studies have been carried out in all Amazonia (Elisabetsky & Posey 1994, this volume). Clearly, we need to assign priorities. A map for Amazonian conservation priorities has been composed. It might be useful to designate areas that are more in need of study as a guide to all these students who are looking for thesis projects.

Dagne: There is more to the contribution from Mexico than was presented here. *Dioscorea* is an elegant example of a contribution from Mexico.

Lozoya: Dioscorea composita was never used in traditional medicine. It was discovered by Dr Marker, in the 1940s, by serendipity. His genius was to show that the saponin, diosgenin, isolated from the root of this plant could be chemically modified and from that molecule the steroidal hormones could be obtained. In 1975 the Mexican government decided to protect *D. composita* as a natural resource. This was blocked by the international pharmaceutical companies. The same companies that today talk about protecting the 'biodiversity' of the tropical rain forest. Today, Mexico imports steroid hormones because the big companies never accepted the government's conditions for exploitation of this Mexican resource.

Balick: In Belize, just a little south of you, *Dioscorea* is a very important medicine amongst the Maya. Traditionally, they have always used it to 'build the blood'.

Berlin: Dioscorea floribunda is recognized among the highland Maya as a contraceptive.

Lozoya: D. composita was never considered a medicinal Indian plant. Marker observed an indigenous practice very common on the east coast, which is to poison the fish using saponins from the root of *Dioscorea*. Marker observed the foam on the water and deduced that there was a high concentration of certain saponins. He took the root and isolated the saponins. That's the real story, the rest is legend. Today, Mexican Indians from Chiapas use some Compositae roots as a 'contraceptive' because they see this on television, thanks to our academics telling the romantic story of the origin of the contraceptive pill. We are producing the 'tradition', not they, in this case.

Martin: Oaxaca is one area of Mexico where further ethnobotanical study is required. Very few groups have been studied in Oaxaca and of those, only one or two communities have been the subject of a systematic appraisal of the plants being used. There has been no systematic study of the ethnobotany of the Mazatec, the Zogue or other major indigenous groups. Nor has there been any salvage ethnobotany of the Ixcatec people or other ethnic communities that are disappearing.

It is true that more ethnobotanical studies in Oaxaca might not find any new drugs, but ethnobotany is not solely at the service of searching for new drugs. There are many reasons to do ethnobotany, such as to serve conservation and community health programmes. So we probably don't need any more anthropologists in southern Mexico, but we certainly need to have people from the communities themselves continuing to do their own ethnobotany.

Lozoya: Over the last 20 years I have read almost all the ethnobotanical studies performed in Mexico; generally, their medical interpretation is useless. For example: they are called 'Medicinal plants of community X'. In the first section they give the historical background and the social problems of the community. The second section offers a list by Latin name of the plants used by that community; the list usually gives the popular name of the plant, the part of the plant used, but then, the medical use is described in one or two words by examples: 'for worms' means nothing medically or pharmacologically. This type of information has been reported for many years.

In the last 20 years, hundreds of such studies have been performed by Mexicans or by foreign researchers with the same superficial grade of information. When you give that information to a pharmacologist, he will ask: how much of the plant do you have to give to the patient? Is the leaf useful when collected in the morning or at night? What symptoms are presented by the patient? This information is not available from the classical ethnobotanical field studies. There is no real medical information about the use of the plant when collected by botanists.

What are we going to do with the information from those studies? Use them to protect the biodiversity of the area? To prepare a herbal *vade-mecum* (pharmacopoeia) for Indian communities? To create 'popular' books of ethnobotanical information—a 'cure yourself guide' with these natural remedies? The rural areas of Mexico do not need to develop knowledge of traditional medicine; they need access to the medicine represented by potable water, antibiotics, better nutrition and vaccinations—to combine both medical cultures. Indian communities are not interested in new uses of medicinal plants; they already know and use them. It is the people of Mexico City who are interested in the use of medicinal plants because this is part of the 'natural way of life' currently in fashion.

Martin: You say that local people need better nutrition. The indigenous people that I work with in Oaxaca consume up to 40 different types of edible greens, which provide very good nutrition. What do you want to do, provide them with manufactured food, or promote the use of plants they use at the current time? Ethnobotany can resolve some of these problems.

Berlin: I'm surprised you say the rural people of Mexico are not interested in herbal medicine. You are a co-author on a multivolume encyclopaedia, of which I am a principal author, that deals with the medical ethnobiology of the highland Maya, which will probably be several thousand pages long. These are real data that reflect an active, functioning ethnomedical system based on herbal medications.

Lozoya: We have been working with Mexican rural people, including the Maya, for many years. Certainly, they use traditional medicine; they have a huge indigenous pharmacopoeia. This will soon be printed, after 500 years of existence, by you and by us; this is something of which I am very proud. But for the 21st century, those communities in my country need real development. You and I have learned a lot from indigenous practices, but are those people solving their health problems? No. Do they require other things, not only plants? Yes. They need food, nutrition and potable water. We have worked in the past to give the communities basic groups of plant remedies to solve their health problems. Herbal remedies for the poor and high tech medicine for the rich is not the solution. I prefer a new approach: herbal remedies for the rich people of the cities and high tech medicine for sanitation and nutrition for the poor.

I believe that from the encyclopaedia of the Maya several fantastic new remedies can be developed in the next 20 years. But we need to use those remedies in a different strategy and introduce them into the social security hospitals of Mexican modern medicine. We need to take, for instance, 100 patients and, with the information from the encyclopaedia, perform the first clinically controlled study to understand how one herbal tea works. Then, with the clinical information obtained, we can develop a modern 'herbal remedy' produced in the country by agroindustrial means and perhaps export it to the USA and Europe—they will love it!

Elisabetsky: Xavier, although I can't agree with you entirely, I do identify with many of your concerns. The Brazilian position on doing clinical trials with traditional medicines as they are prepared by traditional people is that it is unethical not to study these, because millions of people are ingesting them. We ought, for instance, to see if there are psoralens in the preparations they are using, because these are carcinogenic.

We tried to do clinical trials in Brazil with 20 plants. One lesson was that because clinical trials are limited to acute administration of certain doses, they might result in false negatives. Cohort studies of people who are already using a medicine in a traditional form assessed by medical doctors could be a better way to evaluate these medicines.

We have an obligation to add Western scientific value to traditional medicines. If we do not study plants and prepare better formulations, people will continue to wake up at 3 o'clock in the morning when they are sick, go to a clinic and hope they will see a doctor for five minutes. It has been suggested that they then receive phytotherapy, but that does not add anything to the picture. It is our responsibility to solve our own problems. We have to find an appropriate methodology, if the existing one is not appropriate.

Lozoya: I am proposing a different methodology with the same tools. I want just to change the order of the steps. I am proposing to start with clinical studies of the original herbal remedy. If the Food & Drug Administration (FDA) does not allow US scientists to do this, it's a pity. In Mexico we can do this because we know those herbs, we use them, culturally we have the people's acceptance. We are already using this methodology. We are selecting plants using existing ethnobotanical information and we are introducing them directly into clinical studies. If the ethnobotanist says: 'This plant is useful for certain skin diseases', we take a patient, we apply the popular remedy and three weeks later, according to clinical results, we can say whether plant X really is useful. If it is, plant X goes to the pharmacology department and the chemical department for basic studies to understand how it works. How much information are they going to obtain from those basic scientific investigations and how long it will take? We will give them 10 years to answer those questions. But meanwhile, the plant will continue to be used in the clinics. If plant X is not effective, we take it out and do not perform any other type of investigation. So, we are using clinical studies as a screening procedure in selecting medicinal plants for new drug development. In the USA, is this position considered immoral? I am not sure. For us, this is the clinical validation or conformation of a very old practice of our popular culture. The general legislation of the USA does not allow experiments on human beings. We are not doing 'experiments', we are confirming by modern scientific procedures what already occurs in practice. The problem is that for US scientists such 'medicine' does not exist because culturally they do not accept it. So don't talk about ethics when we are addressing our problems; this attitude to me is ethnocentric. For countries like Mexico, the only possible way to develop new drugs is with this methodology. The Chinese have demonstrated that it works!

Dagne: You use ethnobotanical information, select a plant and then you extract. The minute you deal with extracts there is a modern input. Do you have the right to give an extract to a patient without going through the standard protocols of the day?

I don't think we should be mystified about studying plant materials of proven use in the ethnomedical setting. If a plant material is useful in ethnomedicine, we can certainly subject it to all aspects of scientific investigation. Every informed patient is going to ask: what does this plant material contain? We should be in a position to give the answer. With modern facilities this can be done in a very short time. There is some sense of urgency, but we should not exaggerate this. We do not oppose the ethnic communities using these medicines, but our acculturated communities demand a lot more justification than simply the use of a plant material by ethnic groups as a rational for using them.

Lozoya: I believe that if Western society does not modify its approach to the selection of plants to develop new drugs, the Chinese, Japanese and Koreans will overtake in the next 10 years. They are developing new drugs from a tremendous amount of clinical information that they have produced in the last 10 years. We have no clinical information about how our plants work in traditional medicine. I am proposing to reverse the model. It's a scientific approach. First we use clinical evaluation, under the supervision of ethical committees, with the appropriate protocol, in double-blind studies, with the appropriate herbal preparation and the appropriate physicians. We require rapid screening of all the plants collected in Mexico to select the effective ones for study by chemists and pharmacologists to develop new remedies.

Balick: Two years ago the National Institutes of Health chartered the Office of Alternative Medicine as part of the NIH structure. I've been privileged to be on the panel that advises that office. People on the panel are concerned that the double-blind clinical trial methodology is not appropriate for alternative medicine, including herbal medicine. How do you use a placebo or do a controlled trial for acupuncture? They have decided on something called field trials, instead of double-blind trials. Proof is obtained by a panel of medical professionals and then the process goes on from there. Interestingly, there was a request for proposals for grants for alternative medical systems, including traditional medicine, and in the first round over 450 proposals were received. Unfortunately, only six could be funded with the money that was available, but it shows a real interest in these things in the United States. David Eisenberg of Harvard Medical School found that one-third of the population in the USA was using traditional medicines and alternative medicines (Eisenberg et al 1993). Make no mistake—there is *great* interest in traditional medicine within Western society.

Farnsworth: Of those 450 grant proposals, only 60 were concerned with medicinal plants. The others were on alternative lifestyles, diet and nutrition, acupuncture, electromagnetic forces, etc. The purpose was to get traditional healers working with Western-style scientists to design and implement preliminary studies.

Balick: Many projects in the final selection were mind–body-type studies. My point was that this is a beginning. The US government has never recognized this sort of paradigm before; I think these new activities are promising.

Cragg: I really applaud the efforts of both Maurice Iwu and Xavier Lozoya in this concept of developing crude plant drugs clinically for the use of the local people. This is a good way to address the health needs of people in those countries, who cannot afford the high-cost, Western-style drugs.

I find the criticism of the FDA a little tedious. The medicinal plant community is always accusing the FDA of denying the American public access to these treatments. The FDA is there for a particular purpose, that is to protect the American public. There are good reasons for this; thalidomide is the classic example. The work being done in Nigeria to introduce quality control is admirable. I assume that in Mexico there is tight quality control of these plant preparations, no doubt in China and India too. It is up to the medicinal plant community in the United States to address this issue, as these other countries are doing, and to have a tight form of control. My feeling is that the FDA will then be more receptive. But to imply that the FDA is denying the US citizens access to these materials for economic reasons is totally wrong and is insulting to the mission of that agency, which is to protect the American public.

Iwu: The FDA is a very efficient agency and has been a safety shield for the whole world. My point is that if in a free society I am allowed to use Western technology by choice, including guns that may kill me, cigarettes and alcohol that are injurious to my health, should the Americans be denied the choice of using remedies that are available elsewhere? They are rational people, they should be able to choose. If we offer them hocus-pocus medicine, they won't buy it. The problem is not only with America. We cannot send processed timber to Japan; we can only sell them the raw product. The EC allows only European drugs as phytomedicines. We in the Third World should be able to develop our own pharmaceutical industry. There is a global market; it should be global in all its ramifications. There is a strong but subtle trade barrier against Third World phytomedicines.

Albers-Schönberg: Dr Lozoya placed the investigation of the mechanism of action almost at the end of the agenda in the development of a drug; it belongs at the very beginning. The FDA asks us in the US to develop not only effective but also very safe drugs. In response to this demand, the pharmaceutical sciences have made tremendous progress in the last 30 or 40 years, progress which we cannot simply reverse. Of the 55 drugs that Paul Cox mentioned (Cox 1994, this volume), how many would gain approval today? We have learned how to use them, with great caution, but we would dearly like to have better ones.

Today, we are trying to understand a disease as well as we possibly can, then choose the biochemical step at which we think we can effectively and safely intervene. This puts the mechanism of action at the very top of the development process. A good example is Mevacor for the control of cholesterol levels, which was discovered by deliberately targeting the rate-controlling step in the biosynthesis. A carefully chosen mechanism of action is the best—but by no means absolute—safeguard against unpleasant surprises in clinical trials. I wonder how many unacceptable side-effects would turn up in clinical trials of an ethnomedicine.

Lozoya: Recently, in the last 6–7 months, our government was asked to discuss with the FDA some practical problems that the United States is facing. These are problems in health and the use of medicinal plants in the southern United States, where the increase in population is producing a very important market for traditional medicines and herbal remedies. This has induced the government in Mexico to create legislation for these types of remedies. Are we facing the reality of the social and economical problems, or are we just creating schedules of work for our understanding of science? I think the first point is more important.

I'm not against the existing procedures and methodologies for clinical studies. I am against a dogmatic position that does not accept changes in approaches and thinking in the study of medicinal plants. I do not believe that all the purified, active compounds produced by the pharmaceutical industry are safe. In medicine, there are several examples where, several years later, we discovered that our remedies were not as safe as we thought. Of course, this is the production of knowledge. I believe that it's necessary, scientifically, to maintain an open mind towards alternative ways of producing herbal drugs. One way is that, depending on the disease, depending on the plant and depending on the scientific knowledge, we can develop different types of drug and produce then in different ways. It's necessary also to recognize that there are enormous political and economical factors influencing this area of research.

This is the end of the 20th century. We have to recognize that scientists in Africa and Latin America are not different in training and abilities from those in the USA and Europe. Scientists in the so-called 'Third World' countries are no longer interested in being suppliers of information and raw materials for 'developed' societies! We cannot continue sending extracts to laboratories in the US and Europe for screening for activity against cancer or AIDS, for example, with a promise that if something successful results our institutions will benefit. We are capable of a more open academic and technological interaction now and of equal contributions to laboratory development.

Finally, if the health authorities of the United States or Canada will not consider the use of medicinal plants in clinical studies, without basic pharmacological, chemical and toxicological information, good for them! We can do this. Perhaps, some clinical studies can be performed in our countries and some high tech studies can be performed in other places, but I think there is room for everybody.

Schwartsmann: I disagree totally with what Dr Lozoya just said. I don't think that we in the Third World are just sending compounds to be tested elsewhere. We have to find ways in the Third World to be creative and have the wisdom and experience to take best advantage of the cooperation with developed countries.

In contrast to what Dr Lozoya stated, I believe that a scientific approach is very important in this area. Consider, for example, giving digitalis to patients as a pill or as a crude plant extract. You have a much better chance of giving the right amount of the active component, if you give a purified compound, because you know more accurately the amount you have in the preparation. I have doubts about how consistently you can administer doses when you are giving a plant extract.

Secondly, when you look at the development of plant-derived anticancer drugs, the initial use was not always the most valuable one. For instance, the initial use of vinca alkaloids was as antidiabetic agents. These are now improving survival in many illnesses as antitumour agents. If we hadn't done research on the pharmacology of the derivatives, we would never have developed analogues that we can give more safely to people. We gain many advantages by doing sophisticated studies on the mechanism of action of these compounds.

Cox: I support Dr Schwartsmann's comment. When Withering initially used digitalis he had a lot of toxicity problems. He then used dried leaves, still had standardization problems and eventually moved to water infusions. Something that plagued him constantly was trying to get a standardized dosage.

Iwu: The issue is the fundamental question of what constitutes medicine. Is something a medicine only when it consists of one single molecule? I'm from a discipline where we have no problems standardizing mixtures. We standardized our belladonna tincture preparations using one of its components. Digitalis is taken by preparing the leaf, there is no isolation involved. *Rauvolfia* leaves can be standardized for human use by assessing the reserpine content.

We in the Third World need to liberate ourselves from the European Cartesian model. Science is no longer black or white, there is a grey area. We are asking you to follow us and explore this grey area. We should not treat the body as a machine; traditional medicine does not believe in this concept. I'm not really saying that we shouldn't pursue some pure science. Some of the universities in the Third World countries are very good and doing 'respected' scientific work. We can combine Western and traditional medicine. For some diseases, it's better to use a drug with a precise, dramatic effect. For most common diseases, simple remedies are effective. The biggest disease in terms of the number of people affected is malaria. Is it morally right that we ignore these diseases scientifically? Our primary obligation is to provide health care for mankind. If 80% of mankind does not benefit from ethnobotanical studies, they become colonial exploitation.

McChesney: There is, perhaps, some division of philosophy here which even extended discussion may not bridge. Our programme, we hope, falls in that grey area. We are an academic institute for research on tropical diseases. We are interested scientifically in all these diseases, but realistically we are interested in diseases for which we can get money to support studies.

Elisabetsky: We usually put things versus one another. There is a fallacy in thinking that Western medicine is 100% precise, safe and scientific. It is not. We treat leishmaniasis with glucantime because we don't have anything better, but it is a poison. This is not very different from a person in the middle of nowhere using a plant because that's the best they can get there. The message of this symposium should be that these differences and limits must be recognized.

Dr Lozoya does have a point. If we present to an institution, like the Overseas Development Agency, anything that is oriented towards low-cost drugs and we do not provide information or extracts, we don't get funds. Even extremely bright people in the Third World find it difficult to get funds. The plea from this symposium should be that we recognize the diverse ways of dealing with health and the need for better studies of traditional medicines and their development. In my courses in pharmacology, I tell my students that the *sine qua non* for a good drug is that it can be ingested by the patient. I would love to give vincristine and vinblastine to everybody, but we cannot afford it. As Maurice Iwu said, if 80% of the population cannot take these kinds of drugs, we have to develop other drugs.

Jain: In India 70–80% of drugs on the market in Ayurvedic, Yunani or Siddha medicine have evolved by methods that have been worked out (as Dr Lozoya and Professor Iwu described) by direct clinical testing and evaluation, without detailed pharmacological or chemical studies. Several institutions working on indigenous systems are now paying attention to laboratory research, but mainly to satisfy people who find fault with Ayurvedic medicine. The Indian Standards Institution tries to set standards for several of the raw drugs, but for only a very small portion of the indigenous pharmacopoeia. They have set standards for the spices used in medicine and for several of the aromatic plants used in the drugs. But before a drug is marketed, it does go through a process of testing and has to pass certification by the drug controller.

Albers-Schönberg: I certainly agree with Dr Lozoya on the tremendous urgency of the health problems in the Third World. But, we have to separate two distinct issues that seem to be getting confused here: how can Western science help lessen the health problems of millions of people in the developing world and should we allow chemically undefined traditional medicines into our 'Western' system because they don't seem to have any obvious side-effects?

The first question presents a dilemma. Should we abandon our own safety-conscious standards when it comes to the developing world, in order to be able to respond faster? Merck has provided Mectisan for the treatment of river blindness only after rigorous testing. But drugs for which so many lucky circumstances come together as for Mectisan are very hard to find. We should devise an effective mechanism by which traditional medicines can be evaluated by modern science. The same approach might be taken for the production of traditional medicines by applying the agricultural and forestry knowledge of the developed world. These are also real chances for meaningful technology transfer.

The second question is much less urgent. My initial answer would be no, except in some real emergency. As I said before, we should not turn back the scientific clock. Scientific progress has given us a longer life expectancy and will see to it that those gained years will be rewarding and productive ones.

As Elaine Elisabetsky has said, we have to try to understand each other better and work together. Criticizing the system of the developed countries doesn't help.

Lozoya: I agree completely that we should respect our differences and also that we should work together. I have the impression that we require a new terminology to be agreed here among us. In the last 20–25 years, our increased understanding of plant-derived drugs has allowed them to be assigned to different categories. All these plant-derived drugs are necessary, all are useful and all are important.

I shall give two examples of the present commercial reality in Mexico. In January 1994 a new piece of legislation will come into operation in Mexico dealing with plant-derived drugs. This will allow better control and understanding of the utility of medicinal plants in modern society. In Mexico, the authorities recognize that there are at least three types of plant drugs. One is the commonly used 'crude' dried plants sold in local markets and used mainly by people who understand and accept traditional medical concepts and beliefs. Such a category of products is recognized and controlled by the health authorities in Mexico because it represents an important resource for millions of people. Legislation for this category is confined to some restrictions on well-known toxic, hallucinogenic or similar species and with hygienic conditions for the place of sale.

The second category is semi-processed plant materials called 'herbal *products*'. These include dried leaves, pieces of roots and powdered materials used mainly for the preparation of teas, decoctions and infusions. They are for oral administration only and are presented as commercial products in boxes, tea bags, etc. Legislation concerning these products is very precise: producers must be registered officially and have permission to sell.

The third is called 'herbal *remedies*' and corresponds to natural products extracted from plants and presented in pharmaceutical form, e.g. as tablets or pills. More detailed information is required about their medicinal properties, doses, toxicological data and way of use before their sale is permitted.

The important point is that the first two groups of products are now recognized officially in Mexico as plant drugs belonging to a specific category not related to the previously existing categories of pharmaceutical drugs or food.

It's very important for researchers in this field, from ethnobotany all the way down to pharmacology, biotechnology and phytochemistry, to accept that the ideas of natural products from the 1940s and 1950s have changed. The classical isolation of a pure compound probably followed by synthesis is not the only way to develop a pharmaceutical drug. That pathway was and is very useful, but it is not applicable in all cases and in all circumstances. In my opinion, that approach is particularly important for traditional chemical/pharmaceutical Western companies. It is one way, but there are others.

Some *Opuntia* cacti in Mexico are important components of traditional herbal remedies given as co-adjuvants in the treatment of diabetes. The fresh stems of the plant are used to prepare a drink every day to be administered orally. You can imagine the practical difficulties associated with keeping and using a cactus in such a form for medicinal use on a large scale. We scientists were not able to produce a herbal remedy from this plant, because the active component is unstable, it does not last for 24 hours. So in this case, it is imperative to isolate the pure compound, perhaps to synthesize it or to modify the molecule and find new pharmaceutical approaches to develop this as a drug. On the other hand, to cure acute diarrhoea, it is enough to drink a cup of guava leaf tea. We know the active compound and its mechanism of action on intestinal peristaltism. We have done the phytochemistry and the toxicology, but the product is several times more active as a herbal remedy than as a pure substance. So we have promoted the production of a commercial guava leaf tea herbal remedy for the treatment of intestinal disorders. This has the advantages of low toxicity, easy preparation, cultural acceptance and low-cost production.

We must understand what type of disease or health problems we are going to tackle, in which country and for which reason. Scientists have to be open minded and accept that there is not only one methodological approach to validation and promotion of plant drug remedies. In this sense, we must accept that the Western approach of concentrating on pure compounds in the study of medicinal plants, especially in the USA, Canada and the UK, has been unsuccessful in the last three decades compared with other countries. Of course, one could say that finding one good drug of this type every 20 years is sufficient to protect the investment of a company via exploitation of the patent and control of the market. In such a case, we are talking about a certain type of business, not about pharmacological science.

Schwartsmann: As a clinician, I have some difficulties with that idea. In my opinion there are two approaches. One is to go as deep as you can in terms of finding the actual mechanism of action and the active principle of your plant product. Then, even if you don't have a very successful agent, you may have a lead that you could modify with help from a chemist or by looking at the structure–activity relationship.

The alternative is to take the whole plant, for example as a tea, and test it in the clinic. This has to be done in a proper scientific way. As an example, for four centuries, people in Europe used to treat several infections by bleeding patients. Then a surgeon in the 19th century did a simple experiment and assigned 100 patients to two groups; one was bled, the other was not. He saw no difference. Then this treatment, that had been used for 400 years, simply vanished.

Iwu: Isn't medical evaluation acceptable as clinical evidence rather than actually testing? Can't this be done retrospectively? Can't you evaluate efficacy by monitoring the effect on the people using the product?

Schwartsmann: Yes, you can do it.

Iwu: Exactly, so, why can't you re-formulate it and give it back to them?

Schwartsmann: You also have to address the problem of the placebo effect. You have to use some methodological tools to eliminate a chance effect of finding activity. Placebo means in Latin 'I please you', so it's very appealing.

Dagne: One gets the impression from our earlier discussions that not discovering new drugs would constitute a failure of ethnobotany. If ethnobotanical research leads to drug discovery, all well and good, but this is not the only criterion for success in this field. We should also realize that an immense amount of knowledge is being generated. It has been said that traditional medicine is declining. I disagree; I believe there is more interest in traditional medicine now than there was a few decades ago. Nevertheless, there is still a lot to be done to popularize traditional medicine, particularly with young people. Some aspects of traditional medicine should be introduced into the curricula of departments of medicine, chemistry, pharmacy and pharmacology. Even in Third World countries, where in many cases the majority of the population relies on traditional medicine, there is hardly any coverage of the subject in university courses. Some of the reservations emanate from the difficulties encountered in substantiating the claims of traditional medicine. This attitude will gradually change as more and more reliable bioassay systems are introduced.

Prance: I should like to thank everyone for a most lively debate this morning. There are clearly different approaches that exist and we mustn't rely on any one method, there are many different methods that will contribute.

It's been important to see the challenge of some of the developing world health problems and the challenge to address diseases that tend to be neglected, like malaria, leishmaniasis and other common tropical diseases. I feel that we'll get nearer to this if we work closely together with the developed and the developing world.

References

Cox PA 1994 The ethnobotanical approach to drug discovery: strengths and limitations. In: Ethnobotany and the search for new drugs. Wiley, Chichester (Ciba Found Symp 185) p 25–41

Eisenberg D, Kessler RC, Foster C, Norlock FE, Calkins DR, Delbanco TL 1993 Unconventional medicines in the United States. N Engl J Med 328:246–252

Elisabetsky E, Posey DA 1994 Ethnopharmacological search for antiviral compounds: treatment of gastrointestinal disorders by Kayapó medical specialists. In: Ethnobotany and the search for new drugs. Wiley, Chichester (Ciba Found Symp 185) p 77–94

Ethnobotany and research on medicinal plants in India

S. K. Jain

National Botanical Research Institute, Lucknow 226001, India

Abstract. Vast ethnobotanical knowledge exists in India from ancient time. Since the 1950s the study of ethnobotany has intensified; 10 books and 300 papers have been published. Our work over four decades, both in the field and literary studies, has resulted in a dictionary of Indian folk-medicine and ethnobotany that includes 2532 plants. India has about 45 000 plant species; medicinal properties have been assigned to several thousand. About 2000 figure frequently in the literature; indigenous systems commonly employ 500. Despite early (4500–1500 BC) origins and a long history of usage, in the last two centuries Ayurveda has received little official support and hence less attention from good medical practitioners and researchers. Much work is now being done on the botany, pharmacognosy, chemistry, pharmacology and biotechnology of herbal drugs. The value of ethnomedicine has been realized; work is being done on psychoactive plants, household remedies and plants sold by street drug vendors. Statistical methods are being used to assess the credibility of claims. Some recent work in drug development relates to species of *Commiphora* (used as a hypolipidaemic agent), *Picrorhiza* (which is hepatoprotective), *Bacopa* (used as a brain tonic), *Curcuma* (antiinflammatory) and *Asclepias* (cardiotonic). A scrutiny of folk claims found 203 plants for evaluation. Less well known ethnomedicines have been identified that are used to treat intestinal, joint, liver and skin diseases.

1994 Ethnobotany and the search for new drugs. Wiley, Chichester (Ciba Foundation Symposium 185) p 153–168

Ethnobotany is the best word to define the experience of the first humans, who observed birds and animals and tested leaves, fruits and tubers for their ability to satisfy hunger or heal wounds. The term Ethnobotany was coined by Harshberger in 1895 in the *Philadelphia Evening Telegraph*, but the discipline has existed for ages. Various definitions and concepts have been assigned to the word Ethnobotany, but it is now almost universally taken as the total direct relationship between humans and plants. This relationship is usually advantageous to both, humans and plants, but there are relationships beneficial to only one or even harmful to one (Jain 1989). Indian epics, ancient literature and folklore are replete with references to plants.

The Indian scene

India has all the three elements that contribute to the ethnobotanical richness of an area—floristic diversity, ethnic diversity and a rich tradition. India has a great variety of climatic and physiographical conditions: from the cold, arid inner valleys in the far northern Himalaya to the warm and humid Western Ghats; from hot, dry areas of western Rajasthan to the wettest spot in the world (Cherapunji) in the east; from the lofty Himalaya and the Gangetic plains in the north, across the plateau in the peninsular region to a long coastline and the islands. This has provided a home to about 65 000 kinds of animals and 45 000 plants, of which about one-third are flowering plants. It is estimated that 5000 species of flowering plants are endemic.

The unique and diverse flora is matched by some 400 different ethnic groups at varying levels of acculturation. Some are now assimilated into modern society with formal higher education and elite urban vocations, but many are still, at least partially, food gatherers and hunters.

Relationships between humans and plants have been studied directly or indirectly by a variety of scholars and from very different approaches—material, economic, ecological, religious, social and anthropological.

Historical review of ethnobotanical work

Papers by Janaki-Ammal (1956) and Jain (1963, 1965) triggered intensive ethnobotanical studies in many institutions. In the early 1980s, the Department of Environment of the Government of India funded an All-India Coordinated Research Project on Ethnobiology. This work was started in over a dozen institutions under the Botanical Survey of India (BSI), the Council of Scientific and Industrial Research (CSIR), in some universities and in other laboratories. A large area was covered in this quick survey and significant ethnobotanical data were recorded. Much of this work has been published in Indian and international journals. This prompted many institutions to start post-graduate programmes in ethnobotany. In the last 35 years, some 10 books and 300 papers based on research in India have been published and about two dozen doctoral theses written. Though most of the work relates to actual field research, some deals purely with literary or herbarium research. Papers relating to medicinal plants outnumber others. This literature has been reviewed by Jain et al (1984) and Mudgal (1989).

The establishment of a Society of Ethnobotanists in 1982 and of the journal *Ethnobotany* in 1989 placed the discipline on a firm footing in India. Both the society and the journal have international membership and contributions.

Some unorthodox but very useful aspects of ethnobotany have also been studied. Sinha (1988) made a study of the street drug vendors who actually wander around to sell plant, animal and mineral products, particularly aphrodisiacs and cures for sexually transmitted diseases. Dixit (1991) studied household remedies in a district of Rajasthan in western India.

Author's contributions

My own literary, field and herbarium research in ethnobotany started in 1951. The field work was done in several states of India, particularly in Maharashtra, Rajasthan, Gujarat, Madhya Pradesh, Orissa, Bihar, West Bengal and north-eastern states.

Quantification and credibility of folk medicine

Prescriptions in folk medicine are usually and obviously based on the individual's personal experience with patients. In the modern sense there seems little scientific or experimental basis to these claims. The use of a plant for the same purpose in several societies or regions has been taken as one criterion for greater credibility (Saklani 1992, Jain & Saklani 1992). Johns et al (1990) and Phillips & Gentry (1993) proposed statistical methods for the quantification of consensus on such remedies. A similar approach was attempted by Varghese et al (1993) among the Kharia tribe in India. They considered three types of disease—gastroenteritis, malaria and joint diseases. The factors considered were the prevalence of the disease, the abundance or occurrence of the plant, the degree of choice (1–4) possible from available plants, the reported effectiveness (degree 1–4) when the plant was used singly or in a mixture, and the number of people reporting its use for that disease.

Comparative and deductive studies

About seven years ago it was thought that considering (i) the great ethnic diversity (tribal peoples form 7.5% of the Indian population), (ii) phytogeographical variations, (iii) the rich flora, (iv) the voluminous literature based on intensive field data covering 35–40 years, and (v) the growing interest of medicinal chemists, pharmacologists and pharmaceutical companies in new sources of drugs, all the available knowledge concerning ethnobotany, particularly ethnomedicine, would be brought together. Data on over 2500 plants were extracted from nearly 400 publications (mostly primary sources) and published in the form of a dictionary (Jain 1991). Fresh field work in suitable locations and the recording of new folk claims, or verification of old ones, has been a continuing activity of our research group.

We realized that a considerable part of indigenous knowledge is already well documented, widely known and has passed into wide usage and into texts; some of it has also been subjected to modern evaluation. Yet, some of this knowledge existed only in folk traditions. The data were compared with major publications

TABLE 1   Plants used in Indian ethnomedicine for intestinal diseases (diarrhoea and dysentery)

Acacia catechu	Combretum roxburghii	Ludwigia adscendens
Acalypha alnifolia	Crinum pratense	Micromelum minutum
Achyranthes bidentata	Delphinium vestitum	Morinda angustifolia
Alnus nepalensis	Flacourtia indica	Nelsonia canescens
Begonia palmata	Glossogyne bidens	Neptunia triquetra
Blumea fistulosa	Hedyotis scandens	Picrasma javanica
Boehmeria macrophylla	Hybanthus enneaspermus	Pyrrosia adnescens
Boerhaavia diffusa	Hymenodictyon orixense	Sterculia villosa
Careya arborea	Indigofera linnaei	Tectaria coadunata
Chromolaena odorata	Knema linifolia	Tragia involucrata

TABLE 2   Plants used in Indian ethnomedicine for joint diseases (gout and rheumatism)

Ailanthus excelsa	Chlorophytum arundinaceum	Hymenodictyon orixense
Argemone mexicana	Chrysanthemum pyrethroides	Laportea interrupta
Aristolochia tagala	Clerodendrum colebrookianum	Lindera pulcherrima
Barringtonia acutangula	Datura innoxia	Orthosiphon rubicundus
Biophytum sensitivum	Dillenia pentagyna	Pholidota imbricata
Capparis sepiaria	Fagopyrum esculentum	Polygala arvensis
Cassia auriculata	Gerbera piloselloides	Skimmia laureola
Cassia tora	Holarrhena antidysenterica	Strychnos nuxvomica

and about 300 lesser-known remedies, based on 203 species of plants, were selected. An effort was made to determine from secondary or tertiary sources (Rastogi & Mehrotra 1990, 1991, 1993) the present status of knowledge about the chemistry and biological activity of these species. This work (Jain et al 1991) provides the botanical name, the family name, more common local names, a brief botanical description, the distribution, the flowering and fruiting periods, the folk claims with details of dosage, etc. where these are available, the source of information, any earlier known medicinal uses, notes on the chemistry, biological activity or pharmacology, and the name of any allied species known to have similar properties. The 300 prescriptions related mainly to 50 categories of ailments. Numerous remedies were reported for intestinal diseases (Table 1), joint diseases (Table 2), liver complaints (Table 3) and skin disorders (Table 4). At least 25 of these claims are supported by similar properties in another species of that genus.

Medicinal plants research in India

Historical

Written records of the use of plants for curing human or animal diseases in India can be traced back to the earliest (4500–1600 BC) scripture of the Hindus,

TABLE 3   Plants used in Indian ethnomedicine for liver complaints

Achyranthes porphyristachya	*Combretum pilosum*
Alcea rosea	*Costus speciosus*
Allamanda cathartica	*Cyperus rotundus*
Barringtonia acutangula	*Euphorbia ligularia*
Bauhinia purpurea	*Gouania tiliaefolia*
Begonia palmata	*Hedyotis scandens*
Berberis kumaonensis	*Helminthostachys zeylanica*
Bergenia ligulata	*Hymenodictyon orixense*
Betula utilis	*Leea alata*
Cissampelos pareira	*Lygodium flexuosum*
Cochlospermum religiosum	

the *Rigveda*. The juice of the legendary plant 'Soma' (the identity of which is controversial, but it is widely believed to be a mushroom, *Amanita muscaria*) is mentioned as *Oshadhi*, meaning a heat-producer. While exploring the therapeutic uses of 'Soma', the Indo-Aryans discovered more plants with medicinal properties and the word *Oshadhi* acquired a wider meaning to include all medicines.

The Vedic Aryans were familiar with about 100 medicinal plants; several additional plants were described in a later work *Atharva veda*. This was followed by monumental ancient treatises on the subject, like *Charak Samhita* (1000–800 BC), *Sushrut Samhita* (800–700 BC) and Vagbhatta's *Astanga Hridaya*. The Yunani system, which originated in Greece in about 400 BC, came to India through Arab physicians, who accompanied Mogul invaders. Its popularity declined with the fall of the Moguls, but it had already been partly amalgamated with the Indian system and came to be known as *Yunani-Tibb*. The Siddha system, with a recorded history from about 2000 BC, is believed to have originated from Lord Shiva and to have been passed on through his wife Parvati to a number of disciples. Its use became more common in Dravidian civilization. The Siddha system comprises some 1000 biological products.

The texts of each of these three systems dealt with herbs used in that system only; books in English, written between the 18th century and today usually include plants from all three systems.

The progress of Hindu medicine in India declined with invasions by and the influence of the Greeks, Scythians, Huns, Moguls and Europeans. During the British rule, there was further intermingling; new medicinal plants were also introduced. Mitra & Jain (1991) have summarized the English literature produced in the 19th century in the form of catalogues, dispensatories, pharmacopoeias and descriptions or illustrations of plants with notes on medicinal uses, which culminated in Watt's (1889–1896) six-volume *Dictionary of Economic Products*. Timely publication of Hooker's *Flora* (1872–1896) helped the understanding of the identity and occurrence of some 15 000 taxa of higher plants in India.

TABLE 4   Plants used in Indian ethnomedicine for skin diseases

Aerva lanata	*Calotropis procera*	*Martynia annua*
Ampelocissus barbata	*Clematis buchananiana*	*Premna barbata*
Anogeissus latifolia	*Cheilanthes farinosa*	*Quercus leucotrichophora*
Arisaema jacquemontii	*Euphorbia nivulia*	*Ranunculus arvensis*
Artemisia japonica	*Euphorbia uniflora*	*Rhamnus triquetra*
Blepharispermum subsessile	*Flacourtia indica*	*Skimmia laureola*
Blumea laciniata	*Holarrhena antidysenterica*	*Strychnos nuxvomica*
Boehmeria macrophylla	*Holoptelia integrifolia*	
Caesalpinia pulcherrima	*Manihot esculenta*	

The first half of the 20th century witnessed an initial awareness of and organized research in various aspects of medicinal plants. This is illustrated by, for example, the establishment in 1921 of a School of Tropical Medicine in Calcutta, the study of poisonous plants under the auspices of the Indian Council of Agricultural Research (ICAR) and the establishment in 1941 of a Drug Research Laboratory at Jammu. In that year, the ICAR established a Medicinal Plants Committee to finance agricultural research in State Agriculture and Horticulture Departments and in Directorates of Medicinal Plants in West Bengal and Tamil Nadu. This led to the All-India Coordinated Research Project on Medicinal and Aromatic Plants with nine centres in different parts of India.

With independence of the country in 1947, interest in herbal medicine was revived and major agencies for medicobotanical research were established or strengthened. To improve understanding of the taxonomy and distribution of the indigenous flora, the BSI (which had its headquarters in Calcutta), the Forest Research Institute (FRI, Dehradun), the National Botanical Research Institute (NBRI, Lucknow) and some universities initiated field work in underexplored regions. Numerous publications appeared on enumeration, accounts of the flora, taxonomic revisions, pharmacognosy, phytochemistry and active principles or biological activity of Indian plants.

The discovery of reserpine from *Rauvolfia* spp. (Fig. 1) and later of vinblastine from *Catharanthus roseus*, diosgenin from *Dioscorea* spp. and solasodine from *Solanum* spp. boosted interest in herbal drugs. Some major publications on medicinal plants were also brought out, such as the *Indian Pharmaceutical Codex* (Mukerji 1953), the *Glossary of Indian Medicinal Plants* (Chopra et al 1956, 1969) and Chopra's *Indigenous Drugs of India* (Chopra et al 1958).

The ethnobiological basis of research on medicinal plants

All the initial knowledge about the curative properties of plants (as also of animal or mineral products) was the outcome of early humans' observations on birds and animals, combined with accidental use of plant products or individual experience through trial and error. This knowledge descended through oral

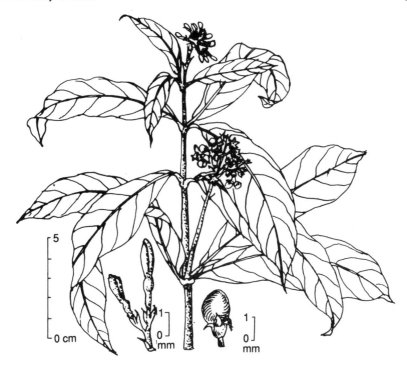

FIG. 1. *Rauvolfia serpentina* (snake root). A decoction of the roots is used as a hypotensive and sedative agent, believed to be useful in anxiety, insomnia, stress and epilepsy.

communication from teacher to taught or in group discussions during religious festivals, pilgrimages and fairs. A part of such knowledge gained greater credibility through acquiring a reputation for efficacy, multilocational use, or some kind of trial on patients. Some of this preferred knowledge went into texts and eventually into various systems such as the Ayurvedic, Yunani or Siddha, or other 'systems' outside India. A 'system' is here considered to satisfy four requirements: (i) written texts, (ii) organized training institutions, (iii) recognized degrees of training and qualification, and (iv) practise of the system only by registered practitioners.

With advances in experimental methods in phytochemistry and pharmacology, in the course of time, several of the folk medicines were tested for active principles and biological activity; the successful ones were added to Indian pharmacopoeias. Jain & DeFilipps (1991) reviewed some of this significant work.

A large mass of knowledge, however, continued to survive outside the organized systems. At a specific point in time, only this could be termed as folk medicine or ethnomedicine.

Recent work

In 1964, the Ministry of Health formulated a composite Drugs Research Scheme. It provided the basis eventually for three autonomous Central Councils for Research in Ayurveda and Siddha (CCRAS), in Yunani (CCRUM) and in Homeopathy (CCRH).

During the last few years, several elaborate or succinct, and often illustrated, accounts of medicinal plants of India have been brought out by Jain (1985) Kapoor (1990), Jain & DeFilipps (1991), Chatterjee & Pakrashi (1991), Thakur et al (1992), Husain et al (1992) and Husain (1993).

The botanical distribution of medicinal plants of India

The majority of known medicinal plants belong to the following families of flowering plants: the Acanthaceae, Apiaceae, Apocynaceae, Asclepiadacae, Asteraceae, Brassicaceae, Caesalpiniaceae, Convolvulaceae, Cucurbitaceae, Euphorbiaceae, Fabaceae, Lamiaceae, Liliaceae, Malvaceae, Mimosaceae, Poaceae, Polygonaceae, Ranunculaceae, Rosaceae, Rubiaceae, Rutaceae, Scrophulariaceae, Solanaceae, Verbenaceae and Zingiberaceae.

Major diseases (on the basis of remedies)

The diseases or ailments for the treatment of which a large number of medicinal plants (over 75 species) are used are asthma, boils, bronchitis, coughs and colds, cuts and wounds, digestive disorders, eye diseases, fevers, intestinal diseases (cholera, diarrhoea, dysentery, etc), joint diseases, leprosy, liver complaints, piles, skin diseases, snake bite, sores, urinary diseases and venereal diseases.

Over 75 species of plants have been attributed the property of being one or more of the following: abortifacient, antihelminthic, astringent, carminative, demulcent, emetic, expectorant, laxative and purgative, stimulant and tonic.

Pharmacognosy

The work on the pharmacognosy of medicinal plants in India has been reviewed by Mitra (1985) and Mehrotra (1986).

Phytochemistry and pharmacology

The phytochemical and pharmacological work on medicinal plants in India during the last 25 years was recently reviewed in a three-volume work compiled at the Central Drug Research Institute (CDRI) (Rastogi & Mehrotra 1990–1993). These books are supplements to Chopra et al (1956, 1969); they provide data in a succinct form and readily available references.

TABLE 5 **Clinical tests of herbal drugs at the Regional Research Institute of Unani Medicine, Aligarh, India**

Disease treated	Plant	No. patients treated	Fully cured	Partially cured	Not cured	Abandoned
Vitiligo	Ammi majus	2292	509	1358	275	150
	Ruta graveolens	400	30	168	96	99
Malaria	Caesalpinia crista	48	16	32	0	0
Tropical pulmonary eosinophilia	Iris sp.	148	88	17	8	35

Results cited from V. K. Singh & S. Hashmi (1993), with permission (Regional Research Institute of Unani Medicine, Aligarh. Research activities and achievements (1970–1992). Government of India, New Delhi).

Biological screening of Indian flora

Screening plants for biological activity was initiated about 25 years ago, mainly at the CDRI, Lucknow. Initially, any plant material was tested, but with the accumulation of more data on ethnomedicobotany in the last 15 years, screening has been directed towards ethnomedicinal plants. Some 3500 plant samples (about 3000 species) have been tested and the results published in a series of papers (Dar et al 1968–1990).

Sustained work on biological screening and follow-up pharmacological, toxicity and clinical tests have focused attention on several plants. These include: *Commiphora mukul* (Burseraceae) for hypolipidaemic activity; *Picrorrhiza kurrooa* (Scrophulariaceae) for hepatoprotective activity; *Bacopa monnieri* (Scrophulariaceae) and *Centella asiatica* (Apiaceae) as a brain and nerve tonic; *Curcuma longa* (Zingiberaceae) for antiinflammatory activity; *Asclepias curassavica* (Asclepiadaceae) as a cardiotonic and *Boswellia serrata* (Burseraceae) as an antiinflammatory agent.

Clinical tests of herbal remedies for vitiligo, malaria and tropical pulmonary eosinophilia have shown promising results (Table 5).

Psychoactive plants

Recently, we have started field and literary work on hallucinogenic and related plants; we have listed about 400 plants reported to possess these properties. This work is continuing.

Conservation

The demand for herbal drugs in India is met mostly from natural populations; this is now causing much concern. Shortage also leads to adulteration. Cultivation

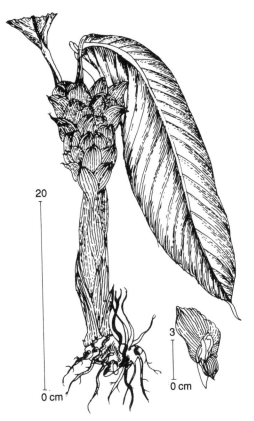

FIG. 2. *Curcuma longa* (turmeric). The rhizomes are boiled in water or fried in oil, then powdered and taken with honey (and some other ingredients) to treat influenza, respiratory and skin diseases. The powdered rhizome is taken as a carminative and antiinflammatory agent; it is also applied locally as an antiseptic.

activity is confined to only a few species in genera such as *Aloe, Ammi, Atropa, Cassia, Catharanthus, Cephaelis, Cinchona, Costus, Cuminum, Curcuma* (Fig. 2), *Cymbopogon, Digitalis, Dioscorea, Emblica, Ephedra, Glycyrrhiza, Hyoscyamus, Mentha, Ocimum, Papaver, Piper, Plantago, Podophyllum, Psoralea, Rauvolfia, Ricinus, Sesamum, Trigonella* and *Withania*.

Conservation of genetic diversity through the collection of germplasm is now occurring at some centres of CSIR and ICAR, also in some universities. Attempts are also being made to preserve and multiply germplasm using tissue culture.

Some botanical gardens in India are devoted chiefly to medicinal plants, such as the experimental gardens of the Central Institute of Medicinal and Aromatic Plants, of the BSI and the Central Councils for Research in Ayurveda, Siddha and Yunani and of their regional sub-stations, particularly the Ayurvedic

Medicinal Plants Garden at Pune (near Bombay). Yet, all these holdings put together may not represent even 30–40% of the known medicinal flora of India. The diversity in climate and in flora defies very rich living collections at any one location.

There is, therefore, a need for a network of herbal gardens in the country for the purpose of conservation, education and research in the development of new drugs.

Conclusion

India has a rich tradition of herbal medicine. The floristic diversity provides a wide choice of species. Ethnobotanical work has brought several thousand folk claims about the medicinal properties of plants into the scientific literature; many of these need more concerted effort for drug development. Promotion of indigenous medicines will also help the people from whom the traditional knowledge was obtained.

Acknowledgements

I am grateful to the CSIR and to Dr P. V. Sane, Director, NBRI for help and facilities, and to Drs R. Gupta, U. Shome and R. Mitra for useful suggestions.

References

Chatterjee A, Pakrashi SC (eds) 1991 The treatise on Indian medicinal plants, vol 1. Council of Scientific and Industrial Research, New Delhi

Chopra RN, Nayar SL, Chopra IC 1956 Glossary of Indian medicinal plants. Council of Scientific and Industrial Research, New Delhi

Chopra RN, Chopra IC, Handa KL, Kapur LD 1958 Chopra's indigenous drugs of India. UN Dhur & Sons, Calcutta, Bengal

Chopra RN, Chopra IC, Verma BS 1969 Supplement to glossary of Indian medicinal plants. Council of Scientific and Industrial Research, New Delhi

Dar ML, Dhar MM, Dhawan BN, Mehrotra BN, Aswal BS, Bhakuni DS, Ved Prakash, Goel AK, Jain S, Srimal RC 1968–1990 Screening of Indian plants for biological activity. I–XIV. Indian J Exp Biol, vols 6–28 (a series of papers with various author combinations)

Dixit 1991 An ethnobotanical survey of ingredients of domestic remedies in use in Ajmer district (Rajasthan). PhD thesis, University of Rajasthan, Jaipur, Rajasthan

Hooker JD 1872–1896 Flora of British India. L Reeve & Co, Kent, UK, 7 vols

Husain A 1993 Medicinal plants and their cultivation. Central Institute of Medicinal and Aromatic Plants, Lucknow, Uttar Pradesh

Husain A, Virmani OP, Popli SP et al 1992 Dictionary of Indian medicinal plants. Central Institute of Medicinal and Aromatic plants, Lucknow, Uttar Pradesh

Jain SK 1963 Studies in Indian ethnobotany. Plants used in medicine by tribals of Madhya Pradesh. Bull Reg Res Lab Jammu 1:126–128

Jain SK 1965 Medicinal plantlore of the tribals of Bastar. Econ Bot 19:236–250

Jain SK 1985 Medicinal plants of India. National Book Trust, New Delhi

Jain SK 1989 Ethnobotany: an interdisciplinary science for holistic approach to man-plant relationships. In: Jain SK (ed) Methods and approaches in ethnobotany. Society of Ethnobotany, Lucknow, Uttar Pradesh (Proc II Train Course, Lucknow, 1989) p 9–12

Jain SK 1991 Dictionary of Indian folkmedicine and ethnobotany. Deep Publications, New Delhi

Jain SK, DeFilipps RA 1991 Medicinal plants of India. Reference Publications, Algonac

Jain SK, Saklani A 1992 Cross-cultural ethnobotanical studies in northeast India. Ethnobotany 4:25–38

Jain SK, Mudgal V, Banerjee DK, Guha A, Pal DC, Das D 1984 Bibliography of ethnobotany. Botanical Survey of India, Calcutta, Bengal

Jain SK, Sinha BK, Gupta RC 1991 Notable plants in ethnomedicine of India. Deep Publications, New Delhi

Janaki-Ammal EK 1956 Introduction to the subsistence economy of India. In: William LT Jr (ed) Man's role in changing the face of the earth. University of Chicago Press, Chicagao, IL, p 324–335

Johns T, Kokwaro JO, Kimanani E 1990 Herbal remedies of the Luo of Siaya District, Kenya: establishment of quantitative criteria for consensuses. Econ Bot 44:369–381

Kapoor LD 1990 CRC handbook of Ayurvedic medicinal plants. Wolfe Medical Publishers, London

Mehrotra BN 1986 Research in pharmacognosy on Indian medicinal plants. In: Dhawan BN (ed) Current resaerch on medicinal plants in India. Indian National Science Academy, New Delhi, p 6–20

Mitra R 1985 Bibliography on pharmacognosy of medicinal plants. National Botanical Research Institute, Lucknow, Uttar Pradesh

Mitra R, Jain SK 1991 Medicinal plants research in India. A review. Ethnobotany 3:65–77

Mudgal V 1989 Literature on ethnobotany—Indian and foreign. In: Jain SK (ed) Methods and approaches in ethnobotany. Society for Ethnobotany, Lucknow, Uttar Pradesh (Proc II Train Course, Lucknow 1988) p 42–57

Mukerji B 1953 Indian pharmaceutical codex I—indigenous drugs. Council of Scientific and Industrial Research, New Delhi

Phillips O, Gentry AH 1993 The useful plants of Tambopata, Peru. I. Statistical hypothesis tests with a new quantitative technique. Econ Bot 47:15–32

Rastogi RP, Mehrotra BN 1990, 1991, 1993 Compendium of Indian medicinal plants. Central Drug Research Institute, Lucknow, Uttar Pradesh, 3 vols

Saklani A 1992 Cross-cultural ethnobotanical studies among the tribes of northeastern India. PhD thesis, Garhwal University, Srinagar, Uttar Pradesh

Sinha RK 1988 Survey of medicinal plants and plant products sold by the herbal vendors and their socio-economic study. PhD thesis, University of Rajasthan, Jaipur, Rajasthan

Thakur RS, Puri HS, Husain A 1992 Major medicinal plants of India. Central Institute of Medicinal and Aromatic Plants, Lucknow, Uttar Pradesh

Varghese E, Jain SK, Bose N 1993 A quantitative approach to establish the efficacy of herbal remedies: a case study on the Kharias. Ethnobotany 5:149–154

Watt G 1889–1896 A dictionary of the economic products of India. WH Allen, London, 6 vols

DISCUSSION

Posey: What's the relationship between plants categorized as spices and those categorized as medicines?

Jain: Most housewives in India know which spices are good for health. They use a lot of turmeric, cumin, fennel and cloves in food in various preparations. They would quickly say that these spices are good for digestion or good for stopping dyspepsia.

Balée: In these very early descriptions of medicinal uses of plants, such as the *Rigveda* and *Arthava veda*, is there a heavy emphasis on the materia medica and less on other possible uses of the plants, including food plants and recipes?

Jain: No. I referred only to the materia medica. The Vedas are supposed to contain all the knowledge. There are profuse references to plants as food, as dyes and so on.

Cragg: With the popularity of these preparations, is there a threat to the existence of some of these plants? Are organizations like the Central Institute of Medicinal and Aromatic Plants in Lucknow handling the cultivation?

Jain: There is a threat to several of these species. In many cases, pharmaceutical companies are already using substitutes because the original species are not easily available. There is a major programme of the Indian Council of Agricultural Research to cultivate medicinal plants. There is a project coordinator working on the cultivation of these plants and some small farms have been established in different parts of the country.

Barton: I understand there is a law governing the sale of medicinal plants in India—that there are special provisions in your pharmaceutical law.

Jain: For sale of the ingredients within the country, there are no regulations. Products manufactured by a pharmaceutical company have to pass certification by the Drug Controller. But many products that have not gone through the development process and certification, etc, do find a way to market. The Drug Controller cannot physically regulate everywhere these are sold, particularly in the remote areas.

Dagne: To what extent have the medicinal plants that are sold in the market been studied and their effects authenticated?

Jain: The street vendors who sometimes sell these drugs are unqualified people, they are unregistered practitioners. Each of these vendors collects about 25–30 plant drugs, as roots or leaves or whatever. Their customers are those who are superstitious or those who have no means of buying standard preparations or expensive drugs.

Farnsworth: What percentage of the plants described in these publications are used singly or in mixtures?

Jain: I could not say the exact percentage, but the majority are given as mixtures.

McChesney: Are the components of these mixtures prepared individually and then mixed, or are they mixed together and the preparation made from the mixture?

Jain: Different preparations are prepared in different ways. The large majority of the practitioners use the Ayurvedic system and prepare their own medicine.

They grow the plants in their back yard, or house garden, from which they prepare the ingredients. Only a few things are obtained from other sources. The mixtures are prepared in a variety of ways; they are boiled, decanted, filtered, powdered, made into pastes, etc. But, except for one or two groups of Ayurvedic medicines that are called *Asavás*, which are fermented, few preparations have long-lasting preservative qualities, most have to be prepared fresh.

Farnsworth: In many of the medicines used in India, black pepper is a constituent. C. K. Atal at Jammu showed that piperine, which is the major constituent in black pepper, enhances the bioavailability of many drugs, including ampicillin and even synthetic drugs. It increases their absorption substantially. This may explain why pepper is a frequent component of medicinal mixtures.

Cox: Some of the remedies contain extracts from multiple plant species. Is part of the reason to try to prevent side-effects? For example, if you have a plant that treats diarrhoea but has some other side-effect, do they add a second plant to try and ameliorate the side-effect?

Jain: It sounds reasonable, but I do not think the Ayurvedic medicines cause side-effects. None of us has lived long enough in any of these communities, where these medicines are being used, to observe the progress of the patient and whether there are any side-effects. Our information is based on the reports from these people.

Elvin-Lewis: Several years ago, I had the honour to be sent by the WHO to India to help my dental colleagues understand and evaluate the therapeutic rationale of their plants. In doing so, I was able to 'crack the code' in understanding their uses of Ayurvedic mixtures.

I looked at the plants that were used in oral hygiene and to treat a variety of oral infections. I didn't simply make a list of the plants used in each remedy, but looked very carefully at the therapeutic rationale behind their inclusion in a formulation. Most of these plants had been well studied chemically and therefore it was possible to determine the types of bioreactivities one might expect. I discovered that in some formulations there might be some redundancy of components with similar types of activities, such as mouth fresheners, antibiotics or emollients. Also, when known toxic plants were included, these were often at concentrations akin to homeopathic levels or at least well below the level associated with toxicity. Through the combination of many plants containing substances of low potency, as the Ayurvedic system has done, one can benefit from complementary actions and evoke efficacy with low toxicity.

My colleagues, especially at St Georges Dental College, Lucknow, have addressed these situations with plant extracts that have been formulated into dentifrices. They have studied the clinical activity of these and have shown that they have antiplaque and ameliorative capacities against gingivitis. In the context of Western dentistry, we are referring to related classes of compounds that are present as a mixture and that possess various degrees of antibiotic and antiinflammatory activities. How then can we ensure reproducible therapeutic success when using partially purified extracts and herbal mixtures rather than purified compounds? It may be necessary to achieve some form of quality control by regulating the amounts of significant bioreactive compounds within them—not unlike techniques applied to wine or tea blending.

Further, I believe that if one wants to study Ayurvedic medicine, the therapeutic rationale of the mixtures will be better understood if one takes the time to pull apart the constituents and understand them first, so you can ascertain how they complement each other. This is not to say that the mixtures are not creating new types of compounds that in themselves are valuable, as has been found in a Chinese remedy for hepatitis.

Jain: In India, some pharmacologists have started to re-evaluate Ayurvedic medicines. Two pharmacologists, Dr Sharadini Dahanukar and Urmila Thatte of Bombay, have published a book called *Ayurveda revisited* (Dahanukar & Thatte 1989). In this, they mention the Ayurvedic medicines that they have evaluated using pharmacological tests.

Martin: I would like to ask about the search for new drugs within the Ayurvedic tradition. We know that, for example, in the history of Greek medicine, Dioscorides in the 1st century AD wrote a pharmacopoeia which was essentially unquestioned for 1600 years. It even inhibited research on medicinal plants in different parts of Europe. In the Ayurvedic tradition is there a constant process of reviewing new medicinal plants and incorporating them into the pharmacopoeia? Or are the 526 species that were recorded in 800–700 BC still the ones that are used primarily?

Jain: At present, about 2000 plants feature in Ayurvedic books; of these some 500 species are more commonly employed and they are used in the majority of Ayurvedic prescriptions. With regard to new additions, there are now several research institutions in Ayurveda. Their scientists are working on pharmacological and/or clinical research problems; their clinical research leads to the incorporation of new plants into Ayurvedic medicine.

Lewis: What is the frequency of the use of different systems in India— Ayurveda, Unani and Western medicine?

Jain: Western medicine is most common in the urban areas, towns of about 100 000 people and more. There, I would think 75% of the people would be receiving Western medicine and only 15% indigenous systems of medicine. As you move to towns and villages of fewer than 10 000 people, more people use indigenous medicines. India is about 80% rual. Therefore, 75–80% of the population in India still receives traditional medicine or Western medicine of very simple 'first aid' level.

Lewis: Do people go to different types of practitioner, if something doesn't work in one case?

Jain: Yes; in very many cases. But, there are certainly many (hundreds of thousands) very loyal people who will not take Western medicine at all or anything outside Ayurveda or homeopathy. For chronic diseases, very many finally go to Ayurveda.

Cox: Could Ayurvedic medicine be compared to a pan-cultural medicine like Japanese *kampah* medicine? Has it displaced some indigenous medicines? For example, do the people in the Gar desert of India have indigenous medicine that's different from Ayurvedic?

Jain: Wherever there is an organized system of indigenous medicine available, whether it is an Ayurvedic doctor or a Yunani or Siddha doctor, people go to them. Folk medicine or ethnomedicine is in vogue only in remote and usually hilly places where none of the organized systems are available.

Elisabetsky: I'm amazed how many practitioners of herbal medicine there are now in the United States of America. How do you feel about these?

Balick: Many of the Ayurvedic practitioners in the United States are Indian and others have been trained by the Indian physicians now living in United States. There are various degrees of sophistication in their work and there is, in some cases, dilution of the original intent or philosophy of Ayurveda, which relies primarily on diet, exercise and meditation to maintain a proper balance in life.

Jain: Many people all over the world are being attracted to natural products and so there is a revival of interest in Ayurveda.

Farnsworth: There is a Maharishi Medical University in the United States located in Iowa. They train Western-qualified practitioners in Ayurvedic medicine and this has become quite widespread. In the NIH study section on alternative medicine, of the 60 proposals that we had, 20–25% involved Ayurvedic medicine.

Reference

Dahanukar S, Thatte U 1989 Ayurveda revisited. South Asia Books, Columbia, MO

Ethnopharmacological investigation of Chinese medicinal plants

Xiao, Pei-Gen

Institute of Medicinal Plant Development, Chinese Academy of Medical Sciences, WHO Collaborating Centre for Traditional Medicine, Beijing, China

Abstract. Chinese medicinal plants, which have been used by the Chinese people for centuries, are still being widely used today within the framework of health-care services. A recent nationwide survey has shown there are 7295 species of plants used medicinally in China. An ethnopharmacological investigation included families containing over 100 medicinal plant species in Chinese medical flora, as well as genera containing over 15 medicinal species with a high proportion of therapeutic members. Medicinal plants are an important source for new drug development in China. Up to now about 200 new drugs have been developed directly or indirectly from Chinese medicinal plants. Nearly one half of the new drugs are from a single Chinese medicinal plant, or its active principle(s) and synthetic derivatives, or active fraction(s) or even the total extract. More than half are from composite prescriptions.

1994 Ethnobotany and the search for new drugs. Wiley, Chichester (Ciba Foundation Symposium 185) p 169–177

Chinese medicinal plants, which have been used by the Chinese people for centuries, are still being widely used today within the framework of health-care services. The production of plant-based medicaments and related products is worth around US$5800 million per annum. Traditional Chinese drugs (which are mostly derived from medicinal plants) make up 45% of the market, Western-style drugs around 55%. The importance of Chinese medicinal plants is reflected in the latest *Pharmacopoeia of the People's Republic of China* (in two volumes, 1990 edn.), of which the first volume is totally devoted to traditional Chinese drugs and their varied preparations with a total of 784 items. All these facts constitute a very good basis for an ethnopharmacological investigation.

General status of medicinal plant use

The latest nationwide survey has shown that there are 7295 species of plants used medicinally in China. Of these species, 281 belong to the thallophytes, 39 to the bryophytes, 433 to the pteridophytes, 73 to the gymnosperms; most,

TABLE 1   Families containing over 100 medicinal plant species used in Chinese medicine

Family	Medical species/ medical genera	Examples of genera with important medical uses
Asteraceae	570/135	Artemisia, Senecio, Achillea, Aster, Atractylodes, Eupatorium
Fabaceae	531/112	Astragalus, Sophora, Acacia, Cassia, Caragara
Laminaceae	253/50	Salvia, Scutellaria, Mentha, Nepeta, Elsholzia, Thymus, Rabdosia
Liliaceae	217/43	Allium, Polygonatum, Asparagus, Fritillaria, Lilium, Smilax, Veratrum
Orchidaceae	198/57	Orchis, Gastrodia, Dendrobium, Habenaria, Liparis
Rosaceae	155/34	Crataegus, Rosa, Prunus, Potentilla, Rubus, Chaenomeles
Rubiaceae	141/39	Rubia, Uncaria, Galium, Gardenia, Hedyotis, Mussaenda
Apiaceae	140/44	Angelica, Ligusticum, Heracleum, Peucedanum, Bupleurum, Ferula
Euphorbiaceae	134/36	Phyllanthus, Euphorbia, Croton, Mallotus
Scrophulariaceae	132/38	Scrophularia, Rehmannia, Veronica
Saxifragaceae	107/21	Hydrangea, Chrysosplenum, Bergenia, Saxifraga
Asclepiadaceae	102/33	Cynanchum, Tylophora, Hoya
Papaveraceae	100/14	Papaver, Corydalis
Ranunculaceae	210/30	Aconitum, Delphinium, Coptis, Thalictrum, Clematis, Anemone

however, are angiosperms, of which 5499 are dicotyledons and 970 are monocotyledons.

This knowledge provided the basis for the publication of two series of books more recently: *A Pictorial Encyclopaedia of Chinese Medicinal Herbs* (Xiao 1988–1990) and *New Compendium of Chinese Materia Medica* (Wu et al 1988–1991).

Ethnopharmacological investigation

From the 7295 species of Chinese medicinal plants, the plant families which contain over 100 medicinal plant species are listed in Table 1. The genera that contain over 15 medicinal plant species and a high proportion of species with therapeutic properties are listed in Table 2.

During the nationwide survey, 35 000 items of ethnopharmacological information on Chinese medicinal plants were collected. These were subsequently analysed systematically by calculating the following statistical coefficients (Xiao et al 1986, 1989).

TABLE 2 Genera containing over 15 medicinal species and a high proportion of species with therapeutic properties

Genus	Medical species/ total species	Main ethnopharmacological properties
Aconitum	46/167	Anodyne, antirheumatic, dispels internal cold, an arrow poison
Aralia	17/30	General tonic, antirheumatic
Aristolochia	20/42	Antipyretic, antidotary, for various infections, treatment of abdominal distension with pain
Artemisia	43/180	Antipyretic, antidotary, antiinflammatory, carminative
Berberis	60/200	Antipyretic, antidotary, for various infections, antidysenteric
Clematis	28/108	Diuretic, antirheumatic, promotes blood circulation, anodyne
Codonopsis	21/48	General tonic, invigorates the function of digestive system, relieves weakness
Corydalis	68/200	Anodyne, promotes blood circulation, antipyretic, antidotary
Cynanchum	23/45	Antipyretic, promotes blood circulation, antirheumatic
Delphinium	24/113	Anodyne, antirheumatic, antiinflammatory
Dendrobium	25/60	Tonic, for febrile diseases with thirst and dry mouth, dry cough and chronic tidal fever
Fritillaria	17/22	Antitussive, expectorant
Gentiana	27/230	Antipyretic, antidotary, antiinflammatory, antirheumatic, diuretic, stomachic
Hypericum	16/50	Antimicrobial, antidotary, treats hepatitis, promotes blood circulation, emmenagogue
Ilex	22/160	Antiinflammatory, antipyretic, antidotary, for the treatment of cardiovascular illness
Lysimachia	37/120	Promotes blood circulation, emmenagogue, antirheumatic
Polygala	17/40	Antitussive, general tonic, relieves weakness, tranquilizer
Polygonum	38/120	Astringent, antidysenteric, haemostatic, antipyretic, antidotary
Rabdosia	22/90	Antimicrobial, anticancer, antipyretic, antidotary, antirheumatic, antiinflammatory
Rubus	44/150	Astringent, haemostatic, antiinflammatory
Salvia	29/75	Promotes blood circulation, used to treat coronary heart diseases, antipyretic, antidotary
Scutellaria	17/55	Antipyretic, antidotary, promotes blood circulation, haemostatic
Swertia	15/70	Bitter tonic, antipyretic, antidotary, stomachic
Thalictrum	30/67	Antipyretic, antidotary, antidysenteric, antimicrobial

1) The family medicinal coefficient (α_f) = medicinal genera/total genera, e.g. Magnoliaceae $\alpha_f = 4/11 = 0.36$
2) The general medicinal coefficient (α_g) = medicinal species/total species, e.g. Magnolia $\alpha_g = 10/30 = 0.33$
3) The coefficient of traditional therapeutic usage within a genus (TRI) = $\dfrac{C_1^2}{C_2} \times 100$

   C_1 represents cards of a certain traditional usage, e.g. antipyretic. (Each card represents one species as it is used in a certain locality.) C_2 represents total cards of those species with this traditional usage.
4) The coefficient of the extent of traditional therapeutic usage within a genus (β) = Sp_1/Sp_2

   Sp_1 represents the total number of species with a certain traditional usage; Sp_2 represents the total number of species of this genus.

According to our experience, TRI of greater than or equal to 300 can be treated as significant, worthy of further investigation. The following are a few examples of the coefficients for certain genera:

Cimicifuga. $\alpha_g = 0.75$; antipyretic and antidote TRI = 1370, $\beta = 0.71$. Our studies revealed that *Cimicifuga* possesses potential antimutagenic and anticancer properties.

Aconitum. $\alpha_g = 0.75$; antipyretic and antidote TRI = 5582, $\beta = 64$; anodyne TRI = 439, $\beta = 19$. Lappaconitine (a compound derived from this genus) has been developed as a new drug with analgesic activity.

Houttuynia. $\alpha_g = 1.0$; antipyretic and antidote TRI = 2235, $\beta = 100$. Sodium houttuyfonate has been developed as a new drug with antimicrobial activity.

Ethnopharmacology and new drug development

Medicinal plants are also an important source for new drug development in China. Up to 1992, about 200 new drugs had been developed directly or indirectly from Chinese medicinal plants (Xiao 1992). Nearly one half of the new drugs are from a single plant; the rest are from composite prescriptions. New drugs derived from a single plant include:

1) Semisynthetic or totally synthetic compounds based on active principles isolated from traditional Chinese drugs, e.g. biphenyl dimethoxy dicarboxylate, which lowers levels of serum glutamic-pyruvic transaminase), changroline (antiarrhythmic), indirubin (for treatment of chronic myelocytic leukaemia) and ligustrazine (used to treat occlusive cerebral blood vessel diseases).
2) Active principles from medicinal plants, e.g. agromophol (a taenifuge), artemisinin (antimalarial), bao-gong-teng A (used to treat glaucoma), huperzine A (used to improve memory and to treat myasthenia gravis) and scutellarin (for the treatment of occlusive cerebral blood vessel diseases).

3) Active fractions from a medicinal plant, e.g. spores from *Ganoderma lucidum* (used to treat myotonia) and the root or total isoflavonoids from *Pueraria lobata* (for the treatment of cardiovascular illness).
4) Extracts from a medicinal plant, e.g. a fermentation extract from *Cordyceps sinensis* (an immunomodulator), a pollen extract from *Typha angustata* (used to treat hyperlipidaemia) and root extracts from *Tripterygium wilfordii* (used to treat rheumatoid arthritis) and from *Salvia miltiorrhiza* (used in the treatment of cardiovascular illness).

Concluding remarks

Medical practice has taught us to appreciate that ethnopharmacological information is an important source for new drug development. Because medicinal plants have been used for centuries and tested by millions of people, there has been ample opportunity to find satisfactory medical agents. Ethnopharmacological investigation, as theoretical research, will provide more useful clues for new drug development as well as lead to better use of resources.

References

Wu ZY, Zhou TY, Xiao PG (eds) 1988–1991 New compendium of Chinese materia medica. Shanghai Science and Technology Publisher, Shanghai, vols 1–3
Xiao PG (ed) 1988–1990 A pictorial encyclopaedia of Chinese medicinal herbs. The Commercial Press (Hong Kong), Hong Kong; People's Medical Publishing House, Beijing, vols 1–10
Xiao PG 1992 Traditional medicine and new drug development. Herba Pol 38:141–147
Xiao PG, Wang LW, Lu SJ, Qiu GS, Sun J 1986 Statistical analysis of the ethnopharmacologic data based on Chinese medicinal plants by electronic computer. I. Magnoliidae. Chin J Integr Tradit West Med 6:253–256
Xiao PG, Wang LW, Qiu GS, Sun J 1989 Statistical analysis of the ethnopharmacologic data based on Chinese medicinal plants by electronic computer. II. Hamamelidae and Caryophylidae. Chin J Integr Tradit West Med 9:429–432

DISCUSSION

Iwu: I'm fascinated by what is happening in China; this is a model for most Third World countries. During a visit to China, I was really impressed by the amount of institutional support from the government. In most Third World countries, the academics are on one side and the government is on the other side.

Bohlin: You mentioned *Aconitum*; how do you handle the toxicity of the alkaloids. Do you mix it with other plant extracts?

Xiao: *Aconitum* is known to be very toxic. People in the rural areas use the root for the treatment of serious rheumatalgia. They rub the root around the rim of the cup, but only three times. If you go round 10 times, the patient will be poisoned. In our traditional medicine, we have to detoxify the *Aconitum* root by processing. The processing hydrolyses the toxic alkaloids and their toxicity decreases dramatically. The prepared *Aconitum* root is called *Fu-zi*; it is used to relieve pain by dispelling cold. It is also a good cardiotonic.

Lozoya: It's always a pleasure to see the way China is solving this long-lasting problem. You use four categories of plant products. What is the difference between categories (3) and (4)—between active fractions and extracts?

Xiao: The extracts are in a much more crude form.

Lozoya: So for these extracts you have no knowledge of the active component. In those cases, is standardization achieved by assaying another substance?

Xiao: Yes. We control by measuring some major component, such as flavonoids.

Peter: Do you have any formally approved drugs which consist of active fractions or crude extracts? If so, do you always know the active ingredients and can you quantitate them?

Xiao: Yes, in many cases. For instance, we know that in *Radix puerariae* the isoflavonoid is active. But for economic reasons, we extract the total isoflavonoids and make a tablet form. If we extract only a single active component, we have to control and evaluate the active ingredient qualitatively and quantitatively. In many cases, we just make the total extracts.

Peter: But in this way you have solved the problem of standardization.

Xiao: Yes. For most of these preparations, when they are developed as a new drug, you have to offer detailed documentation, then you can get approval.

Lewis: In the category of crude extracts, do you have a quality control?

Xiao: We control the quality in several ways. One is that we have to fix the pharmaceutical techniques and say how many ingredients are present in the crude drugs. Each crude drug is very well evaluated in order to guarantee its quality. We can also carry out qualitative analysis for some of the ingredients, for example, flavonoids by means of thin-layer chromatography. These assays are done mainly at the primary stage of drug development, on the crude extract form. We are willing to clarify the active principles later on, but even if we have not determined these, we still use the extract. For instance, we have developed an injection called 'Jisheng–Injection', made from the spores of *Ganoderma lucidum*, but we don't yet know the active ingredient.

Farnsworth: You mentioned one of my favourite plants, *Tripterygium wilfordii*. It is little known in the West, most of the relevant publications are in China. An extract of the root wood from this plant has been clinically tested and standardized for treatment of severe arthritis and lupus. The only way to treat these in the West is with very powerful drugs, such as methotrexate and steroids. The active principle is known. *T. wilfordii* contains about 30 very unusual diterpenes; it's the only plant in the history of the world that ever produced a diterpene with a methyl group in the 3 position. These diterpenes were first discovered by Professor Kupchan in the United States in the cancer programme at NCI. They have anticancer properties but these were never pursued, presumably because of the low yield. There are occasional side-effects but none is irreversible. These extracts are standardized in terms of the two major diterpenes. These two compounds are also effective in rodents as male contraceptives at concentrations of 30 ng/kg orally. I have never seen any plant-derived pure chemical compound active at that dose level. The Chinese have patented this. In this case, it is probably not commercially feasible to isolate the active compounds, because they're only present in very low yields. There is a good case for, in some situations, using a standardized crude extract.

McChesney: Professor Xiao, one of the important activities that you referred to was immune modulation or immune stimulation. How do you measure that in a clinical setting, so as to be able to develop a pharmaceutical preparation, either a crude plant extract or a purified substance?

Xiao: It is difficult; we measure many different things (Xiao 1993).

Lozoya: In what dosage form are Chinese medicines usually given?

Xiao: The most common form of administration of traditional drugs at the moment is tablets, capsules and granules—rather like instant coffee. But in traditional Chinese medicine, another common form is as a powder mixed with honey.

Balée: Many new drugs have been developed on the basis of the traditional medicine of which you speak. What percentage of those have already been described in the Chinese written materia medica? What percentage of those plants come from new field research that you and your colleagues have done?

Xiao: Of the 5136 medicinal plants that I mentioned, around 1000 are already recognized in the Chinese written materia medica. More recently, in a nationwide investigation we found 7295 species of plants used medicinally. So up to now, the total number of Chinese medicinal plants is around 8000, because a proportion have not yet been identified. Only 10% of these are in popular use, which means around 800 species.

Balée: Do anthropologists help you in this work, in the search for new drugs in China and in research in traditional medicine?

Xiao: Yes, but only occasionally.

Balée: Which part of China is the most important source of new drugs from traditional medicine?

Xiao: It's difficult to say. It depends on the target. Even today, many pharmaceutical factories ask us to find new chemical structures from medicinal plants. The chemical content of the plants, even of a single species, varies depending on the environment—there is ecological variation. For instance, if we obtained samples of a plant from the Tibetan high plateau, they might contain higher concentrations of a particular ingredient than samples of the same plant from other regions.

McChesney: You said that about half the Chinese population prefers Western-style medicine and about half prefers traditional medicine. Have you done a geographical analysis of those data. Is the half that prefers Western-style medicine an urban population and the half that prefers Chinese style a rural population?

Xiao: I think the distributions of the people who prefer each sort of medicine are almost the same. For instance, I am a Western-style scientist; I can go to the hospital for Western-style treatment, but sometimes I prefer a traditional style of treatment, for example for chronic conditions.

Cragg: I know that for cancer treatment, people are using combinations of Western drugs and traditional Chinese medicines, for instance at the cancer hospital in Beijing. Is this the case for other diseases too?

Xiao: Yes. In many cases, we integrate these two kinds of medicine in clinics, because several Western-style drugs cause some side-effects. We often combine these with Chinese drugs, particularly those that are good for the body—in cancer treatment we use immunostimulatory agents.

McChesney: With the great reliance on the use of plant preparations in traditional medicine, and in some ways what we might call the Westernization of that into convenient dosage forms, such as pills, are those almost entirely prepared from cultivated plants or are the plants still largely gathered from the wild?

Xiao: Most of the plants used in traditional medicine are cultivated.

McChesney: Have they been in cultivation for a long time?

Xiao: Yes, but only few of them. Most of the traditional Chinese drugs have been cultivated recently, particularly in the last 30 years.

Elvin-Lewis: In Chinese medicine, many of the drugs are prepared by mixing plant extracts together. This also occurs in Ayurvedic medicine. I would propose that in the future when we have to deal with these crude mixtures or crude extracts from plants, we learn a lesson from the tea scientists and the wine scientists, who have become expert in blending for flavour. We may eventually have a whole group of new scientists, who are expert in blending for efficacy.

Craveiro: Are there any essential oils used in traditional medicine in China?

Xiao: There are many, particularly from the plant families Apiaceae, Asteraceae, Zingiberaceae and Laminaceae.

Prance: In China, are you looking at any plants from other parts of the world or are you confining your efforts to the plants that are already in China?

Xiao: We are always learning from other countries. Recently, we introduced American ginseng from the United States. This is good for general health as a tonic. In some ways it is better than Korean ginseng! We have also introduced *Rauvolfia* for the treatment of hypertension. We have started to investigate taxol thoroughly as an anticancer drug from the United States.

Lewis: What do you think about the efficacy of ginseng? Is either Chinese and American preferred over the other?

Xiao: Yes. Chinese ginseng is the same species as Korean ginseng, but Korean ginseng right now is more popular worldwide. American ginseng is another species. In China we prefer to use American ginseng. According to the theory of our traditional medicine, the property of Korean ginseng is too hot. Some people take this remedy and it causes nose-bleeds. Other people take it at night and they feel too excited to sleep. American ginseng is much more mild; it has an action we call *Yin*-tonic, for replenishing the vital essence. It is beneficial for patients with symptoms of *Yin*-deficiency. We are developing American ginseng as an antiageing agent combined with some other ingredients. This is being tested in clinical trials; all the parameters show very good results.

Reference

Xiao PG 1993 Immunological aspects of Chinese medicinal plants as anti-ageing drugs. J Ethnopharmacol 38:167–175

Ethnobotany and drug discovery: the experience of the US National Cancer Institute

Gordon M. Cragg, Michael R. Boyd, John H. Cardellina II, David J. Newman, Kenneth M. Snader and *Thomas G. McCloud

Developmental Therapeutics Program, Division of Cancer Treatment, National Cancer Institute, Bethesda, Maryland 20892 and *Program Resources, Inc., NCI-Frederick Cancer Research and Development Center, P.O. Box B, Frederick, MD 21701, USA

Abstract. Between 1960 and 1981 the National Cancer Institute (NCI) screened 114 000 extracts of 35 000 plants, mainly collected in temperate regions. Of the three clinically active anticancer drugs so far discovered in that programme, none was isolated from a plant collected on an ethnobotanical basis, though various *Taxus* species, which are the source of taxol, are reported to have been used medicinally. Since 1986, the NCI has focused its collections in tropical and subtropical regions worldwide; collections cover a broad taxonomic range, though priority is given to medicinal plants when relevant information is available. As of August 1993, 21 881 extracts derived from over 10 500 samples had been tested in a screen for activity against the human immunodeficiency virus (HIV); 2320 of these extracts were of medicinal plant origin. Approximately 18% of both the total number of extracts and the medicinal plant-derived extracts showed significant anti-HIV activity; in each instance about 90% of the active extracts were aqueous. The activity of the aqueous extracts has been attributed mainly to the presence of polysaccharides or tannins. Four plant-derived compounds are in preclinical development at the NCI; only one of the four source plants, obtained from a non-contract source, was collected on an ethnobotanical basis. At this stage the results indicate that the current NCI collection policy offers the best chances for the discovery and development of agents for the treatment of AIDS (acquired immune deficiency syndrome) and cancer.

1994 Ethnobotany and the search for new drugs. Wiley, Chichester (Ciba Foundation Symposium 185) p 178–196

The mission of the United States National Cancer Institute (NCI) established in 1937 is 'to provide for, foster, and aid in coordinating research related to cancer'. An important goal of such a mission is the discovery and development of new agents with chemotherapeutic value in the fight against cancer; the responsibility for achieving this goal rests with the Developmental Therapeutics

Program (DTP), a major component of the Division of Cancer Treatment. With the emergence of AIDS (acquired immune deficiency syndrome), followed by its establishment globally in the 1980s, the NCI added to its mission the discovery and preclinical development of novel agents against the human immunodeficiency virus (HIV). The DTP developed a high-flux *in vitro* screen which has provided a national resource permitting scientists from organizations worldwide to submit materials for anti-HIV testing. The value of natural products as sources of potential, novel chemotherapeutic agents was recognized at the inception of the NCI drug discovery effort in 1955 and an extensive programme for the testing of thousands of materials of animal, microbial and plant origin has been operating for the past 35 years.

Plant acquisition programme: 1960–1982

The long history of the use of plants in sophisticated traditional medicine systems and in folkloric medicine makes this a particularly valuable area for study and the importance of medicinal plants in the development of a broad range of drugs used in modern medicine is clear (Farnsworth et al 1985). This was one of several factors that prompted the NCI to initiate a systematic programme for the collection and screening of plants for antitumour activity, because many claims have been made for the beneficial action of plants in the treatment of cancer. Over 3000 species of plants with reported antitumour properties were catalogued in the very thorough review published by the late Jonathan Hartwell (1982). Of greater importance, however, was the development of the efficacious anticancer agents, vinblastine and vincristine, isolated from *Catharanthus roseus*, and of podophyllotoxin, isolated from *Podophyllum peltatum* or *Podophyllum emodii*, which was later semisynthetically converted into etoposide and teniposide (Cragg et al 1993).

In 1960, the NCI established a collaborative programme with the United States Department of Agriculture (USDA) for the collection and screening of plants, mainly from temperate regions. The strategy adopted was to collect a broad taxonomic range of plants selected largely at random, with the exclusion of those plants too small or uncommon to permit adequate sampling (approximately 0.5 kg per sample). As screening progressed, a greater emphasis was placed on the collection of certain families (e.g. Apocynaceae, Rutaceae, Simaroubaceae) shown to be good sources of active extracts, while the collection of families providing few active extracts (e.g. Graminae) became less important. During 1960–1982, over 114 000 extracts of some 35 000 plants were tested against a range of rodent tumour systems. The details of the procurement strategy and the distribution of preclinical activity observed in these plant collections have been reviewed (Perdue 1976, Barclay & Perdue 1976).

Anticancer drug discovery: 1960–1982

As stated above, the strategy adopted by the USDA in its plant collections was largely one of random selection of a broad taxonomic range as opposed to selection based on medicinal use. The ethnobotanical approach was not used because, despite the many claims made for the anticancer properties of plants (Hartwell 1982), it is unlikely that cancer, as a specific disease entity, would be readily diagnosed in traditional medicine or folklore. This applies, in particular, to the slow-growing tumours of internal organs (e.g. lung, colon, ovarian, prostate) which are major killers in modern society. Furthermore, cancer is mainly a disease of older populations and when one considers the limited life expectancy in earlier times or of most societies reliant on herbal medicine, the likelihood of observing or diagnosing many cancers would be minimal. A major exception here is external tumours and many of the references in Hartwell's catalogue are to the treatment of skin cancers, warts, external growths and swellings, tumours of the oral cavity and genital tumours, as well as undefined 'cancers' or 'tumours'.

Despite the adoption of the so-called random collection approach, a retrospective analysis of the source plants of extracts that showed activity in the preclinical animal screens indicated that collections guided by knowledge of traditional medicinal use and poisonous properties yielded at least 50% more active extracts than did random ones (Spjut & Perdue 1976). It was noted, however, that several of the earlier test systems used (e.g. Sarcoma 180 and Walker 256 carcinosarcoma) were particularly sensitive to ubiquitous plant constituents, such as tannins, phytosterols and saponins—compounds not considered as potentially useful candidates for drug development. Thus the conclusions from the analysis by Spjut & Perdue are questionable in terms of drug development.

This collection and screening programme led to the discovery and characterization of many novel agents belonging to a wide variety of chemical classes (Hartwell 1976, Suffness & Douros 1979). Few of these agents, however, satisfied the stringent preclinical development requirements and advanced to human clinical trials. In assessing the value of medicinal plants in the discovery and development of anticancer drugs, it is instructive to consider those agents showing significant activity in clinical trials or general clinical use.

Currently, five plant-derived anticancer drugs (Fig. 1) are in clinical use. Of these, only taxol was discovered in the NCI programme, but the NCI played a substantial role in the clinical development of the other four. The best known of these agents are the so-called vinca alkaloids, vinblastine and vincristine, isolated from the Madagascan periwinkle, *Catharanthus roseus*. These alkaloids first became available in the 1960s; they are now used extensively in the treatment of a wide variety of cancers, often in combination with other agents (Neuss & Neuss 1990). *C. roseus* was used by various cultures for the treatment of

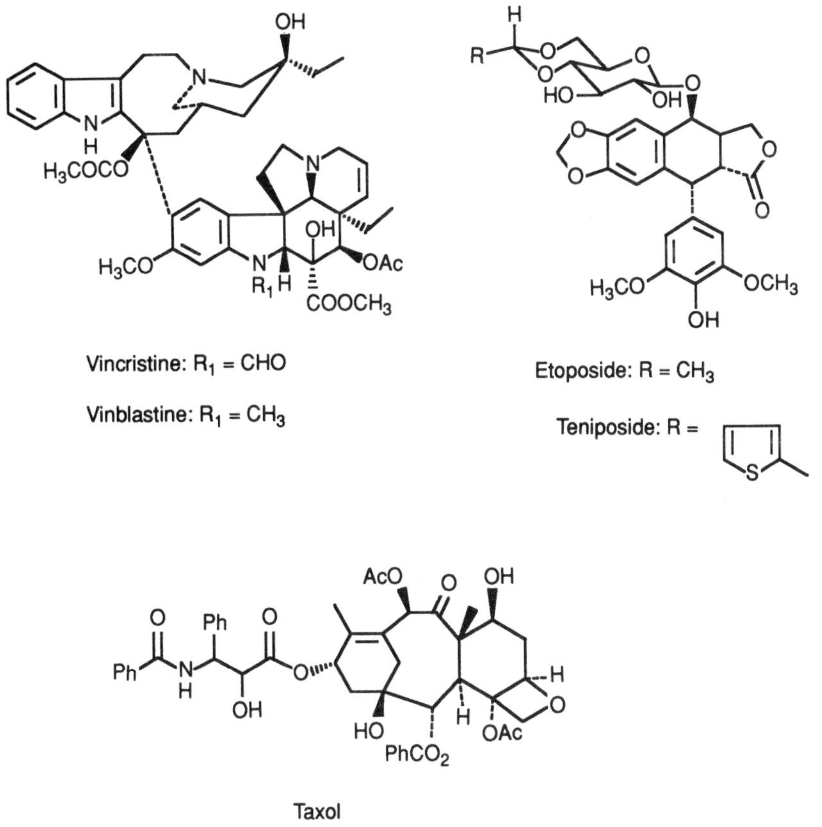

Vincristine: R_1 = CHO

Vinblastine: R_1 = CH₃

Etoposide: R = CH₃

Teniposide: R =

Taxol

FIG. 1. Clinically used plant-derived anticancer agents.

diabetes and these agents were discovered during an investigation of the plant as a source of potential antihypoglycaemic agents. Their discovery, therefore, may be indirectly attributed to the observation of an unrelated medicinal use of the source plant.

Podophyllotoxin was isolated from the roots of various species of the genus *Podophyllum*. These plants possess a long history of medicinal use by early American and Asian cultures, including in the treatment of skin cancers and warts (Hartwell 1982). Clinical trials of various derivatives of podophyllotoxin were discontinued owing to severe toxic side-effects, but two semisynthetic derivatives of epipodophyllotoxin, an isomer of podophyllotoxin, have been developed into the clinically effective agents etoposide and teniposide (Cragg et al 1993).

Taxol initially was isolated from the bark of *Taxus brevifolia*, collected in the US Pacific Northwest as part of the USDA random collection programme.

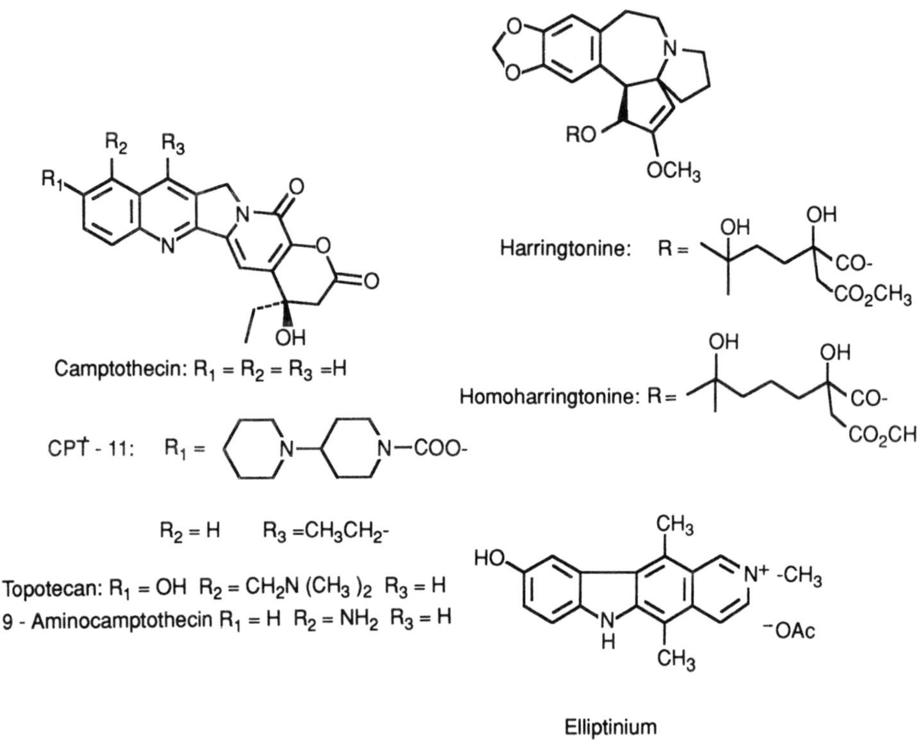

Camptothecin: $R_1 = R_2 = R_3 = H$

CPT - 11: $R_1 =$

$R_2 = H$ $R_3 = CH_3CH_2-$

Topotecan: $R_1 = OH$ $R_2 = CH_2N(CH_3)_2$ $R_3 = H$
9 - Aminocamptothecin $R_1 = H$ $R_2 = NH_2$ $R_3 = H$

Harringtonine: $R =$

Homoharringtonine: $R =$

Elliptinium

FIG. 2.   Plant-derived anticancer agents in advanced clinical trials.

The use of various parts of *T. brevifolia* and other *Taxus* species (e.g. *canadensis*, *baccata*) by several Native American tribes for the treatment of a variety of non-cancerous conditions has been reported (Moerman 1986). The leaves and fruits of *T. baccata* are likewise used in the Ayurvedic traditional medicine system (Kapoor 1990). There is one report of the use of *T. baccata* in India for the treatment of cancer (Hartwell 1982); the reported preparation is clarified butter, which implies use in the treatment of external conditions. Taxol appears to occur in the leaves of all *Taxus* species together with key precursors (baccatins) that can be readily converted synthetically to active taxol analogues, such as taxotere; taxotere is currently in clinical trials.

Other plant-derived anticancer agents are currently in clinical trials (Fig. 2) (Cragg et al 1993). Topotecan and CPT-11, both semisynthetically derived from camptothecin, are showing considerable promise; another derivative, 9-aminocamptothecin, has recently entered trials. Camptothecin is isolated from *Camptotheca acuminata*, a tree used extensively in China as an ornamental and as a source of firewood (Perdue et al 1970). Extracts of *C. acuminata* were investigated in the 1950s as a source of steroidal sapogenins for conversion into cortisone; they were later randomly selected for antitumour screening. The active extracts yielded camptothecin and several other active analogues. Camptothecin (as its sodium salt) was advanced to clinical trials, but was dropped because of severe bladder toxicity. Chang & But (1987a) report that the Chinese use *Xishu*, derived from the fruit, root, bark, stem bark or leaf of *C. acuminata*, as an antineoplastic agent, but it is not clear whether it was used in the traditional Chinese medical system or is the result of recent pharmacological studies in the USA and China. The first reported antitumour activity of extracts came from research workers connected with the NCI screening programme (Wall et al 1966).

Homoharringtonine, isolated from the Chinese tree, *Cephalotaxus harringtonia* var. *drupacea* (Sieb or Zucci) and several other species of this genus, is in clinical trials and has shown some efficacy against various leukaemias. Chang & But (1987b) report studies of the Chinese preparation, *Sanjianshan*, derived from the bark and leafy branch of *Cephalotaxus fortunei*, but in this instance it appears that the antitumour activity was first observed in the NCI screening programme (Powell et al 1972). In fact, no medicinal properties are reported by Chang & But for this particular preparation. Elliptinium, a derivative of ellipticine, has been in clinical trials in France for over 10 years and has shown some activity in the treatment of several cancers. Ellipticine is isolated from species of several genera of the Apocynaceae family, including *Bleekeria vitensis* A. C. Sm. It exhibited significant activity against a number of animal tumour models, but development towards clinical trials was abandoned owing to severe toxic effects observed in animal studies. *B. vitensis* is reported as a Fijian medicinal plant having anticancer properties (Cambie 1986).

The emphasis on the collection of families and genera yielding relatively greater numbers of active extracts in the USDA collection programme resulted in the repeated isolation of known agents or analogues possessing similar activity. The failure to isolate novel agents, particularly the failure to isolate agents effective against the resistant solid tumour disease-types, led to the termination of the programme in 1982. This apparent failure might have been due more to the limitations of the primary *in vivo* mouse leukaemia screens used for most of the collection period than to a deficiency of Nature.

Plant acquisition and screening programme: 1986–present

In 1985 a decision was made to develop a new *in vitro* screening strategy involving the use of cell lines from human solid tumours (Boyd 1989, 1993). This decision was accompanied by the implementation, in 1986, of a new NCI natural products programme involving new collection, extraction and isolation components. The development of the *in vitro* anti-HIV screen in 1987 (Boyd 1988) provided further impetus for the revitalization of the NCI's focus on natural products.

Since September 1986, contracts have been held by the Missouri Botanical Garden, New York Botanical Garden and the University of Illinois at Chicago (assisted by the Arnold Arboretum at Harvard University and the Bishop Museum) for the collection of plants from Africa and Madagascar, Central and South America, and South-East Asia, respectively. As with the earlier USDA collection programme, the present collections encompass a broad taxonomic diversity, but collectors are instructed to give priority to the collection of medicinal plants when reliable information on their use is available.

The dried plant samples are shipped by air freight to the NCI Natural Products Repository (NPR) in Frederick, Maryland, where they are stored at $-20\,°C$ for at least 48 hours to minimize the survival of plant pests and pathogens. After grinding, the plant samples are extracted sequentially with a 1:1 mixture of methanol:dichloromethane then water to give organic solvent and aqueous extracts. All the extracts are assigned discrete NCI extract numbers and returned to the repository for storage at $-20\,°C$ until requested for screening or further investigation.

Extracts are tested *in vitro* for selective cytotoxicity against panels of human cancer cell lines representing major disease types, including leukaemia, melanoma, lung, colon, central nervous system, ovarian and renal cancers (Boyd 1989). *In vitro* anti-HIV activity is determined by measuring the survival of virally infected human lymphoblastoid cells in the presence or absence of the extracts (Boyd 1988). Extracts showing significant selective cytotoxicity or anti-HIV activity are subjected to bioassay-guided fractionation by NCI chemists and biologists to isolate the pure chemical constituents responsible for the observed activity.

Recent developments in the NCI drug discovery programme

Potential anticancer agents

The screen now employed by the NCI involves 60 human cancer cell lines. The extensive research required in the development of this screen delayed the implementation of routine testing until 1990. The nature and limited capacity of the screen necessitate initial testing of extracts at a single high concentration ($100\,\mu g/ml$) followed by multidose testing of extracts exhibiting cytotoxicity in the single-dose assay. Over 30 000 plant extracts have been tested in the single-dose assay, but only relatively few of the selected active extracts have been subjected to multidose testing. Experience thus far indicates that fewer than 1% of the extracts show some degree of selective cytotoxicity. Interesting patterns of differential cytotoxicity have been associated with known classes of compounds, such as cardenolides, cucurbitacins, lignans and quassinoids (Cardellina et al 1993a). Several new leads have been discovered and are being investigated further, but the limited data available do not permit meaningful analysis at this stage.

TABLE 1 Primary *in vitro* test results of medicinal[a] and non-medicinal plant extracts for activity against human immunodeficiency virus

Type of extract	Medicinal plants			Non-medicinal plants		
	Total	No. active	% active	Total	No. active	% active
Aqueous	1093	386	35.3	9 111	3060	33.6
Organic	1227	38	3.1	10 450	457	4.4

[a]Medicinal uses include applications on humans and animals.

Potential anti-HIV agents

Between September 1986 and August 1993, close to 37 500 plant samples were acquired for the Natural Products Repository in Frederick and over 26 000 were extracted to give approximately 52 000 extracts. Analysis of the anti-HIV test results for sets of medicinal and non-medicinal plant extracts (Table 1) indicates that there is little difference between the level of activity observed for the aqueous extracts (35.3% versus 33.6%) and organic extracts (3.1% versus 4.4%) of the two sets. About 90% of active extracts are aqueous in each case. All active aqueous extracts are tested for the presence of polysaccharides by diluting an aqueous solution of the extract with an equal volume of ethanol (Beutler et al 1993); the precipitate and filtrate, together with a sample of the original extract, are tested for anti-HIV activity. Concentration of the activity in the precipitate indicates the presence of polysaccharide material and such samples are dropped from further consideration. Active filtrates are tested for the presence of tannins by being passed through a small polyamide column (Cardellina et al 1993b). Irreversible adsorption of the active material onto the column indicates the presence of tannins and signals no further interest in the sample. Concentration of the activity in the eluate, however, indicates the presence of compounds of potential interest and these extracts are subjected to bioassay-guided fractionation procedures. Active polar organic solvent extracts are also subjected to the polyamide process prior to proceeding with further fractionation. The elimination of aqueous extracts containing polysaccharides and/or tannins (often referred to as dereplication) is still proceeding and thus far only 10 (including one from a medicinal plant) of 1500 extracts processed have qualified for further investigation.

Of the 495 active organic extracts, 129 have been dropped from further consideration after dereplication. The bioassay-guided fractionation of 30 has been completed; the remainder are undergoing either dereplication or advanced bioassay-guided fractionation. Four novel plant-derived agents with *in vitro* anti-HIV activity have been isolated and selected for preclinical development. The dimeric alkaloid, michellamine B (Fig. 3), has been isolated from the leaves of the tropical vine, *Ancistrocladus korupensis*, collected in the rain forest regions of south-western Cameroon (Manfredi et al 1991). Michellamine B shows

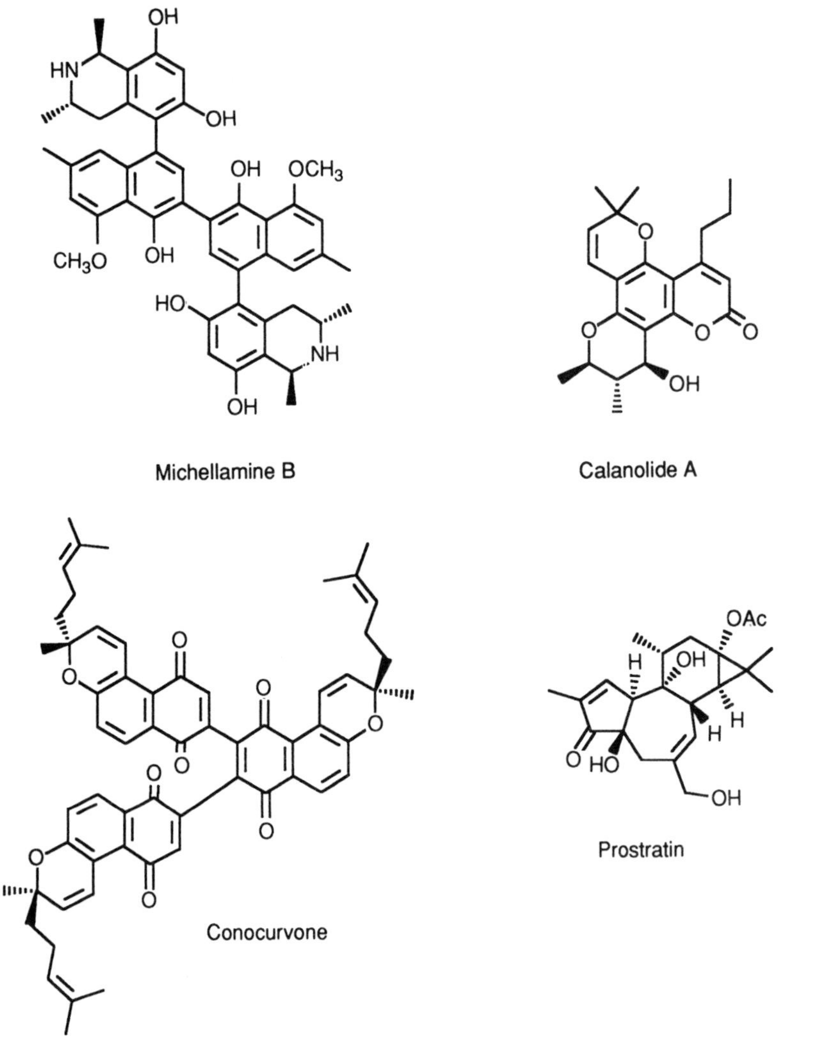

Michellamine B

Calanolide A

Conocurvone

Prostratin

FIG. 3.   Plant-derived anti-HIV agents in preclinical development.

in vitro activity against both HIV-1 and HIV-2; it is in advanced preclinical development. *A. korupensis* is a new species of the genus *Ancistrocladus* and has no history of medicinal use by the local population. While the aerial parts of *Ancistrocladus abbreviatus* are reported to be used in the treatment of measles and fever (Iwu 1993), it is significant that extracts of this and other *Ancistrocladus* species do not exhibit anti-HIV activity.

Calanolide A is a novel coumarin isolated from the leaves and twigs of the tree, *Calophyllum lanigerum* Miq. var. *austrocariaceum* (T.C. Whitmore), collected in the rain forest regions of Sarawak, Malaysia (Kashman et al 1992). It shows potent activity against HIV-1 and several strains of the virus resistant to agents such as AZT (3'-azido-3'-deoxythymidine), but not against HIV-2; it is in early preclinical development. Repeated collections of the source species have thus far failed to yield further supplies of calanolide A, but another species, *Calophyllum teysmanii* var. *inophylloide*, collected in the same region, has yielded a related active compound, costatolide, which is being considered for preclinical development. While there have been reports of the use of *Calophyllum* species, including *Calophyllum inophyllum*, for the treatment of tumours in the Philippines and New Caledonia (Hartwell 1982), antiviral activity has not been reported.

Conocurvone, a novel trimeric naphthoquinone (Decosterd et al 1993) isolated from a *Conospermum* species endemic to Western Australia shows potent *in vitro* activity against HIV-1 and is in early preclinical development. As far as can be determined, no medicinal uses for this genus have been reported.

Prostratin, isolated from the stem wood of the Western Samoan tree, *Homalanthus nutans* (Forster), is the only example thus far of a potential anti-HIV agent discovered on the basis of ethnobotanical plant selection (Gustafson et al 1992). The source tree is used in Western Samoa for the treatment of a variety of diseases, including yellow fever: an extract was provided to the NCI by Professor Paul Cox. The significance of this discovery is discussed by Cox (1994, this volume).

Intellectual property rights: issues of international collaboration and compensation

The NCI and its collection contractors have collected plants from over 25 countries. They recognize the tremendous value of these natural resources to the source countries and the significant contributions being made by scientists and, in some instances, traditional healers in these countries to the performance of the NCI collection and drug discovery programmes. Through its Letter of Collection (Letter of Intent), the NCI has unequivocally stated its intent to deal with source countries in a fair and equitable manner; agreements based on this Letter have been signed with organizations or Government agencies in seven countries and negotiations with several other countries are in progress. Space does not permit a detailed discussion of these policies, but the NCI has already demonstrated its willingness to collaborate with source country scientists through undertaking joint research projects in NCI facilities with scientists from 10 countries. It is interacting closely with scientists and Government agencies in Australia, Cameroon and Sarawak in pursuing the development of the potential anti-HIV agents, conocurvone, michellamine B and calanolide A, respectively.

While the issue of compensation has not, as yet, arisen, the NCI will require licensees of its patented discoveries (e.g. michellamine B) to negotiate mutually acceptable terms of compensation (e.g. percentage of royalties from drug sales) directly with the relevant source country organizations.

Conclusions

In the investigation of plants as sources of potential novel anticancer, and more recently, anti-HIV drugs, the NCI has adopted a broadly based taxonomic approach to plant collections, as opposed to one based strictly on the collection of medicinal plants. Analysis of the clinically effective anticancer agents discovered from plants collected in the USDA programme (1960–1982) indicates that the majority were isolated either from plants having a reported history of medicinal use, sometimes in the treatment of cancerous conditions, or from plants belonging to genera containing species with reported medicinal uses. These observations support the collection and screening of plants on the basis of reported medicinal uses and/or the use of an ethnobotanical strategy. Strict adherence to medicinal plant collections, however, might well have missed important discoveries, such as camptothecin and homoharringtonine. While none of the potential anti-HIV agents discovered in the ongoing programme has advanced to clinical trials, of the four agents in preclinical development, only prostratin was discovered on the basis of ethnobotanical evidence.

The chances of discovery of novel clinically effective anticancer and antiviral drugs are very low; from the early NCI programme, they may be estimated at about one in 10 000 species screened. The urgent need for novel agents to treat cancer and AIDS requires that as many sources as possible be searched; the authors feel that the real possibility of missing the discovery of a clinically effective agent by focusing only on medicinal plant collections justifies the current NCI policy of broadly based taxonomic plant collections. However, the value of traditional medicine in the discovery of anticancer agents has been demonstrated and the NCI policy does require that priority be given to the collection of medicinal plants when reliable information is available.

Acknowledgements

The NCI gratefully acknowledges the collaboration and support of the many individuals and organizations worldwide that make these programmes possible. From the collection of organisms in over 25 countries to the clinical trials of new drugs, and the studies of occurrence and prevention, this is truly an international effort in the fight against the scourges of AIDS and cancer. The NCI recognizes the indispensable contributions being made through the provision of valuable natural resources, expertise, knowledge and skills; through policies of collaboration and compensation, as stated in the Letter of Collection, the NCI wishes to assure participating countries of its intentions to deal with them in a fair and equitable manner.

References

Barclay AS, Perdue RE 1976 Distribution of anticancer activity in higher plants. Cancer Treat Rep 60:1081–1113

Beutler JA, McKee TC, Fuller RW et al 1993 Frequent occurrence of HIV-inhibitory sulphated polysaccharides in marine invertebrates. Antiviral Chem & Chemother 4:167–172

Boyd MR 1988 Strategies for identification of new agents for the treatment of AIDS: a national programme to facilitate the discovery and preclinical development of new drug candidates for clinical evaluation. In: Devita VT, Hellman S, Rosenberg SA (eds) AIDS: etiology, diagnosis, treatment, and prevention. Lippincott, Philadelphia, PA, p 305–319

Boyd MR 1989 Status of the NCI preclinical antitumor drug discovery screen. In: Devita VT, Hellman S, Rosenberg SA (eds) Principles and practice of oncology updates. Lippincott, Philadelphia, PA, vol 3:1–12

Boyd MR 1993 The future of new drug development. In: Niederhuber J (ed) Current therapy in oncology. Introduction to cancer therapy. BC Decker, Philadelphia, PA, p 11–22

Cambie RC 1986 Fijian medicinal plants. In: Steiner RP (ed) Folk medicine. American Chemical Society, Washington, DC, p 67–89

Cardellina JH II, Gustafson KR, Beutler JA et al 1993a National Cancer Institute intramural research on human immunodeficiency virus inhibitory and antitumor plant natural products. In: Kinghorn AD, Balandrin MF (eds) Human medicinal agents from plants. American Chemical Society, Washington, DC (Am Chem Soc Symp Ser 534) p 218–227

Cardellina JH II, Munro MHG, Fuller RW et al 1993b A chemical screening strategy for the dereplication and prioritization of HIV-inhibitory aqueous natural products extracts. J Nat Prod 56:1123–1129

Chang H-M, But PP-H 1987a Pharmacology and applications of Chinese materia medica. World Scientific Publishing, Singapore, vol 2:1167–1172

Chang H-M, But PP-H 1987b Pharmacology and applications of Chinese materia medica. World Scientific Publishing, Singapore, vol 1:54–61

Cox PA 1994 The ethnobotanical approach to drug discovery: strengths and limitations. In: Ethnobotany and the search for new drugs. Wiley, Chichester (Ciba Found Symp 185) p 25–41

Cragg GM, Boyd MR, Cardellina JH II et al 1993 Role of plants in the National Cancer Institute drug discovery and development program. In: Kinghorn AD, Balandrin MF (eds) Human medicinal agents from plants. American Chemical Society, Washington, DC (Am Chem Soc Symp Ser 534) p 80–95

Decosterd LA, Parsons IC, Gustafson KR et al 1993 Structure, absolute stereochemistry, and synthesis of conocurvone, a potent, novel HIV-inhibitory naphthoquinone trimer from a *Conospermum* sp. J Am Chem Soc 115:6673–6679

Farnsworth NR, Akerele O, Bingel AS, Soejarto DD, Guo Z 1985 Medicinal plants in therapy. Bull WHO 63:965–981

Gustafson KR, Cardellina JH II, McMahon JB et al 1992 A nonpromoting phorbol from the Samoan medicinal plant *Homalanthus nutans* inhibits cell killing by HIV-1. J Med Chem 35:1978–1986

Hartwell JL 1976 Types of anticancer agents isolated from plants. Cancer Treat Rep 60:1031–1067

Hartwell JL 1982 Plants used against cancer. A survey. Quarterman Publications, Lawrence, MA

Iwu MM 1993 Handbook of African medicinal plants. CRC Press, Boca Raton, FL, p 13

Cragg: The NCI has a compound repository of over 400 000 compounds. These are being screened.

Dagne: Would it be possible for the NCI to publish or make available the list of plant materials that have been screened, even if the results were negative? This would save a lot of people from repeating these screens. For instance, from Ethiopia and indeed from many parts of Africa, there was massive collection of plants in the 1960s. I know that some of the results have been published, but you may also have extensive unpublished data at the NCI that could be of interest to the countries from where the plants were collected.

Cragg: We are not distributing publicly the data on the plants collected in the latest programme, which started in 1986, because these plants are still being screened by other organizations which have access to our repository and there are obligations to the source countries. Data on the earlier collections could be made available.

Farnsworth: The organization that preceded the National Cancer Institute was called the Cancer Chemotherapy National Service Centre. They used to publish in supplemental issues of *Cancer Research* all of their negative screening data.

Dagne: But in the early days, in the 1960s, there were no collaborating institutions.

You mentioned calanolide A. If this were developed through partial synthesis, chemical modifications and so forth, would that eliminate the obligation to the country in which the plant was originally found?

Cragg: This is a good question. If calanolide A is developed as the effective agent, even if it has been synthesized, the source country will still benefit from whatever percentage royalties are negotiated between the country and the successful pharmaceutical company. Of course, they won't benefit from the use of their natural resources as a source of that compound. They might, because even though a relatively simple synthesis of calanolide A has been published, once this is scaled up it may not be economically viable compared with the natural source.

Calanolide A has been synthesized independently by a pharmaceutical company. We beat them to patent application, but they had made an independent discovery of very similar compounds showing activity in a different screen. They used a reverse transcriptase screen as opposed to our cell-based screen. I'm certain that the company will try to synthesize active analogues. If one of these is developed rather than the original compound, they may consider avoiding the NCI patent, in which case, the source country and NCI would not derive any benefit whatsoever.

I would like to put in a word of caution here to countries where we are collecting. We've lost time in negotiations about providing source material for costatolide, which is the compound related to calanolide A. This is proving to be very valuable time lost now. If we had had that material nine months ago, we could have been pursuing costatolide vigorously; we might even have had it into toxicology testing and on its way to possible clinical trials. But we've been held up and if an alternative compound to calanolide A is developed by synthesis, NCI might not even develop one of these derivatives. That would be a sad loss for the source country. I understand that one wants to derive maximum benefit in negotiations, but it is important to appreciate the urgency of advancing these compounds as quickly as possible.

Posey: What are the political or ethical rules within the pharmaceutical industry that govern circumventing patents?

Lewis: We all know patent circumvention is possible via analogues or isomers, but we also know the origin of the material to which the changes are made. We have a tremendous moral responsibility to prevent such circumvention so that the country of origin is compensated in some way for the initial genetic information.

Cragg: Our legal office will be watchful regarding this issue. The Director, Dr Thomas Mays, really has his heart in this programme; he's very conscious of wanting to protect the rights of source countries and so forth. But it's a possibility and you cannot blame a company for wanting to get round patent restrictions.

Lewis: I do! To get around, to actually do it on purpose—I think that is morally wrong. There should always be some recognition of the origin of the material.

Cragg: In this case, the company had an independent natural source, from another area in the world, so I don't think we can claim that.

Albers-Schönberg: A natural product patent today contains so much detail that anyone can, in a short time, compete effectively with some significant, non-trivial improvement. This kind of competition is an everyday fact of life among the pharmaceutical companies of the entire developed world. It forces one, as the original discoverer, to try to make such improvements oneself and to protect these improvements with patents in a great number of countries. For instance, Merck, under its agreement with Costa Rica's INBio, will pay agreed royalties on any marketed product that traces back to a plant or other sample that the company first received from INBio. But if some other company genuinely improves on this product and gains the market, then not only Costa Rica but also Merck will be out of luck. It would not only be extremely difficult to prohibit such competition—where do you draw the line between what is allowed and what is not? Who would police it?—it would also be contrary to the original idea behind the patent system. This was first formally introduced by a very liberally minded person in what was then a developing country—Thomas Jefferson, in 1790. The idea was to stimulate creativity and inventiveness by giving inventors the rights to their inventions, but only for a limited time. At the same time, it challenges the inventor by allowing improvements to supersede a patent. I would encourage all countries that have not yet joined the international patent agreements to do so. It is the best protection.

An important aspect of the patent system in this context is the following. By challenging a patent in court, one can force the owner of the patent to document, step by step, the history of the invention. This makes it impossible to isolate an interesting compound from, let's say, a Nigerian plant, that one has received under an agreement with Nigeria, and then to search for the same plant or a close relative in one's own country and claim it as the first source.

Peter: When a company synthesizes analogues of a biologically active substance, the primary objective is always to find out whether the activity and other pharmacologically important parameters can be improved; the objective is not to circumvent a patent.

McChesney: Most natural product agents, once they have been purified from their native mixtures, do not make good pharmaceuticals. They are interesting drugs, but not good pharmaceuticals. They may not formulate well; they may not be bioavailable; they may not have the appropriate stability; they may have toxic side-effects as single agents. Often an analogue programme is designed to overcome those issues, more than to get around a particular patent. This impacts very dramatically this issue of compensation. How much do the appropriate levels of compensation change as you go down that development track? Are they the same if the molecule is only a pharmacophore representation of the original lead that has been altered through years of chemistry? Or is it some fraction of the original negotiated amount?

Prance: We at Kew have negotiated several contracts with pharmaceutical companies, both in medicines and insecticides. Basically, the level of royalty is higher if the product comes directly from a plant, less if it is an analogue. Several agreements that I have seen are like that. This is alright if the original company is making the analogue. If another company is making the analogue, the situation is more complicated.

Iwu: The problem of patent circumvention was one of the things that nearly broke up the Rio Convention on Biodiversity in 1992. At what point do you stop the right to ownership? Someone said it is like if you sell a bicycle and the owner decides to use that bicycle as taxi, do you collect money each time the buyer collects a fee from a passenger?

There is an even more difficult problem: if a company licenses a product, how much access does the host country have to the patented product? The NCI's Letter of Intent addresses this very well; it allows both the NCI and the source country to have access. Usually, this is not practicable. If there is a licence between a Third World country and a drug company, the company owns the rights to the product. If you want to get the product, you have to buy it from them. I think it was India that made the point at the UN Convention on Environment Development in Rio that if a product originates from a country and is patented, the source country should not be denied access to that product. These are the problems in biodiversity prospecting, but I suppose we will have to sort them out as we go along.

Cragg: As my colleague, Tom Mays says, this is a constantly evolving process. I think our first test case might be with michellamine B in Cameroon—if it or some analogue shows some activity in clinical trials. The NCI is strongly committed to doing its best for the source country.

Barton: You were away ahead of the everyone else in terms of providing compensation to the source nation. Can you tell us a little about the decision-making process that led to that?

Cragg: We owe a lot to our botanical colleagues— Mike Balick, the New York Botanical Garden, and the staff of Missouri Botanic Garden and the University of Illinois at Chicago. They emphasized to us at the start of the programme in 1986 that this is very important. We would be collecting in many countries and there should be policies for compensation and for collaborating with these countries in various aspects of drug development. We also had some very useful dialogue and negotiations with Elaine Elisabetsky and her colleagues at one stage when we were considering a Brazilian collaboration.

King: John, you said that the NCI is way ahead of the field in compensation for source countries. What criteria are you using to make that statement?

Barton: I'm certainly not making comparisons with the tradition of the scientific community, but rather with the tradition of the pharmaceutical community. In terms of any transaction structure, such as the Merck–INBio agreement, NCI was the first.

Dagne: The issue of collection is becoming extremely controversial. We hear of companies and research institutes, particularly from the North, who make collections through small companies or individuals, usually by paying for materials. When this comes out in the open, it elicits many accusations and counter-accusations. Now, in particular after the Rio Convention, to what extent should there be a clear understanding of how collections ought to be made? How does one choose a partner in the respective countries?

Cragg: We rely on our collection contractors, such as Mike Balick, who know the countries well and have established collaborations there. These are often with national institutions, such as botanical gardens; examples are the Bogor Herbarium in Indonesia or the Botany Department of the Philippines National Museum. We would not deal with individuals; we deal with reputable organizations at every stage.

Rubin: You made reference to pharmaceutical prospecting: there is the example of INBio and I think the new round of contracts for International Cooperative Biodiversity Groups will inspire more source countries to develop a greater capacity to do large-scale collections. What do you see as the future for such collections? Do you think there will be many of these? Do you think private participants—major pharmaceutical companies—will become involved in these types of collections?

I think in the new round of contracts, there will be much greater technology transfer involved in the large-scale collections to benefit the source countries.

There will be, in the first instance, royalties paid on any drugs that become commercialized. There might be a higher scale of royalties where ethnobotanical knowledge has been used. Given this background, what sort of future is there for this type of prospecting?

Cragg: We've had numerous enquiries from many of the countries where we are working about setting up their own screening programmes. We are happy to help transfer technologies. In the future, the collections, the extractions and some preliminary screening might be done in the source country. Then, depending on the expertise and facilities available in that country, they might proceed further with some isolation work.

We invite source country scientists to our labs where they participate in the extraction and isolation of active agents and gain experience with the procedures and equipment used by NCI. We can provide them with some basic materials so that they may return to their countries and carry out natural product collections and extractions themselves. I feel that this is the emerging trend, towards value-added products.

Posey: Is the NCI involved in the Human Genome Diversity Project, which aims to sample DNA from indigenous people and others around the world? What is the situation if, for example, they find human genetic material that has natural resistance to cancer? Apparently, patents are being sought for genetic material taken from two Guarani Indian women in Panama, who have exceptional natural immune systems.

Barton: The key scientist in the Human Genome Diversity Project is L. Cavalli-Sforza; his key legal advisor is Hank Greely. The basic scientific idea is to take samples of genetic material from a variety of human groups that may be on the verge of disappearing in many cases. The aim is to learn something about human evolution, comparing this with the evolution of language, etc. It is basically a scientific concern. Hank Greely is seeking to formulate the patent policy of this programme to be responsive to the obvious concerns. They want to distribute the material on the assumption that no patents will be derived from it.

I don't think that the case you mentioned is associated with this programme. There was some material taken from Panama. A patent application was submitted, I believe by the Centers for Disease Control (CDC). There was much criticism and the CDC decided not to pursue the patent application for that reason.

Reference

Lewis WH, Elvin-Lewis MP 1994 Basic, quantitative and experimental research phases of future ethnobotany with reference to the medicinal plants of South America. In: Ethnobotany and the search for new drugs. Wiley, Chichester (Ciba Found Symp 185) p 60–76

From shaman to human clinical trials: the role of industry in ethnobotany, conservation and community reciprocity

Steven R. King and Michael S. Tempesta

Ethnobotany and Conservation, Shaman Pharmaceuticals Inc., 213 East Grand Avenue, South San Francisco, CA 94080-4812, USA

Abstract. Shaman Pharmaceuticals is a development-stage company engaged in developing traditional pharmaceuticals identified through a discovery process focused on isolating active compounds from tropical plants with a history of medicinal use. This process has resulted in two products in clinical trials: Provir™, an oral product for the treatment of respiratory viral infections, and Virend™, a topical antiviral product for the treatment of herpes. Shaman's drug-discovery process targets specific plants that have been used for medicinal purposes by native peoples. By integrating the sciences of ethnobotany, medicine and plant natural product chemistry, Shaman has been able to achieve time and cost savings for the identification of active compounds and preclinical development of its initial products. Numerous drugs have entered the international pharmacopoeia via the study of ethnobotany and traditional medicine. Two important elements of this approach to drug discovery are the percentage of plants that show activity against specific viral pathogens and the correlation between the folk therapeutic classification of plants used and the percentage of those plants that have shown activity in our antiviral screens. Conservation and direct reciprocity to indigenous communities are important features of Shaman Pharmaceuticals' drug-discovery process.

1994 Ethnobotany and the search for new drugs. Wiley, Chichester (Ciba Foundation symposium 185) p 197–213

The title of the symposium, 'Ethnobotany and the search for new drugs' provides an excellent opportunity for an analysis of the importance of ethnobotany, indigenous knowledge and tropical forests in drug discovery. Papers presented at the earlier Ciba Foundation Symposium entitled 'Bioactive compounds from plants' confirmed that a number of leading scientists in the areas of ethnobotany, pharmacognosy and ethnobiology view ethnobotany as an important methodology for discovering new bioactive molecules (Balick 1990, Farnsworth 1990, Cox 1990). However, it is also apparent that ethnobotany represents a rapidly evolving multidisciplinary approach that requires continual monitoring and modification to achieve the most successful results. A. B. Cunningham (1993) has highlighted the strong links between the conservation of biological diversity, the knowledge of indigenous peoples and the interests of the global community in developing novel therapeutic agents for a wide variety of diseases.

In this paper, several perspectives are integrated and the utility of this approach and its implications for both drug discovery and maintenance of human health in diverse regions of the world are addressed. Evidence is provided that supports the importance of ethnobotanical investigation as a source for drug discovery and highlights indigenous knowledge as a key element in local health care by traditional healers and indigenous people throughout the world. Information is presented on the antiviral activity of plants collected with a range of ethnobotanical indications, including antiviral, antifungal, antiinfective and central nervous system treatments. The importance of unifying various approaches to drug discovery by combining the disciplines of ethnobotany, ethnopharmacology, chemotaxonomy, medicine and pharmacognosy with bioactivity-directed isolation and structural elucidation of bioactive molecules is described. Recent work by a variety of scientists in these disciplines has indicated that the ethnobotanical approach has merit for drug discovery (Soejarto 1993, Gentry 1993, Hudson 1990). We want to stress the importance of simultaneously integrating all of these approaches to maximize our ability to discover new therapeutic molecules from plants. We also want to validate the importance of local ethnomedical practices with plants that have a high level of *in vitro* activity against a number of well-known and widespread viral infections.

Finally, we describe some of the activities that we consider to be important models for establishing reciprocity with local communities of tropical forest peoples. Many of these communities are in a constant struggle to acquire land rights, adequate food and health care. We suggest that the success of ethnobotanical research programmes and drug development will be increasingly and intimately linked to the welfare and management of natural forest habitats by local and indigenous people.

Brief history and methods

Shaman Pharmaceuticals Inc. began operations in 1990. Utilizing an ethnobotanical approach to the collection of tropical medicinal plant species, we have been successful in bringing two products into human clinical trials within 24 months of the company's founding. These two products are Provir™, an oral product for the treatment of respiratory viral infections, and Virend™, a topical antiviral product for the treatment of herpes. In the initial selection of therapeutic targets, as well as the addition of new ones, four key conditions have to be met for the isolation of successful drugs. First, there have to be incidence and phytotreatment of the disease among indigenous peoples. Second, there have to be *in vitro* and/or *in vivo* models of the human disease that not only allow initial screening, but also are amenable to efficient bioassay-driven isolation of active compounds. Third, there must be clearly defined clinical endpoints that allow unambiguous determination of efficacy. Finally, there must be a sizeable market that warrants the expense of discovery and development. Our drug-discovery process targets plants that have been used medicinally by native people for specific viral and fungal infections, central nervous system disorders and diabetic conditions.

The first phase in our discovery process of new therapeutic agents from tropical forests involves a multidisciplinary field research programme. The most logical starting point for assigning priority within plant collections is the knowledge of the local and indigenous people who live in and around tropical forests. Our methodology comprises the following steps. Before any research expedition, we prepare a full regional study on the epidemiology, traditional medicine, culture and ecology of the people and the environment in which they live. Information on the plants known to be utilized in any given area is assembled by searching several international databases on ethnomedicine, medicinal activities of plants and any known chemistry of plants with such activity. We also search for data from international and national hospitals in remote areas and treatment programmes that work with local and native people. All of this information is synthesized and integrated into our field research programme. The most highly ranked plant candidates are reported to our integrated field research teams which consist of an ethnobotanist and a Western-trained medical doctor.

Once in the field, ethnobotanists and medical doctors work together on a highly focused data collection strategy. The goal of this method is to identify and collect plants with the highest probability of possessing activity in selected areas. This integrated system is aimed at maximizing the probability of new drug discovery. The ethnobotanist and medical doctor prepare case presentations of diseases identified before the trip in epidemiological reports. These brief case descriptions of specific diseases are presented to shamans and traditional healers of tropical forest communities. Photographs of diseases with readily visible signs and symptoms, such as fungal infections or viral skin rashes, are utilized with the case presentations.

Interviews with the informants are conducted very carefully, using no medical terminology for diseases while presenting cases. The focus is on the recognition of common signs and symptoms. Terms such as 'malaria', 'hepatitis' or 'parasites' are not utilized by the interviewing team. A translator for the local language is often necessary to conduct this phase.

We have used translators for Wa, Quichua, Waorani and many other languages without any difficulty, as the case symptoms are designed to be simple but highly specific. Once an informant has recognized and described the same or similar disease condition, the botanical treatment for that condition is recorded. If several independent and reliable informants describe a similar treatment for a disease, the plant is collected.

Additional plant collections and observations are made when the local shaman and/or the Western-trained physician provides health care to the local people. In all cases, the patient and the local healer are asked about the types of plants, if any, used to treat the wound, fever or other condition. In many cases, we have learned of antiinfective plants utilized prior to the arrival of the ethnobotanist/physician team. The plants that look interesting are also collected for analysis. The process of plant collecting involves all the standard methods of field botanists, including the preparation of multiple plant voucher specimens which are deposited in the host country as well as in the United States herbarium. General, bulk field-collecting methods are described well by one of the National Cancer Institute's collecting scientists (Soejarto 1993).

A combination of ethnobotanists, medical doctors and local indigenous expertise has significantly enhanced our ability to collect plants and associated medical data with a much greater degree of confidence in their role as bioactive therapeutic agents in humans. This methodology is constantly being tested and improved through our field research expeditions and has yielded a number of compounds under preclinical development at this time.

Antiviral activity of plants according to primary ethnobotanical use

In this section, we shall illustrate a discrete analysis of ethnobotanical use and correlation of the indicated activity in antiviral screens. Recent papers have indicated that in the case of activity against the human immunodeficiency virus (HIV), ethnobotanical information has provided a statistically higher proportion of activity than have randomly collected plants (Balick 1990, Soejarto 1993, but see Balick 1994, this volume). This is interesting, but it is important to bear in mind that shamans and traditional healers have not historically treated or faced HIV and therefore have not had a long time to develop compounds for the treatment of infection with it. Nonetheless, many plant compounds are being utilized by people throughout the world to combat this virus. The information we are presenting is more specifically oriented towards viral diseases that have been present for several hundred years and in many cases several thousand years.

We have tested the antiviral activity of 207 different plants (Table 1). All of these plants were collected with ethnomedicinal indications of antiviral, antifungal or antiinfective activity or effects related to disorders of the central nervous system. We have classified primary ethnobotanical use according to the following criteria. Antiviral activity is graded as weak, good or strong (Table 2). Antiviral plants are plants with a primary use for fever, coughs, jaundice/hepatitis, herpes, warts, conjunctivitis, diarrhoea (especially without blood) and gastrointestinal problems. Antifungal plants are those for which the primary use is for *Candida* (e.g. thrush, vaginal yeast infections), dermatophytes

TABLE 1 Summary of antiviral activity of 207 plants used in ethnomedicine

Primary ethnobotanical use	No activity	Weak activity	Good activity	Strong activity	Total tested
AV	14	9	5	28	56
AV & AI	17	9	4	14	44
AV & AF	4	0	1	5	10
AV & CNS	3	2	0	6	11
AV & AI & AF	2	1	0	0	3
AV & AI & CNS	2	1	0	2	5
AV & AF & CNS	2	0	0	3	5
AI	10	1	2	4	17
AI & AF	6	1	1	2	10
AI & CNS	1	0	1	1	3
AF	7	4	1	15	27
AF & CNS	4	0	1	1	6
CNS	3	2	0	5	10
TOTAL	75	30	16	86	207

CNS, used to treat disorders of the central nervous system; AF, antifungal; AI, antiinfective; AV, antiviral. The criteria used to rank antiviral activity are given in Table 2.

and skin fungal infections (e.g. *Tinea capita*, *Tinea pedis*, etc). Antiinfective plants are plants with a primary use for wound healing, infected eyes, burns, boils or unspecified 'skin diseases', 'skin problems', 'skin afflictions' and 'skin rashes'. Analgesic and other plants associated with central nervous system disorders are those with a primary use for pain, headache, dental caries, scorpion sting, ray sting, snakebite, epilepsy, mental illness, to aid sleeping, as a fish poison, arrow poison, hallucinogen or as snuff.

Table 1 shows that 50% of the plants that were utilized primarily to treat only diseases likely to be caused by viral agents do act directly as antiviral agents. The other groups of plants are correlated more closely to the 'powerful' plant category, i.e. they are used for several different ailments or for unrelated

TABLE 2 Criteria used to rank antiviral activity

No activity	Weak activity	Good activity	Strong activity
Toxic or $ED_{50} > 600 \, \mu g/ml$	$ED_{50} > 100 \, \mu g/ml$ and $SI \geqslant 2$	$50 \leqslant ED_{50} \leqslant 100 \, \mu g/ml$ and $SI > 5$	$ED_{50} < 50 \, \mu g/ml$ and $SI > 9$

The viruses against which the plants were tested were herpes simplex 1, influenza A and respiratory syncytial virus. SI, selective index.

ailments. These plants tend to be more evenly distributed across the spectrum of strong, good or weak levels of activity in the primary assays. Overall, 42% of all plants screened for antiviral activity show strong activity, 22% show good to weak activity and 36% show no activity at all.

Fifty-five per cent of the plants that had a primary ethnobotanical use solely as antifungal treatments also showed strong antiviral activity. This supports our observations so far that plants utilized for fungal or viral infections are more potent plants. In some cases, antifungal and antiviral plants are used for both indications as generally powerful antiinfective plants. We would suggest that papers reporting on the correlation between ethnomedicinally collected plants and their activity should also indicate what the ethnomedical uses of the plants are. This will help distinguish 'medicinal collections' from plants that are targeted to match the therapeutic areas for which they are being screened. A similar analysis of our antifungal assay results is in preparation. Another trend we have observed is that the strong antiviral activity is located mainly in the more polar extracts, which end up yielding, in addition to the well known polymeric polysaccharides and tannins, very active alkaloids, quinones, terpenoids, flavonoids and phenolic glycosides, among others.

Therefore, we would suggest that the interest in developing antiviral drugs, an area that is greatly in need of new therapeutic agents, can be enhanced and facilitated by focusing expressly on the use of plants by shamans and traditional healers to treat viral *and* fungal diseases.

Implications for traditional medicine and local health care

It is important to remember that the management of local health-care problems by traditional healers and shamans utilizes a great number of plants in addition to other therapies. The fact that 64% of these 207 plants show antiviral activity *in vitro* strongly suggests that these therapeutic agents are capable of exerting positive direct antiviral effects on the patients treated with such remedies. If only a small percentage of the leads from antiviral plants become successful international therapeutic agents, it does not in any way diminish their utility and importance within the traditional medical systems where they are used. This needs to be emphasized to the communities and the countries where these plants are utilized to help support existing likely efficacious treatments, especially in areas where routine access to Western pharmaceutical products is difficult and/or limited. This has been stated by a number of researchers involved in international research on traditional medicine, but there is always a cry for actual data to support such assertions (WHO 1991). We believe that this type of data can support the role of traditional medicine in local cultures and strengthen the position of organizations such the World Health Organization and Tramil, which continue to stress the importance of local community healers.

Biocultural conservation and community reciprocity

Shaman Pharmaceuticals has two products in human clinical trials and others in preclinical development. We strongly support the integrated approach to the development of novel therapeutic agents, utilizing botany, ethnobotany, medical knowledge, ethnomedicine, pharmacognosy and chemosystematics. Simultaneously, there exists an important mandate to all parties interested in developing therapeutic agents from plants to establish short-, medium- and long-term reciprocal schemes for the return of benefits to the communities and countries in which these genetic resources and indigenous local knowledge reside. The importance of establishing positive long-term collaborative relationships with government research organizations and indigenous people in tropical forest countries cannot be overstated. The international community has now arrived at a critical juncture and has provided the pharmaceutical industry (among others) with an exquisite opportunity to take the lead role in the areas of conserving and safeguarding biological and cultural diversity. Shaman Pharmaceuticals has a collaborative antifungal drug-discovery and development relationship with Eli Lilly and Company. In the course of describing this collaborative agreement to the general public, Eli Lilly became the first international pharmaceutical company to make a public statement regarding the importance of both the biological and cultural diversity of tropical regions of the world. They have embraced publicly the great potential of working with tropical forest plant species and tropical forest peoples. This approach by the management of a pharmaceutical company has set an important precedent for the industry.

The United Nations Conference on Environment Development held in Rio de Janeiro, Brazil in 1992, generated the Convention on Biological Diversity and emphasized the importance of unified global action to protect ecosystems and species within them as part of global priorities. The conference presented an excellent opportunity for vitalizing the links between drug development organizations, local people, national scientists and resource management organizations in the countries that contain the greatest biologically enhanced chemical diversity. These collaborative relationships are essential to successful long-term research, development and conservation activities.

A significant portion of novel chemical entities may be either too structurally complex or too expensive to produce synthetically. For them, large quantities of raw material will be required to bring plant pharmaceuticals to the market place. In the case of Shaman's two lead antiviral products, we are working with four different countries to develop and create long-term sustainable supply and, ultimately, extraction industries. This is an area of great economic interest to the governments and local communities of these countries. Sustainable harvest of new non-timber forest products can contribute to the conservation of biological and cultural diversity by providing alternative income-generating activities to large-scale logging or clearing of the land for livestock (Nepstad & Schwartzman 1990, Plotkin & Famolare 1992).

Collaboration and reciprocity

Shaman Pharmaceuticals has established a policy to provide benefits to the local people, communities and the proper governmental agencies in three ways, using short-, medium- and long-term horizons. We feel that this method is one of the most appropriate and balanced solutions for providing reciprocal benefits to all parties that are due acknowledgment for their contribution to new drug development. We will provide a portion of profits of any and all of our products to all the communities and countries in which we have worked. This spreads out the risk and assures a more rapid return of resources for all of our collaborators. This would include a portion of profits derived from a product that may enter the company via a separate route, including a product licensed from another company.

The second part of our solution is the mechanism. We created, at the formation of this company, a non-profit conservation organization, now called The Healing Forest Conservancy, with an independent board of advisors, which will help determine the appropriate distribution of these resources to communities and governmental organizations. Representatives of the various countries will be included in the process of discussions, research and decision making; all this is independent of Shaman Pharmaceuticals. The Executive Director of The Healing Forest Conservancy, Ms Katy Moran, is currently supporting a number of projects, through outside fund-raising efforts, for the maintenance of biological and cultural diversity in tropical countries. Ms Moran described the goals and activities of the Conservancy at a recent symposium, 'Intellectual Property Rights and Indigenous People' (Moran 1994).

The third part of returning benefits involves, when and where appropriate, the creation of new sustainable natural product supply and extraction industries in the countries in which we work. The details of our research and development work with indigenous federations are described in a separate publication (King 1994). The development of sustainable supply and extraction industries in tropical countries is considered a vital part of economic development by the governments and local people in many countries throughout the world.

Finally, we feel very strongly that it is important to return benefits immediately to any and all the groups with which we work. An organization need only ask the people with whom it works what their immediate needs are and pose the solution in advance to a given research project. We feel that contributing reciprocal benefits based on the requests of local people in the early, middle and later stages of any relationship is the model that should be emulated by all individuals, organizations or corporations studying local people's traditional knowledge.

Details of this approach have been outlined in a separate paper (King & Carlson 1994).

Collaborative relationships require that appropriate technology and resources be exchanged as part of drug development programmes. There must be an emphasis on helping provide measures for the conservation of biological and cultural diversity. This includes methods for improving the appreciation of the cultural dimensions of the use and management of medicinal plant species. The future of drug development based on an ethnobotanical approach clearly lies in relationships with built-in reciprocal benefits and commitments for the provision of a portion of pharmaceutical resources towards surveys of biological diversity, ethnomedicinal uses and conservation activities. The emerging partnership between drug development and the conservation of biological and cultural diversity holds great promise for producing new classes of bioactive molecules. The future is clearly a picture of increasing interdependence with a strong value gained through truly collaborative relationships with countries. One of the key elements to making this a genuinely equitable and reciprocal relationship lies in including local people and communities in the process of discussing, researching and sharing the benefits of the development of novel therapeutic agents.

Acknowledgments

The authors would like to acknowledge the contributions of Lisa Conte, Reimar Bruening, Franco Rutili, Michael Kernan, John Kuo, Shivanand Jolad, Rosa Ubillas, Diana Fort, Charles Limbach, Thomas Carlson, Marilyn Barrett, Martha Gamble, Dennis McKenna, Shaman's Scientific Strategy Team, outside investigators and the entire research, development and administration of Shaman Pharmaceuticals. Special thanks are due to Rowena Richter for her technical and intellectual contribution to this paper and to all of the indigenous and local scientists that we work with in the tropics of Asia, Africa and Latin America, especially Coweûa, Huepe Coba, Tiro Coba and Cesar Gualinga.

References

Balick MJ 1990 Ethnobotany and the identification of therapeutic agents from the rainforest. In: Bioactive compounds from plants. Wiley, Chichester (Ciba Found Symp 154) p 22–39
Balick MJ 1994 Ethnobotany, drug development and biodiversity conservation—exploring the linkages. In: Ethnobotany and the search for new drugs. Wiley, Chichester (Ciba Found Symp 185) p 4–24
Cox PA 1990 Ethnopharmacology and the search for new drugs. In: Bioactive compounds from plants. Wiley, Chichester (Ciba Found Symp 154) p 40–55
Cunningham AB 1993 Ethics, ethnobiological research, and biodiversity. World Wide Fund For Nature (WWF), Gland (WWF/UNESCO/Kew People Plants Initiative)
Farnsworth NR 1990 The role of ethnopharmacology in drug development. In: Bioactive compounds from plants. Wiley, Chichester (Ciba Found Symp 154) p 2–21

King: I would, particularly in the Ecuadorean Amazon. The Shuar or the Quichua are geographically very nearby, but culturally distinct and have larger pharmacopoeias.

Prance: The Yanomani and other tribes that have been isolated also have a small pharmacopoeia.

Balée: It may be that some of these groups have a small pharmacopoeia because they practise a trekking lifestyle. They move around a lot; they don't have very many domesticates; they live in small camps or villages compared to other groups; they have few interlocutors with whom to discuss the biological activity of various plants. These details seem to be associated with a small pharmacopoeia, compared to full-scale horticultural sedentary peoples.

Cox: Cecil Brown did a literature survey on mode of subsistence and ethnotaxonomy, which might be informative (Brown 1985).

Prance: It's amazing how quickly, once an indigenous people does come in contact with Western civilization and gets some of the Western diseases, that their pharmacopoeia begins to increase. Very often they obtain new medicines from neighbouring tribes once they begin to need them. The Yanomani villages that are closest to other tribes and have been the most invaded, have a much bigger pharmacopoeia than the more remote tribes.

Martin: I don't think it is isolation from disease alone, but also cultural isolation and tradition. If there are spiritual modes of healing, you may find more reliance on medicinal herbs. In some communities in Mexico there are spiritualist healers, who in fact use very few medicinal plants. Their few herbal cures cover a very wide range of illnesses.

King: There have been epidemiological studies of Amazonian tribes that showed seropositivity for things like respiratory syncytial virus and herpes simplex before there was routine contact with other tribes or Europeans. These viruses were present in their region, therefore the people may have been exposed, but not necessarily succumbed, to them.

Martin: Steve, you emphasize returning things to the host country, returning things to the community itself. How can Shaman participate in the goal that Maurice Iwu mentioned, of helping countries standardize traditional remedies in a way that can quickly improve the health-care system?

King: We work with quite a few individuals who are equally as effective and aggressive as Dr Iwu in making certain that what they need out of these relationships for their programmes is obtained. With people like him, these relationships work rather well because they make certain that those things happen.

Martin: To what extent are you giving information back to the communities? For example, giving information about the toxicity and the efficacy of their traditional remedies? One model is the Tramil programme in the Carribean, which evaluates herbal remedies that are used commonly by people and tries to give indications of which are effective, which may be toxic and should be discontinued, and those for which there is as yet no information on toxicity or efficacy.

King: The Aguaruna/Huambisa Federation in Northern Peru made part of their provisions that we return data on the progress of our results. I've recently sent them our latest results on toxicity, clinical trials and the efficacy, in a language that is accessible to those who requested the information. In general, we leave the results of our local work right there with the communities, in several cases with school teachers who can then teach them directly.

Rubin: Steve, you and I have spoken before about the $44 million, very successful initial public offering of Shaman. In a very visible and neutral way, this demonstrated the value of standing rain forest and indigenous knowledge. Would it be possible for Shaman to develop a formula for providing an advance on royalties in connection with its different collection efforts around the world? Could that be built into the business plan in some way? A company that relies so heavily on ethnobotanical information, including using this in its marketing, possibly needs to remunerate indigenous peoples in source countries more than do companies that rely simply on random collections.

King: If you look closely at the details of what we do—we are endeavouring more and more to communicate those details—you will see that we have built in that aspect of return to the places where we work, in direct response to their requests and needs. We don't say: here's $4.4 million to share out and spend as you wish. We are working with local programmes of collaborators; they negotiate with us, in the course of our activities, to provide the things they require.

Shaman has to function as a business to be successful. If it doesn't succeed, the rest is a pipedream. Distribution of a large chunk of money all at once would not be the best use of the money *vis-à-vis* the requirements of trying to develop pharmaceuticals. $44 million is an enormous sum, but in the realm of pharmaceutical development and human clinical trials, it is a breath—it goes away rather quickly. If we don't utilize these resources effectively, we're not going to survive.

Iwu: Shaman hasn't yet made any drugs. If we kill the goose that lays the golden egg, we will not get anything. Shaman Pharmaceuticals is unique because it is dominated by scientists. The sooner we have scientists taking over research companies and being able to determine what goes on in the corporate world, the better for us.

Rubin: I understand, but it would set a good precedent for other companies, and a progressive company like Shaman might be in a position to do that.

King: I would like your response regarding the award to Conservation International of $2.7 million, which is essentially the same thing. There is an ethnobiological component to it. Have you something that might teach us and the rest of the group?

Rubin: We at Conservation International are about to receive an International Cooperative Biodiversity Group grant, which is to cover the costs of a collection process. (This programme is a collaboration between the National Institutes of Health, the National Science Foundation and the Agency for International Development.) We are one partner among several; the others include an industrial company, another specialized collection group and an in-country pharmaceutical company. The amount of the grant is not comparable to royalties or to a commercial benefit, in as much as it's covering the cost of engaging in a collection. The negotiations involved an industrial partner, which is where the question of remuneration would come up. The remuneration is going to go to the in-country collaborators, as well as to the university that might be reponsible for research and development with respect to a drug.

Iwu: I don't think the money Conservation International is getting is for plant collection alone; it is for drug development and conservation as a whole. It would be good for organizations like Conservation International to ensure that at least 20% of the funds go to Third World countries.

King: Steve, I agree that setting precedents of the type you suggested would be very valuable. What part of that $2.7 million will actually go into the local country activities and programmes?

Iwu: This is an important issue. In most cases we have institutions in a developed country using Third World countries to get grants, and it's really important to know what percentage of that goes to the Third World country.

Rubin: I would add that where there's involvement of indigenous peoples, there might be an enhanced royalty rate, depending on how extensively their knowledge was used and how helpful their contribution was. This is a new development that will soon be seen in this area. The other possibility is that indigenous peoples or traditional healers could actually be joint patent holders.

King: I am somewhat of the view of Paul Cox that trying to thrust the Western legal paradigm of commercial patent law onto individuals is fraught with some serious problems. Knowledge is generally held by large groups of people and has evolved among groups of people. With the best intentions, compensation to individuals for community knowledge can bring about a great deal of internal disruption and problems within communities. I say that after having sat at a meeting discussing patents with indigenous people. I do like the assertion that the indigenous people are primary participants both in the management and the discovery of these compounds and should be acknowledged within the compensation structure. I am not sure as going as far as Steve Rubin suggested is wise.

Finally, the indigenous people whom we spoke with at a meeting in Peru in spring 1993, said that they were not only interested in financial and other aspects of intellectual property, they were interested in intellectual credit. It should go on the record that 90 pharmaceuticals were derived from their knowledge, that their knowledge has contributed to modern medicine and should be taught in schools and colleges to both native and non-native people. This is something that is not often said and needs to be put in print.

Cox: This is a very interesting discussion about compensation. There are two approaches: one is to involve the local people, the other is to strengthen local institutions. Both are good, but in some unfortunate situations, these two goods are at odds. There are certain countries that actively suppress their indigenous people. It is unclear to me in those cases that the Rio treaty is sufficient to guard the rights of indigenous people. For example, if there were an ethnopharmacological find from the Penan people in Malaysia, would it be appropriate that the Malaysian government, which is essentially in conflict with these people, have sole control over compensation?

Martin: There are disagreements sometimes between indigenous groups and national institutions. Agreements should be made not just between foreign partners and host countries, but also between indigenous groups, national institutions and governmental offices.

Cox: I believe 'indigenous control' may be a way of resolving the conundrum posed to Conservation International (Cox & Elmqvist 1991). If indigenous people are in control of the solution, instead of it being imposed on them from the outside, we avoid a lot of problems. Conservation International should put indigenous people on their board and let them control the budget and decide which Western consultants to pay. This way the indigenous people can direct the organization in a way appropriate to their world view.

We tend, as Westerners, to think solely in economic terms; many indigenous people think in terms of spiritual values that are far more important to them. The issue of credit is crucial in my experience; indigenous people want it made very clear that they are participating.

Indigenous people in my experience in some cultures are very keen that the pharmaceutical products be used to heal people. Healing is very high on their list as a reason for action. The Navajo healers were appalled when I informed them there might be some commercial return from their products. They said, 'We don't sell medicine, do you?' I was astonished by what I regarded as a piercing critique of our own culture. In their view, selling medicine to a sick person is essentially holding somebody hostage. Yet, upon reflection I realized they are right! This is a complex set of issues and it's unclear to me that there is a single, ubiquitous solution to be imposed by Westerners on all indigenous people. I would like to hear more the voices of indigenous people in determining these solution sets.

Lewis: I agree that our interaction with the indigenous community is critical and that compensation in some form must be provided. There is another aspect that we must consider, namely, the genetic resource of the country being examined. Is there not also a responsibility to the country of origin, through perhaps the academic institutions? In our programme we have arranged compensation with an academic institution and a museum.

King: This is built in to Shaman's programme. Government groups don't have much trouble becoming involved in the benefits of intellectual property.

The government of the country, the Ministry of Agriculture, the local museum of natural history and governmental departments of natural resources of Peru and Ecuador are recipients of infrastructure and building materials. Scientists within the institutions perform research as part of a joint scientific programme. This is a very big part of every project we have. The part that gets left out is the local communities of the indigenous people—you're absolutely right.

Jain: In India there are two institutions working with indigenous people. They are organizing a small training programme for the children of the indigenous people. In one region in Kashale in Maharashtra, near Bombay, young boys and girls sit in meetings with their elders and discuss uses of herbs. They are also taught about uses of plants from outside their villages. Therefore, something gets gradually added to their own pharmacopoeia. The other such group is in the state of Bihar. Shaman or other companies could help these indigenous people by encouraging such situations where knowledge can be exchanged and enhanced.

Balick: When we do general taxonomic surveys or floristic surveys in countries, we sign Letters of Intent with the local institutions. In the case of ethnobotanical surveys, we have also signed such letters with indigenous federations, such as the AWA Federation in Ecudaor and the Carib council in Dominica.

The Belize Association of Traditional Healers voted at one of their meetings on how they would like to receive compensation. Four levels were proposed—compensation to the individual who gave the information, his or her community, the Healers Association or the government. They voted by a margin of 10:1 for the benefits from information they had provided to go to the Healers Association; this will be the case when this particular Letter of Intent is signed. It probably will be in collaboration with the government, but a component will return to the traditional healers.

The Belize Association of Traditional Healers approached the Healing Forest Conservancy, a non-profit foundation established by Shaman Pharmaceuticals, for money to demarcate an ethnobiomedical forest reserve; the money was forthcoming. This was before there was an interest in plant collection on the part of the company. It is an example of benefits that come with no intentions.

In our dealings with the NCI, we've always put in a budget for what are called up-front benefits and NCI has always very graciously agreed to fund them. This has been the same with other contracts we have, such as with Pfizer Inc., to collect within the USA. There are modest up-front benefits, such as for development of infrastructure and money for local institutions that we collaborate with.

From the NCI funds given to us for direct collecting costs, a budget line is included that contains as much as 10% of our collecting budget for the benefit of local institutes and communities. This is not 20% or 40%, but at least it's a start.

Rubin: We have an up-front benefit. This would be negotiated with an industrial partner.

Farnsworth: You talk about industry, but there are scientists doing this kind of work who can't possibly go through all this legal hassle. The late Professor Ferdinand Bohlmann in Berlin worked with plants for 40 years; he isolated 10% of all the new structures ever discovered in the history of the world from plants. He mainly collected plants from Brazil and South Africa. I presume he had permission to take those plant samples out of the countries. He had a repository of 10 000 pure compounds. He died and willed these to someone who set up a company. The company is now selling the samples for up to $100 per 10 mg. What about the property rights to compounds isolated by chemists who are interested only in determining structures, chemotaxonomy and publishing papers, with no thoughts about biological activity or drugs?

Elisabetsky: No one has commented on the efficacy of the ethnopharmacological approach. The hit rate of Shaman Pharmaceuticals makes that of every major pharmaceutical company look bad. In a country like Brazil, where you either go ahead killing Indians or sueing anthropologists for trying to prevent this happening, to have an American company investing dollars— hard currency—on traditional medicine is by itself an enormous benefit. This is not easy to quantify but it is a real benefit.

We have suggested that Shaman Pharmaceuticals publish the data showing that some plants are positive in certain screens when, for some reason, Shaman is not interested in developing these products into a drug. Then the source countries can prepare their own formulations, based on this information. This again would be an enormous benefit.

References

Brown CH 1985 Mode of subsistence and folk biological taxonomy. Curr Anthropol 26:43–53
Cox PA, Elmqvist T 1991 Indigenous control: an alternative strategy for the establishment of rainforest preserves. Ambio 20:317–321

Ethnobotany and intellectual property rights

John H. Barton

Law & High Technology Program, Stanford University, Stanford, CA 94305-8610, USA

Abstract. Contemporary intellectual property law permits only the patenting of an identified active principle from a plant, not the plant or folk information relating to medicinal properties of a plant. The most significant rights of indigenous peoples are those deriving from physical control of the plants and the knowledge pertaining to their use. This control can provide the basis for trade secret protection. Such agreements are enforceable in developed nations and should become so in developing nations. There have been recent efforts to strengthen indigenous peoples' rights over genetic resources and relevant folk knowledge but the most far-reaching of these are not yet a part of international law. Pharmaceutical patents combined with trade secrecy can allow firms to develop and market products and ensure that the nation and/or people from which the material or information was derived are properly rewarded. This does not provide protection from competition or with respect to derived knowledge nor does it act retrospectively. At present, rights under the United Nations Convention on Biodiversity are prospective only. These rights belong to the nation and there is little legal pressure for recompense to be shared with indigenous peoples. A uniform agreement that deals in a balanced way with the relative rights of indigenous peoples and of their governments should be developed by non-governmental organizations.

1994 Ethnobotany and the search for new drugs. Wiley, Chichester (Ciba Foundation Symposium 185) p 214–227

The source material and knowledge

It is unlikely that contemporary intellectual property law would permit patenting of a plant solely on the basis of the knowledge that the plant had medicinal properties. The plant itself is not novel. In appropriate circumstances, a claim covering the use of a particular extract from the plant might be valid under contemporary developed world patent law. There may still, however, be a problem of novelty; in at least one case, a patent to a poison obtained by crushing a specific plant was invalidated by previous publications describing indigenous use of the plant, *Dennis v. Pitner*, 106 F.2d 142 (7th. Cir.), *cert den'd* 308 U.S. 606 (1939).

There is not, at this point, any basis for protecting folk information about the medicinal properties of a plant, save for the trade secret approach to be considered below. There have been proposals for copyright-style protection of folklore and these or similar proposals might provide protection for knowledge of medicinals (Jabbour 1983). The World Intellectual Property Organization has developed Model Provisions for National Laws on the Protection of Expressions of Folklore against Illicit Exploitation and other Prejudicial Action (United Nations Commission on Human Rights 1992a), but such proposals for the protection of folklore have never been broadly accepted.

The most significant rights in the hands of the indigenous peoples who may hold genetic resources are those deriving from the peoples' physical control of the plants and of the knowledge of how to use the plants. To the extent that this control is effective, or that users of the material are willing to accept restrictions, it can provide a basis for protection of trade secrets.

Trade secret protection is, in this context, fundamentally an application of contract law. Those who hold the material or information require those who want access to it to accept an agreement. The terms of the agreement are subject to negotiation; in a context such as this, they would include provisions to give the source group a share in any profits as well as provisions against passing on the material to anyone not in a contractual chain. As long as the agreement is enforceable, it can provide protection against misappropriation by the recipient of the material and, by law or by additional agreements, against misappropriation by the employees of the recipient. This approach provides no protection against independent discovery or acquisition of the same information or plant from another group or nation.

In general, such trade secret agreements are enforceable in developed nations. Some developing nations have resisted enforcing these agreements on the grounds that trade secret information may never go into the public domain, but the intellectual property provisions of the Uruguay Round of the General Agreement on Tariffs and Trade will require all participating nations to enforce trade secret agreements (Agreement on Trade-Related Aspects of Intellectual Property Rights, Art 39, MTN/FAII-AIC, 15 December 1993).

Under rapidly emerging international law, the nation within which the medicinal plant is found has control over access to the plant. This is probably the most important substantive provision of the 1992 United Nations Convention on Biodiversity (in force 29 December 1993). Article 15, paragraph 1, states that:

Recognizing the sovereign rights of States over their natural resources, the authority to determine access to genetic resources rests with the national governments and is subject to national legislation.

The implication, of course, is that states will exercise this right by requiring those who seek access to genetic resources to execute material transfer agreements, presumably ensuring that the nation receives a share of the profits. Costa Rica, for example, enacted such a law in 1992 (World Resources Institute 1993).

It is clear from a careful reading of the Convention, including its definitions, that states' authority over genetic resources is prospective only; the Convention gives them no right to seek any rights in materials that have already been exported from the nation under other arrangements. The same is true of the other convention that arguably relates to such resources, the UNESCO Convention on the Illicit Movement of Art Treasures, November 1, 1970 which covers '[r]are collections and specimens of fauna, flora . . . and objects of palaeontological interest' that have been specifically designated by a nation (Article 1).

The Biodiversity Convention gives indigenous peoples no solid rights over genetic resources; this point is a basis for significant criticism of the Convention by the community of non-governmental organizations (NGOs) that deal with indigenous peoples. In Article 8, its provision on *in situ* conservation, the Convention states:

Each Contracting Party shall, as far as possible and as appropriate:

(j) Subject to its national legislation, respect, preserve and maintain knowledge, innovations and practices of indigenous and local communities embodying traditional lifestyles relevant for the conservation and sustainable use of biological diversity and promote their wider application with the approval and involvement of the holders of such knowledge, innovations and practices and encourage the equitable sharing of the benefits arising from the utilization of such knowledge, innovations and practices.

Clearly, this provision envisions that the relative rights of the indigenous peoples and of the national government will be set by national law.

There is also a broader emerging law of treatment of indigenous peoples. In 1957, the International Labour Organization negotiated a Convention Concerning the Protection and Integration of Indigenous and Other Tribal and Semi-Tribal Populations in Independent Countries (Convention 107). This was updated (primarily in response to criticisms that the 1957 Convention was too assimilationist) in 1989 as Convention 169, Convention Concerning Indigenous and Tribal Peoples in Independent Countries (Berman 1988, International Labour Organization 1992). Although these conventions are in force, acceptance has been minimal. In early 1992, four nations (Norway, Mexico, Colombia and Bolivia) had accepted the newer convention and five other nations were considering it (Anaya 1991).

The 1957 Convention, perhaps recognizing that many nations give control of natural resources to the government rather than to the landholder, included relatively strong provisions on indigenous peoples' rights over land and no provisions relating to natural resources. The new Convention, while it does not mention genetic resources explicitly, does deal with 'other resources pertaining to lands', a concept that could include plant genetic resources, in its grant to indigenous peoples of very limited rights over natural resources. Article 15, paragraph 2, states:

In cases in which the State retains the ownership of mineral or sub-surface resources or rights to other resources pertaining to lands, governments shall establish or maintain procedures through which they shall consult these peoples, with a view to ascertaining whether and to what degree their interests would be prejudiced, before undertaking or permitting any programmes for the exploration or exploitation of such resources pertaining to their lands. The peoples concerned shall wherever possible participate in the benefits of such activities, and shall receive fair compensation for any damages which they may sustain as a result of such activities.

This clearly gives no rights to the indigenous peoples with respect to royalties deriving from ethnobotanical materials; it does, however, give them a right to be consulted when a programme is set up.

There has also been a recent effort within the United Nations system toward the strengthening of indigenous peoples' rights. Following a 1981 initiative by the Sub-Commission on Prevention of Discrimination and Protection of Minorities, a Working Group on Indigenous Populations was created. This is a small working group of experts, not a formal international negotiation. It has been working to produce a Draft Universal Declaration on the Rights of Indigenous Peoples, which could serve as the basis for a new convention (United Nations Commission on Human Rights 1992b).

In its 1993 version, the Draft Declaration goes much further than any other document in recognizing indigenous rights in genetic resources:

Indigenous peoples have the right to their traditional medicines and health practices, including the right to the protection of vital medicinal plants, animals, and minerals. (Operative paragraph 22)

Indigenous peoples have the right to special measures to protect, as intellectual property, their sciences, technologies and cultural manifestations, including genetic resources, seeds, medicines, knowledge of the properties of flora and fauna, oral traditions, literatures, designs and visual and performing arts. (Operative paragraph 27)

(United Nations Commission on Human Rights 1993). As noted above, however, this provision is not yet part of international law.

Derived pharmaceutical products

In a 'normal' ethnobotanical development process, material identified, on the basis of information from the indigenous group, as possibly having therapeutic value is collected and various extracts are screened in order to identify the active agents. The molecular structure of these agents is determined and the agents and related chemical substances are evaluated for possible medical application. If an agent survives the entire pharmaceutical review process, it is marketed; it may be produced from the traditional material or it may be produced synthetically.

The agent itself can be patented under United States and probably other law, assuming it has never been isolated or described before. At this time, essentially all developed nations and many developing nations allow patents on pharmaceuticals. The fact of separation of a product from its natural background is enough to satisfy the novelty requirement of US law, *In re Bergy*, 563 F.2d 1031 (CCPA 1977), *remanded sub nom. Parker v. Bergy*, 438 U.S. 902 (1978), *on remand* 596 F.2d 952 (CCPA 1979), *affd. sub nom. Diamond v. Chakrabarty*, 447 U.S. 303 (1980). The same result would apply under Articles 8 and 9 of the Proposal for a [European Communities] Council Directive on the legal protection of biotechnological inventions, COM(88) 496 final, SYN 159, submitted by the Commission on 20 October 1988. Even if the product itself cannot be patented, close analogues that do not occur naturally can be patented in most developed nations (see, for example, the discussion of patenting of neem seed derivatives in Axt et al 1993). Also, subject to similar questions about national rules of novelty, if the active substance is a protein, the protein and the corresponding gene sequence can be patented. It is also possible that specific relevant production processes can be patented.

The combination of trade secrecy and pharmaceutical patents just described is the basis of a plausible—and sometimes followed—agreement pattern. A pharmaceutical firm enters into an agreement with an indigenous peoples' group and/or with that group's national government and agrees to specific terms and conditions in its prospecting for therapeutic genetic resources. From the source nation or group's perspective, this agreement is a material transfer agreement or trade secret agreement. The pharmaceutical firm agrees to patent anything of value that it finds (normally seeking patent protection only in the developed world), to use the patent to protect it and the source against competitors, and to share the profits or royalties according to an agreed formula. This is presumably the pattern of the Merck–INBio agreement (although that agreement has not been made public), it is the implicit pattern of Shaman Pharmaceuticals Inc. (which has set up an affiliated non-profit organization) (Shaman Pharmaceuticals Inc., South San Francisco, CA, undated company summary), it is the pattern underlying the National Cancer Institute's Letter of Intent (revised 9/3/91, appended to Axt 1993) and it is the pattern of the World Resources Institute (1993) draft agreement.

These agreements must, of course, give the recipient appropriate rights to collect materials, learn the folklore and research the material. They must impose obligations on the recipient to evaluate the material, protect the information, obtain patents and share royalties under an agreed formula. Among the other issues that may be resolved are:

Inclusion (or not) of source nation scientists in the developed world research team.
Definition of the precise reach of the rights, e.g. application or not of the agreement to products based on but not the same as the substances found in the plant.

Distinction between products developed under the agreement and products developed on the basis of long-public knowledge about plant-based medicinals.
Possible preferences for production using plant materials derived from the source nation.
Confidentiality pending patent application.
Sub-licensing and protection of source nation interests in recipient collaboration with third parties.
Definition of recipient and manner of payment to recipient.

Evaluation of the current legal situation

Several important points can be made about the current law and the use of a trade secret agreement between an indigenous group and a pharmaceutical firm.
1. There is no protection against competition. If a researcher suspects that a plant is valuable and one nation's terms are quite severe, there is nothing to prevent the researcher from going to another nation in search of the same plant. Similarly, until the product is patented, there is no protection against an independent firm's successful development of a similar pharmaceutical.

Such problems could logically be resolved by new international rights such that one source country could sue another nation arguing reasons the royalty flow should be divided between the two nations or sue an inventor arguing that the innovation really did derive from the nation's genetic resources. It is unlikely that, in general, the developed world will accept the underlying law necessary for such a suit, although, in some cases, such a suit might be based on existing doctrines of misappropriation of a trade secret. But there is nothing to keep developing nations from working together to negotiate standard terms—and probably most users would appreciate such terms, if only to reduce confusion. It would thus be useful to continue the initiative of the World Resources Institute and define not only a model agreement but also a model national law and more detailed precise terms.

2. Under the compromise of the Convention on Biodiversity, the rights of the source nation are prospective only—there is no right to obtain royalties for genetic materials that have already left the nation and there is no right to obtain royalties for folk knowledge at all.

The underlying issue here derives from the function of the payment. An intellectual property provision normally serves as an incentive, e.g. to further conservation of biodiversity and retention of information on folk medicine. On this basis, the payment may be economically justified. If one looks to payment for providing material in the past, however, one is thinking of equitable compensation rather than of shaping behaviour. The interest in compensation is understandable, as is the desire to help indigenous peoples and developing nations. With respect to past transfers, compensation does not create new incentives. More importantly, it is not at all clear that the holding of genetic resources is a particularly useful or efficient criterion for distributing compensation to the poorer persons and nations of the world (Barton & Christensen 1988).

3. There is no protection with respect to derived knowledge. Suppose the most important output of investigation of a particular plant is not a specific patented pharmaceutical but instead a scientific article that leads to many new pharmaceuticals produced by other pharmaceutical firms or outside the scope of the agreement. Should there be any rights in such case?

This is an important fundamental problem. Publication of genome data permits replication of a protein (and probably ultimately of a secondary metabolite whose production is catalysed by a series of proteins) without access to the genetic materials from which the sequences were derived. In short, the international flow of gene sequences may ultimately make control of genetic resources irrelevant. In deciding how to respond to it, the question is one of drawing a rational line between a scientific world of free exchange and a proprietary world of controlled exchange. If the line is too far misplaced, we are all hurt by slowing technological progress. Efforts to exercise control over developments based indirectly on scientific advances are ultimately self-defeating. But we must also recognize that the absence of compensation for published information might reduce incentives to publish. And we would be very unwise to restrict the international flow of genetic information, whether in the name of helping developing nations or of preserving governmental rights over human genome data.

4. At this point, the rights under the Convention on Biodiversity belong to the nation and there is little legal pressure on the nation to share any recompense with its indigenous peoples.

In general, there are both developed and developing nations that are likely to resist a stronger commitment to share the fruits of genetic resources with their indigenous peoples. Although one can argue that the folk wisdom associated with ethnobotanical resources distinguishes these resources from other natural resources, such as oil, there will be too much concern with setting precedents that might affect such other resources. Moreover, there are enormously difficult genuine questions concerning realistic efforts to compensate indigenous peoples. For example, should the compensation be to the individuals providing information or to their community? Should it be distributed as it becomes available or should there be trust arrangements? In an effort to face some of these problems, there have been proposals for the creation of a central collecting group which would spread the benefits of the royalties among many indigenous peoples (Axt et al 1993). There is also experience in the area of providing compensation for other types of resources, such as oil under the Alaska Native Claims Settlement Act, 43 USC §§ 1601 ff.

The best approach at this point is probably to work informally and to explore approaches to protecting indigenous peoples in a model/standard law and agreement. If such models were developed by NGOs and offered on an *ad hoc* basis, some of the political barriers could be avoided because the precedents to national governments would be less strong and the agreements could provide a way to gain experience with alternatives for some of the difficult issues. The World Resources Institute (1993) has proposed a model agreement that reflects careful consideration of the interests of the pharmaceutical firm and a collecting entity; it would be useful to complement this effort by developing proposals to balance the interests of the source nation government and of the indigenous peoples.

References

Anaya S 1991 Indigenous rights norms in contemporary international law. Ariz J Int & Comp Law 8:1–39

Axt J, Corn M, Lee M, Ackerman D 1993 Biotechnology, indigenous peoples and intellectual property rights. Library of Congress, Congressional Research Service, Washington, DC

Barton J, Christensen E 1988 Diversity compensation systems: ways to compensate developing nations for providing genetic materials. In: Kloppenburg J (ed) Seeds and sovereignty. Duke University Press, Durham, NC p 339–355

Berman H 1988 ILO and indigenous peoples: revision of ILO Convention 107. Rev Int Comm of Jurists 41:48–57

International Labour Organization 1992 Seventy-sixth session (1989). 169. Indigenous and Tribal Peoples Convention, 1989. In: International labour conventions and recommendations 1919–1991. International Labour Office, Geneva, vol 1:1436–1477

Jabbour A 1983 Folklore protection and national patrimony: developments and dilemmas in the legal protection of folklore. Copyright Bull 17:10–14

United Nations Commission on Human Rights 1992a Discrimination against indigenous peoples. Intellectual property of indigenous peoples: concise report of the Secretary General. United Nations, Geneva (E/CN.4/Sub.2/1992/30, 6 July 1992)

United Nations Commission on Human Rights 1992b Discrimination against indigenous peoples. Report of the Working Group on Indigenous Populations on its tenth session. United Nations, Geneva (E/CN.4/Sub.2/1993/33, 20 August 1992)

United Nations Commission on Human Rights 1993 Discrimination against indigenous peoples. Draft declaration on the rights of indigenous peoples. United Nations, Geneva (E/CN.4/Sub.2/1993/26, 8 June 1993)

World Resources Institute 1993 Biodiversity prospecting. World Resources Institute, Washington, DC

DISCUSSION

Iwu: In negotiations concerning biodiversity prospecting, we are dealing with three entities—the individual, the ethnic group or community and the nation state. Only two of these are recognized by law; only two can sue and be sued, these are the individual and the state. In most Third World countries, the ethnic communities don't have any legal existence. How does one deal with that kind of situation?

Barton: Assistance to the indigenous peoples is a major role for the non-governmental organizations (NGOs). This is happening; perhaps not in all countries. More and more of these people are getting the protection of the kind of law that one would like to see.

Iwu: In a short-term analysis, a strict patent law such as that here in Brazil may appear to impede biodiversity prospecting. However, this is not exactly true; such a law could be very progressive if the long-term effect on the economic and technological growth of the country is taken into consideration. A good example is India: when the economy of that country was bad and they had a very loose patent law, several multinational companies moved out of India. The absence of the foreign companies allowed the growth of local entrepreneurs. Now that the Indian economy is booming, the foreign companies are going back to India. The point is that if the poor tropical nations can hold on for a short while, even the multinational companies will be forced to re-think their positions on intellectual property rights and patent laws. So we should be taking the Brazilian model as something for the future, although in the short term it may appear as an impediment to the development of new pharmaceuticals from plants.

Barton: A country really has a choice about its patent law. Patent protection for pharmaceuticals will encourage research in the country. It will probably also raise the price of the pharmaceuticals to the public. Many countries—including India when it repatriated its patent law in 1971 from London, and Mexico and Brazil during the 1970s—avoided patent protection for pharmaceuticals because of historical evidence during the 1960s that pharmaceutical companies charged extremely high prices in these countries. By eliminating the patent protection, the prices fell to those of generic compounds. That's the benefit of it. The cost of it is, you get no research. Indeed, there's been essentially no pharmaceutical research in Brazil or India since those decisions were made.

There is an obvious trade-off; does this today look the way the trade-off looked 20 years ago, given that some of these countries now have much stronger scientific capabilities? I think the trade-off looks different in a big country. Mexico is changing its policy. I would urge Brazil and India to do so. The Uruguay round of GATT (General Agreement on Tariffs and Trade) will require that. For a smaller country that doesn't have significant scientific capability, there might still be wisdom in not patenting pharmaceuticals. This is the choice they have to face.

Posey: What is your prediction of the effects of NAFTA (the North American Free Trade Agreement) and GATT on these issues?

Barton: I think NAFTA will clarify whether or not patents actually help a more advanced developing country. If it looks like NAFTA is really working for Mexico, similarly advanced developing countries will probably follow the same kind of patent line.

Craveiro: I have read that 75% of all scientific knowledge is being patented. Of the 25% that is not protected by patents, the major part comes from the developing countries. Is this number correct?

Iwu: I believe that the figure comes from a UNESCO study. If a country doesn't belong to the International Patents Convention, its patents are not respected outside that country. It costs at least $5000 to patent a product in the USA; this is equivalent to the annual salary for a post-doctoral research assistant in most African countries.

Craveiro: This suggests that the present system represents generation of knowledge in developing countries and its use by the developed countries to make products. Professor Barton, you said Brazil should change its patent law to increase the amount of research money in this country. I personally don't believe it's going to happen. A big pharmaceutical company will find it cheaper to do research in their own country, then sell the products in Brazil. We will not get more money if we change the patent law.

Barton: The fact that universities are now being encouraged by the government to patent their research, in the United States and in Europe, has caused a sea change in the conduct of research over the last 10–20 years. Sometimes it's eminently appropriate; much of the time, I think it may have some of the negative effects you are concerned about.

Iwu: Another issue is, where does one draw the line in determining intellectual property rights? Where does the right of ownership actually stop? For example, if someone publishes an ethnobotanical paper on Brazilian folk medicine and in the process reveals the constituents of some local remedies, does this not infringe on the intellectual property rights of the native healer? Should the healer, the community and his nation state be entitled to some compensation? The developed countries insist on royalties for the use of technologies protected by laws that are alien to the forest dwellers. Should the converse not hold, namely that the traditional person's right to ownership be respected according to his own laws? In the computer industry, royalties are paid at all times for the duplication of other peoples' software. Recently, a US silicon chip manufacturer sued a Japanese company for patent infringement and the court ruled for the original owner of the patent. The Japanese claimed unsuccessfully that they had modified the program.

In the development of drugs from natural products, the trick of the trade is not to develop the original compound from Nature, thereby avoiding payment of any royalties to the source country. A very good example concerns the use of non-sugar sweeteners from West African plants. In the late 1960s, research interest was focused on three of these plants: *katemfe* or miraculous fruit (*Thaumatococcus daniellii* of the Marantaceae family), the West African serendipity berry (*Dioscoreophyllum cumminsii*, Menispermaceae) and the miracle berry (*Synsepalum dulcificum*, Sapotaceae). It was shown that the sweet taste of *katemfe* was due to a protein named thaumatin which is 1600 times as sweet as sucrose when compared weight for weight. That investigation, which was sponsored largely by the Tate and Lyle company, to my knowledge was the very first time protein was reported as a sweetener. Later work was directed at splitting the protein molecule into smaller sweet-tasting peptides. Ten years later an American US company announced the accidental discovery of a dipeptide sweetener and has reaped a fortune ever since.

The major question is: should the source country and the local institution that isolated the original protein be paid any compensation?

Barton: In general, for basic research I do not want to see patents. I argue against some of the patents that are granted in the United States on these grounds. If I publish a scientific paper, so many different people may use that result in so many different ways, that it's better for society if the information isn't protected and if the information is freely shared. We want somehow in our scientific system to have a component of research where research incentives work and another component where patent incentives work. The hard problem is where to draw the line. Do I put genetic resources on one side of the line or the other?

Iwu: Why can't you do the same for patent law as you are doing for copyright law? Under copyright law, if you copy more than a certain amount or if you make copies for commercial use, you pay a fee that goes back to the publisher.

Barton: That one gets so complicated very quickly. I have two interests in copyright law and in computer software. One is direct copying, which most of us would agree should be an infringement and I should be able to restrict it. When I sell you the product, you ought to be entitled to make the copies you need to use it and only those copies.

The second set of questions is: you read my program and design a new program on the basis of it. It is not direct copying but an evolution in some way. I think we ought to be pretty free in that, the software industry does not; the courts are divided.

Elisabetsky: When a physicist publishes something about the possibility of creating a new material and that material gets used afterwards for commercial purposes, how is that different? I agree there has to be free exchange of information within the scientific community. People don't publish a paper naming a species in the hope that a pharmaceutical company will make something from this so that they can sue them. This is not the currency of the university. But if someone has purposely done research in the hope of finding a commercial product, this is different.

Barton: It depends on the path between the invention and the commercial product. If I describe a computer circuit, any computer manufacturer can copy that pretty easily. If I do it on public funding, I think it should be put in the public domain. On the other hand, if I'm the first person to discover a pharmaceutical, it takes millions and millions of dollars of investment to turn that pharmaceutical into a product. Therefore, I think I had better patent it to begin with and I should design my patent system to make that incentive possible.

Elisabetsky: So it's different for a researcher who goes to a tribe than for a laboratory scientist or a pharmaceutical company?

Barton: Yes.

King: John, is it correct that you can't file a use patent for some extracted and purified biomedicinals? You stated that without a patent of composition of matter, no pharmaceutical company would develop a compound that already has published uses for the claim they hope to make for its biomedical application.

Barton: I'm still confused on what exactly the law says on use patents, i.e patents on a particular use of a compound. The pharmaceutical firms don't like them, because ultimately there is no restriction on how a doctor uses a product. Once a product is approved by the Food & Drug Administration for a particular use, there's no effective restriction on how it's used. And there is nothing to keep me from trying to find another use for it. I can then sell it and my product ends up being used for your market even though you held the use patent. This is a weak form of protection from the viewpoint of pharmaceutical firms in the US.

Balick: There have been several legal documents that use the term 'indigenous' person. Is this limited to tribal people or does it include non-tribal people?

Barton: You have raised a real problem. The law talks in terms of indigenous with the definition that everybody cites: it talks about traditional territory, separate culture, a desire to be separate from the government. There's a list of factors, but they are clearly oriented towards tribal status. The net result is that something like the International Labour Organization Convention will protect one group of people that's living in the rain forest because they have a tribal organization. Somebody else who moved in because he couldn't get a job in the city and is seeking desperately to support his family doesn't get the protection of the system.

Cox: As an ethnobotanist, in my brief career (I published my first ethnobotanical paper 14 years ago), I've seen a tremendous change in the culture of the science. I entered ethnobotany in the interest of creating public goods. Public goods are commodities that can be consumed but not depleted, such as poetry or musical compositions. My goals in becoming an ethnobotanist were not only to add to the scientific wealth of the world but also to highlight the beauty and intrinsic worth of indigenous cultures.

I'm grateful that NGOs and the private sector are interested in indigenous intellectual property rights. I am uncomfortable with the assumption that we Europeans are the correct individuals and the correct culture to arrive at these formulations. Many of the indigenous cultures I work with are non-monetized; they do not perceive all transactions in economic terms as Western people do. Many of the healers I work with are also interested in creating public goods, namely healing people who are ill. In the formulations coming from the legal community, I hear precious little about the creation of public good, about the rights of people who are ill, and particularly little about the devastation that we may unwillingly wreak on indigenous cultures when we monetize them. In other words, the discussion on intellectual property rights, because it's been framed and conducted by Westerners, could be considered yet another manifestation of Western cultural imperialism. I would be very interested to have my Samoan villagers and the Samoan healers conduct this conversation. I assure you that conversation would be conducted in completely different terms, because the Samoan people see healing as a spiritual calling. They do not charge for healing, since they do not see it as a monetary transaction. They believe that to see healing as a monetary transaction is a travesty of their calling as healers. I would very much like to see a convention of indigenous people and hear their views on this matter, because as an ethnobotanist I am morally bound to regard first and foremost the views of indigenous people.

I am very nervous about NGOs staffed by Western anthropologists and Western lawyers coming down and trying to transact business with indigenous people in the same way as they transact business with a corporation. I would very much like to give the indigenous people voices. Do you think this whole dialogue could be re-framed in indigenous terms?

Barton: Yes. I think it's essential that the kind of discussion we're having be held in conjunction with the indigenous peoples' community.

King: I agree that the entire dialogue has to be re-framed and must include the perspective of the indigenous people in these cultures. This is a challenge. Our entire trade system and intellectual property rights systems, which are not based on indigenous perceptions of the world, may have to change. This may require some lawyers and other people having extended residence or long-term associations with indigenous people to get that perspective.

We at Shaman Pharmaceuticals met with several Peruvian indigenous leaders and their lawyers to discuss what we are doing in detail. The NGOs that are appropriate to these discussions are not the international Western NGOs that Paul Cox referred to; indigenous organizations exist that are quite competent.

Posey: I've heard the words moral rights and human rights. We are talking about basic rights. I don't know if it's clear that patents and trade secrets and these kinds of protection of intellectual property rights are just examples of many strategies to deal with a very complex problem.

I squirm a little after all these years hearing someone who comes from a developing country's elite tell me what *his* people want. In the past 5–6 years, indigenous and traditional peoples have become much better organized and aware of global issues. For example, The Alliance of Indigenous-Tribal Peoples of the Tropical Forests has over 300 group members around the world, many of whom have expert lawyers working on rights and access issues. The International Union for the Conservation of Nature has an Intercommission Task Force on Indigenous Peoples that is focusing on many of the questions raised in this symposium. The Mataatua (Maori) tribes in Whakatane, Aotearoa, New Zealand, in 1993 brought together specialists from 54 indigenous groups from around the world. So, there are plenty of indigenous people who could be sitting at this table with us. If there is ever a follow-up meeting to this, they ought to have significant representation, if not be the majority of participants.

As President of the Global Coalition for Biological and Cultural Diversity, I have developed a covenant dealing with the intellectual, cultural and scientific property rights between a responsible corporation, scientist or scientific institution and an indigenous group. This is an attempt to establish a basic code of ethics and conduct to create equitable partnerships that lead to economic independence for local communities, while providing for the conservation of natural resources. This guides all our work with indigenous peoples and I should like to see it more widely adopted by both individual scientists and organizations involved in similar undertakings.

The primary concern of indigenous peoples is self-determination, which subsumes such basic rights as recognition of and respect for their cultures, societies and languages, as well as ownership of their lands and control over the resources associated with those lands. Intellectual, cultural and scientific property rights are seen as a starting point to defining a more comprehensive category of traditional values, knowledge and resources that have often been exploited without authorization, recognition or compensation. In addition to supporting the above, the Covenant aims to strengthen indigenous cultures through recognition and encouragement of the groups' own values and objectives by helping to find new ways of using biological, ecological and cultural richness through equitable and responsible trade, sourcing, research and development. The idea is to establish long-term relationships built on joint decision making. The Covenant states what is to be protected, particularly sacred property (both material and non-material) and knowledge of both things and processes, such as preparation of useful species. Finally, the Covenant lists the principles to be observed by the various parties, including independent monitors.

I can supply full details of the Covenant to anyone who is interested.

What would be your guidelines on publishing our data, given that in many cases we are taking information that we know to be private and making it public? What are the legal possibilities for making disclaimers in our publications? For example: Information taken from this article is restricted and if any development ever comes out of what you've read or done from this article, you have the moral obligation to compensate the people from whom this information came. Does that have any basis in law?

Barton: Probably not. It may create a moral obligation. This is a problem that I face in the seed context. Should an international gene bank like the International Rice Research Institute send seeds out with a letter that says, this came from country X, you really owe them something if you make a commercial product out of it? This is similar to your notice in the article; it's a little different in that one is providing material and to a narrower audience.

Conservation and ethnobotanical exploration

Gary J. Martin

WWF/UNESCO/Kew People and Plants Initiative, Division of Ecological Sciences, Man and the Biosphere Program, UNESCO, 7, Place de Fontenoy, 75352 Paris, CEDEX 07 SP, France

Abstract. In recent years conservationists have realized that the maintenance of protected areas is closely linked to rural development. As part of their efforts to improve local people's standards of living, they have sought the advice of researchers who work in communities, especially those that border on nature reserves. Ethnobotanists, who are turning their attention to the cultural and ecological crises confronting the regions in which they work, are natural allies in this venture. The joint efforts of conservationists and ethnobotanists are being supported by non-profit organizations, intergovernmental agencies and research institutes. The search for new drugs and other natural products from plants is an important element in this collaboration, but it cannot be divorced from the broader objective of promoting the survival of biological and cultural diversity. Conservationists will support biodiversity prospecting and related efforts only if there is a clear benefit for local communities and protected areas. An example of the concrete actions being taken by conservation agencies is the People and Plants Initiative, a joint effort of the World Wide Fund for Nature, the United Nations Educational, Scientific and Cultural Organization and the Royal Botanic Gardens, Kew. The main objective is to support the work of ethnobotanists in developing countries in studies of sustainable plant use and application of their work to conservation and community development. The initiative provides training workshops and relevant literature; coordinators work in collaboration with local people to create inventories of useful plants and appraise the impact of harvesting specific plant resources in and around protected areas. Phytochemical screening of medicinal plants and preparation of extracts are carried out as part of some projects.

1994 Ethnobotany and the search for new drugs. Wiley, Chichester (Ciba Foundation Symposium 185) p 228–245

In a symposium on ethnobotany and the search for new drugs, a paper on conservation might appear to be an anomaly. A conservationist is often perceived as someone who advocates that local people be kept out of protected areas and that biological resources be preserved rather than exploited. Yet over the last 10 years, ethnobotany and biodiversity conservation have become increasingly intertwined, blending two rich traditions that had been evolving independently since their origin in the 19th century (Ford 1978, McIntosh 1985). Ethnobotanists are becoming aware that research should focus on the ecological and cultural crises facing the world (Toledo 1992), while conservationists are realizing that protection of wildlands depends on the participation of local people (Wells & Brandon 1992). If floral and vegetational surveys provided the impetus to declare the first nature reserves in the late 19th century (McIntosh 1985), it is ethnobotanical surveys that will contribute to conservation efforts in many countries as we reach the end of the 20th century.

The convergence of ethnobotany and conservation

As ethnobotany expanded in the early 20th century, two distinct lines of research became apparent. Natural scientists, who refer to their approach as economic botany, documented local uses of plants, typically organizing the resulting data according to the Western system of plant classification. In addition, they pursued agronomic and phytochemical studies to clarify the potential role that novel useful plants could play in agricultural production, the discovery of new medicines and other aspects of international development.

Social scientists, interested in a systematic appraisal of local people's interaction with the local environment, tended to focus on the folk classifications, social rules and ecological strategies that guide such interaction. Drawing upon the empirical tradition in American anthropology and new techniques devised by linguistic and cognitive anthropologists, ethnoscientists have sought to discover the basic principles which underlay the way that humans perceive the natural world.

One outcome of the interaction between natural and social scientists has been an increased emphasis on applied projects in which there is a direct relationship between indigenous people and academics as co-promoters of research, community development and conservation of biodiversity. Ethnobotanists and other researchers have begun to advocate the return of scientific results to rural people, the application of traditional ecological knowledge in development projects and the promotion of local participation in the management of natural resources (Williams & Baines 1993). While economic botanists continue to look for marketable products in tropical forests and elsewhere, they increasingly focus on how the commercialization of these resources can contribute to resolving the poverty, malnutrition and diminished social status of people in developing countries. And as ethnoscientists continue to study cultural systems, they are aware that local knowledge can play an important role in development programmes.

At the same time, conservation organizations are exploring ways in which the participation of local people can strengthen protected areas and conserve biodiversity (McNeely et al 1990, National Research Council 1992, Shiva et al

1991, World Resources Institute 1992). Today, there are many conservationists who propose to save the world's natural and cultural richness by creating ecological reserves that include rather than exclude communities of traditional agriculturalists, pastoralists or gatherers of forest products. This concept forms the foundation for biosphere reserves and integrated conservation–development projects. The goal of most of the projects is to support rational land use in the buffer zones outside protected boundaries and to increase the local standard of living as a way of diminishing exploitation of biological resources within nature reserves (Wells & Brandon 1992). Conservationists who develop these integrated projects often work with ethnobotanists to find ways in which local people can continue to have access to their traditional resources in buffer zones without damaging the integrity of the protected area. Increasingly, there is interest in promoting the commercial exploitation of selected plants in order to provide income for local communities (Plotkin & Famolare 1992). At the same time, there is concern about the overexploitation of some commercially valuable species that are vulnerable to overharvesting.

Why are ethnobotany and conservation linked?

The debate on the link between ethnoecology and conservation focuses on several interrelated arguments which form the basis of an ideology—an optimistic vision of how local people can contribute to community development. Although some researchers have questioned the validity of this perspective (Diamond 1987, Rambo 1985), it forms the basis for a programme of action that many governmental agencies, indigenous organizations and non-profit groups are seeking to implement (Redford & Padoch 1992).

The basic tenet of this ideology is that the maintenance of protected areas depends on collaboration with local people rather than on a policy of excluding them from the natural habitats near their communities. Hundreds of indigenous groups, ranging in size from a few surviving members of some ethnic minorities to populations of several hundred thousand people, harvest a large number of useful plants from temperate and tropical forests, many of which have been granted protected status or are proposed as ecological reserves (Clay 1988, Davis 1993). In addition, large numbers of non-indigenous agriculturalists have migrated to these pristine areas in search of a better life (Denslow & Padoch 1988).

Some conservationists believe that when local people have access to health care, education, agricultural land and biological resources, they can become allies in the quest for conservation of biodiversity. As part of management plans for specific reserves, they seek to support local efforts towards strengthening cultural identity, fortifying traditional productive activities and improving standards of living, when these actions contribute to the wise use of natural resources.

Many researchers claim that traditional management of the local environment, including techniques of fishing, hunting, farming and gathering wild plant foods, forms a sustainable way of living off of natural bounty without destroying it. They observe that rural inhabitants have developed a profound knowledge of ecological dynamics because forests, savannahs and wetlands have been under their stewardship for hundreds of years. Although the general public is only now discovering the role of forests in buffering the world's climate, controlling erosion, maintaining water tables and recycling nutrients (Ehrlich & Ehrlich 1992), human ecologists have argued that local people are aware of these benefits and sometimes purposefully leave forest cover on watersheds, mountain peaks and other strategic ecological zones in order to maintain agricultural productivity and wild resources. Other observers claim that the close spiritual relationship that local people have formed with Nature provides them with a moral imperative to use natural resources in a sustainable way. They suggest that people raised in Western society can learn a new respect for the forest by following the lead of indigenous people (Davis 1991, von Hildebrand 1992).

Ethnobotanical exploration and conservation

Within this general perspective, there are two specific arguments that explain conservationists' emerging support for ethnobotanical exploration in protected areas and surrounding communities (Nepstad & Swartzman 1993, World Resources Institute 1993). Ethnobotanists are aware that a wealth of species, hidden away in tropical forests, are known only by the local inhabitants who use them for food, medicine, shelter and many other purposes. They have long argued that traditional knowledge of plants can lead researchers to discover new marketable products that will benefit humanity, providing an incentive to explore and preserve natural areas (Schultes 1992).

A related argument is that local people's knowledge of plants allows us to establish that the forest is of more commercial value when left standing than when cut for timber, cattle-raising, commercial agriculture or other ventures (Peters et al 1989). Conservationists argue that the true cost of deforestation is not calculated in the final balance sheet, simply because we ignore the value of many plant resources, including those used for subsistence or traded only in local markets. How can we put a price on medicinal plants that have not yet been analysed by the pharmaceutical industry, food plants not yet seen in the supermarket, or ornamental flowers not found in florists' shops? Resource economists assert that the first step in making such an appraisal is to calculate the current or potential commercial value of plants and animals used by local people.

Even as there is a growing interest in the discovery of useful plants and animals, habitat destruction is causing the rapid disappearance of species of uncalculated value and culture change is provoking the loss of traditional knowledge, sometimes in a generation or two. Because biodiversity conservation, cultural survival and the search for new products are inextricably linked, conservationists argue that a portion of profits from commercialization of forest products and biochemical prospecting must be returned to national institutions and local communities that are directly implicated in the maintenance of natural areas (Johns 1990). Local people who have safeguarded both biological resources and cultural knowledge over many generations have a special right to compensation (Posey 1990).

A critical assessment of the ideology and strategy

From a variety of perspectives, critics are challenging the wisdom that commercialization of forest products will provide an easy alternative to deforestation (Corry 1993, Godoy & Bawa 1993). Some observers have pointed out that markets for non-timber products are relatively unstable (Homma 1992) and that profits often end up in the pockets of middlemen rather than in the hands of local people who are living in close proximity to natural areas (Lescure et al 1992). Ecologists express scepticism about the sustainability of harvesting of many resources as commercial levels of production are attained and as local communities grow in population. Anthropologists observe that if profits are not shared equitably among residents, greater affluence in rural communities may not necessarily lead to general improvements in health, nutrition and education (Dewey 1981). Rapid culture change and further integration into the market economy may lead local people to abandon customary methods of management and obtain new technology for extraction, often replicating the unsustainable use of resources typically associated with large-scale commercial enterprises.

While there are notable examples of local knowledge leading to discovery of new products from the forest, opinions vary on how many species will actually find major success in international or even regional markets (Caballero 1987). Only a small percentage of the world's 300 000 species of plants currently play a role in diet, medicine and industry (Bates 1985). Of the thousands of plants that are being tested for pharmaceutical activity, only a few are likely to be developed into major medications. The search for plants that yield new drugs typically provides few short-term benefits and there is no guarantee that future profits will be shared equitably with the countries and communities where the species are found (Miller 1993). In many tropical countries, the relative benefit of preserving nature in the hope of someday finding a handful of useful products may pale next to the necessity of feeding, clothing and curing people today.

With all of the enthusiasm over the commercialization of plant resources, it is easy to forget that a large number of plants will never be brought to market. Although the vast majority of useful species are unlikely to become sources of products that benefit humanity in general or reap a commercial profit, they are critical to the sustenance of rural communities (Martin 1992).

The People and Plants Initiative

As the debate continues, human rights organizations, such as Cultural Survival, and conservation agencies, such as the World Wide Fund for Nature, Conservation International and the Rainforest Alliance, have been developing programmes to blend ethnobotany with their efforts to preserve cultural and biological diversity. Although each of these programmes merits a detailed discussion, I would like to focus on People and Plants, an initiative on which I have been working since its inception in 1992. As a joint effort of the World Wide Fund for Nature, the United Nations Educational, Scientific and Cultural Organization and the Royal Botanic Gardens, Kew, the initiative has a global reach. Projects are currently being carried out in Africa (including Cameroon, Uganda and Madagascar), Asia (particularly in Malaysia) and in Latin America (with special emphasis on Bolivia, Brazil and Mexico) and the Caribbean. The primary focus is on field activities in biosphere reserves, World Heritage sites and other natural areas in which the partner organizations have been active in the past.

The main objective is to support the work of ethnobotanists in developing countries who are not only studying the use and sustainable harvesting of plant resources but are also seeking to apply the results of their research to conservation and community development. People and Plants coordinators offer training workshops in which participants gain expertise in ethnobotanical methods drawn from various academic disciplines and discuss ethical issues concerning the exploitation of botanical resources in developing countries. On visits to specific field projects, the coordinators work in collaboration with local people, park personnel and researchers to create inventories of useful plants and appraise the impact of harvesting of specific plant resources in and around protected areas. The initiative is also providing local participants with scientific and popular literature, including manuals of practical methods on ethnobotany, resource harvesting and related topics.

Among the fundamental aims of People and Plants is to ensure that local communities and national institutions gain from the conservation and utilization of plant resources. A. B. Cunningham, a People and Plants field coordinator who focuses on Africa, has prepared a set of guidelines to encourage the equitable distribution of benefits from ethnobiological research (Cunningham 1993). The specific way of applying these guidelines in People and Plants projects depends on the needs encountered at each field site. In some places, the focus is on resources that are already harvested for commercial purposes rather than on a search for new sources of drugs. Because they reveal difficulties encountered in harvesting plants sustainably and ensuring benefits for communities, these projects hold a valuable lesson for colleagues who propose to link conservation with ethnobotanical exploration.

In Cameroon, A. B. Cunningham has joined forces with botanist Fonki Mbenkum to assess the harvesting and commercialization of *Prunus africana*, a tree of Afromontane forests which yields the primary ingredient of prostate medicines used in France, Italy and other parts of Europe (Cunningham & Mbenkum 1993). They are discovering that the debarking of these dominant trees, although potentially sustainable, is currently resulting in habitat destruction and a reduction in the amount of the resource available for use in traditional medicine. They recommend various actions, including cultivation in local plantations, preservation of representative wild populations and the setting of sustainable limits to bark harvesting.

In other projects, phytochemical screening of medicinal plants is considered an important element in efforts to build local ethnobotanical expertise and provide community members with information about the plants they use. Kinabalu Park, a protected area in Sabah, Malaysia, is the site of an ethnobotanical inventory which involves local Dusun people, park personnel and researchers from Malaysia and several foreign countries. Although the work is currently focusing on local knowledge of palms there will be a future emphasis on evaluating medicinal plants used in the communities around the park.

Sabah Parks, the state agency which oversees the park, and the *Universiti Malaysia Sarawak*, a newly founded research university in the adjoining state of Sarawak, will provide professors and students from Malaysia and other South-East Asian countries the opportunity to collaborate with Dusun people on ethnopharmacological studies and other research projects. The goal is to support visiting scientists in their quest to investigate local botanical resources, while providing the Dusun people with an assessment of the efficacy and toxicity of local herbal remedies. There are plans to develop a basic phytochemistry laboratory in Kinabalu Park itself, opening the way to initial screening of medicinal plants and the preparation of extracts that can be further analysed in laboratories of various Malaysian universities. The integral participation of park personnel and Dusun community members in the research should contribute to raising their awareness of the potential benefits of ethnobotanical exploration and the value of conserving forested areas even outside of the reserve boundaries. In addition, it should result in a greater appreciation of the complexity of local ecological knowledge, encouraging respect for local healers or *bebelian* who ensure that ethnobotany is a living tradition. If a marketable product were to result from the analysis of local plants, the participants in the initiative— including park authorities, Malaysian university researchers and local people— would be in a relatively strong position to negotiate a share of profits and to ensure that any local production of the resource is sustainable. At present, a formal research agreement is being designed to guarantee that all visiting scientists embrace these goals, thus supporting conservation and development in Sabah.

A similar initiative is underway in the Beni Biosphere Reserve of northern Bolivia. In both *mestizo* and indigenous Chimane villages, park personnel and local people are making an inventory of useful plants which play a role in agriculture, traditional medicine, the manufacture of crafts and other aspects of community life. The national herbarium of Bolivia is coordinating the identification and distribution of all ethnobotanical voucher specimens. Small samples of medicinal plants are sent to the *Instituto Boliviano de Biologia de Altura* and various research institutes of the *Universidad Mayor de San Andrés*. These laboratories have phytochemical screening programmes which focus on parasitic diseases of local importance, such as Chagas' disease and leishmaniasis, that have received little attention from major pharmaceutical companies.

The park staff has joined forces with the *Gran Consejo Chimane*, an indigenous organization, to contribute to the development of Chimane settlements found inside the boundaries of the reserve. As part of a broad literacy and education programme, the results of the ethnobotanical survey will be returned to the communities in the form of popular booklets written in both Chimane and Spanish. Similar materials will be prepared for *mestizo* communities, where residents who are relatively new to the region are requesting advice on the efficacy of locally available medicinal plants, the selection of useful species that can increase the productivity of regenerating forests, and the sustainability of current methods for harvesting plants used in crafts production.

A project that integrates ethnobotany, health care and biodiversity conservation is being carried out in the Manongarivo Special Reserve, which covers some 350 km^2 of forested lands in north-west Madagascar. Nat Quansah, plants officer of the World Wide Fund for Nature—Madagascar and a participant in the People and Plants Initiative, is coordinating two teams, one in the field and one in the laboratory, which are documenting traditional medical practices while seeking to improve the availability of health care for local people. The field team consists of a local plant specialist who has an intimate knowledge of the flora, university-trained ethnobotanists who carry out surveys of medicinal plant use and medical doctors who diagnose diseases and assess the efficacy of local medicinal plants while providing modern medicine to complement the local remedies. The laboratory team consists of pharmacologists who test the pharmacological activity and dosage of medicinal plants used by local communities. There is a constant exchange of information between the two teams, allowing local people to evaluate the results of pharmacological tests and laboratory personnel to interpret their own findings in the light of traditional use of the plant remedies. The central goal of this initiative is to legitimize and promote traditional ways of curing and to make people aware that continued access to herbal medicines is dependent on forest conservation.

Although carried out in diverse cultural and natural locations, these and other projects demonstrate a common approach to the role of ethnobotanical exploration in protected areas. Rather than promoting the discovery and marketing of new products, the emphasis is on reinforcing subsistence use and small-scale commercialization of plants that contribute to the well-being of rural communities, enhancing their ability to participate in conservation initiatives.

In cases where large-scale commercialization of wild plants already exists, the focus is on improving harvesting methods and searching for mechanisms that allow communities to gain an increasing share of the profits. Attention is given to not only the sustainability of plant resources but also the viability of the cultural knowledge that guides local people's management of the environment. All work is carried out by teams of local people, park personnel, researchers and university students as a way of building local expertise in ethnobotany and reinforcing the interaction between communities, national scientific institutions, non-governmental organizations and conservation agencies.

The path ahead

Over the coming years, conservation organizations will continue to focus on the link between ethnobotanical exploration and protection of wildlands. As conservationists become increasingly sophisticated in their understanding of plant resource management and commercialization, there will be additional emphasis on forming multidisciplinary and multicultural teams of researchers who can examine local plant use from various perspectives. Particular support will be given to local promoters of ethnobotany and scientists in the host country who collaborate in the design of resource use studies, because they are in a unique position to apply the results to community development and nature conservation.

Spurred on by the urgency of this conservation issue, there will be a tendency to support rapid participatory ethnobotanical inventories as a first step to guide detailed studies on selected resources. In these scientifically rigorous analyses, attention will be given to posing hypotheses about the link between resource use and conservation, as well as to developing empirical methods to test these ideas.

Conservation organizations—and the general public who support them—will be increasingly perceptive in their assessment of the successes and failures of ethnobotanical exploration. Approval of biodiversity prospecting will be forthcoming only if scientists, governmental officials and commercial enterprises can make a convincing case that these actions benefit natural areas and rural communities.

What course of action will communities and host countries require from outside partners interested in developing new drugs through collaborative ethnobotanical research? In the short term, they will be expected to provide training and supplies for local participants in the research. As work continues over a number of years, the training should be complemented by building local infrastructure, including not only research facilities in protected areas and universities but also health and educational services in communities. Finally, there will be a long-term obligation to return to local collaborators a fair share of any profits earned from the commercialization of products derived from ethnobotanical exploration. These various efforts will be directed towards the common goal of preparing counterparts in developing countries to discover new sources of drugs and to develop commercial products with increasingly limited participation by foreign partners.

Collaborators in local communities will not be simply the beneficiaries of these initiatives, but full partners in the process. They will participate in the design and implementation of the research as well as in the application of the results. In addition to receiving monetary returns, they will expect assistance in analysing and reinforcing their traditional culture, including testing the efficacy of local herbal remedies, measuring the sustainability of forest management practices and designing ways of ensuring that knowledge is passed from one generation to the next.

From the perspective of conservation and community development, the way ahead for ethnobotanical exploration is to follow a path of participatory research guided by explicit research agreements and contracts that define the rights and obligations of all participants at each stage of the project. The potential contribution of ethnobotany to resolving global crises has been well publicized but the world is now waiting for results.

Acknowledgements

The People and Plants Initiative is coordinated by the Biodiversity Unit of the Conservation Policy Division of WWF–International, the Man and the Biosphere Programme of the Division of Ecological Sciences, UNESCO and the Director of the Royal Botanic Gardens, Kew. Financial support for the work in Cameroon has been provided by WWF–International and the Tropical Forestry Program of the Forest Service, United States Department of Agriculture. The initial stages of the Kinabalu Ethnobotany Project were funded by WWF-International, UNESCO and the Tropical Forestry Program; the MacArthur Foundation has approved support from 1994–1996. The work in the Beni Biosphere Reserve is made possible by a Fulbright teaching and research award and by a grant from UNESCO. WWF–International has provided primary support for the work in Madagascar.

References

Bates DM 1985 Plant utilization: patterns and prospects. Econ Bot 39:241–265
Caballero J 1987 Etnobotánica y desarrollo: la búsqueda de nuevos recursos vegetales. In: Toledo VM (ed) Memorias del simposio de etnobotánica del IV congreso latioamericano de botánica. Instituto Colombiano para el Fomento de la Educación, Superior, p 79–96
Clay JW 1988 Indigenous peoples and tropical forests, models of land use and management for Latin America. Cultural Survival, Cambridge MA
Corry S 1993 The rainforest harvest: who reaps the benefit? Ecologist 23:148–153
Cunningham AB 1993 Ethics, ethnobiological research and biodiversity. World Wide Fund for Nature (WWF), Gland (WWF/UNESCO/Kew People Plants Initiative)
Cunningham AB, Mbenkum FT 1993 Sustainability of harvesting of *Prunus africana* bark in Cameroon, a medicinal plant in international trade. UNESCO, Paris (People Plant Work Pap 2)

Davis W 1991 Towards a new synthesis in ethnobotany. In: Rios M, Borgtoft H, Pedersen H (eds) Las plantas y el hombre. Ediciones Abya-Yala, Quito, p 339–358

Davis SH 1993 Indigenous views of land and the environment. World Bank, Washington, DC (World Bank Discuss Pap 188)

Denslow JS, Padoch C 1988 People of the tropical rain forest. University of California Press, Berkeley, CA

Dewey KG 1981 Nutritional consequences of the transformation from subsistence to commercial agriculture in Tubasco, Mexico. Hum Ecol 9:151–188

Diamond JM 1987 The environmentalist myth. Nature 324:9–10

Ehrlich PH, Ehrlich AH 1992 The value of biodiversity. Ambio 21:210–226

Ford RI 1978 Ethnobotany: historical diversity and synthesis. In: Ford RI (ed) The nature and status of ethnobotany. Museum of Anthropology, University of Michigan, Ann Arbor, MI (Anthropol Pap 67) p 33–50

Godoy RA, Bawa KS 1993 The economic value and sustainable harvest of plants and animals from the tropical forest: assumptions, hypotheses and methods. Econ Bot 47:215–219

Homma AKO 1992 The dynamics of extraction in Amazonia: a historical perspective. In: Nepstad DC, Swartzman S (eds) Non-timber products from tropical forests: evaluation of a conservation and development strategy. New York Botanical Garden, New York (Adv Econ Bot vol 9) p 23–31

Johns T 1990 Chemical screening programs in the real world. Chemoecology 1:142–143

Lescure JP, Emperaire L, Pinton F, Renault-Lescure O 1992 Nontimber forest products and extractive reserves in the middle Rio Negro region, Brazil. In: Plotkin M, Famolare L (eds) Sustainable harvest and marketing of rain forest products. Island Press, Washington, DC, p 151–157

McIntosh RP 1985 The background of ecology: concept and theory. Cambridge University Press, Cambridge

McNeely JA, Miller KR, Reid WV, Mittermeier RA, Werner TB 1990 Conserving the world's biological diversity. IUCN/Gland, Switzerland, WRI/CI/WWF-US/The World Bank, Washington, DC

Martin GJ 1992 Searching for plants in peasant markets. In: Plotkin M, Famolare L (eds) Sustainable harvest and marketing of rain forest products. Island Press, Washington, DC, p 212–223

Miller SK 1993 High hopes hanging on a 'useless' vine. New Sci 137:12–13

National Research Council 1992 Conserving biodiversity, a research agenda for development agencies. National Academy Press, Washington, DC

Nepstad DC, Swartzman S (eds) 1992 Non-timber products from tropical forests, Evaluation of a conservation and development strategy. New York Botanical Garden, New York (Adv Econ Bot vol 9)

Peters CM, Gentry AH, Mendelsohn RO 1989 Valuation of an Amazonian rainforest Nature 339:655–656

Plotkin M, Famolare L (eds) 1992 Sustainable harvest and marketing of rain forest products. Island Press, Washington, DC (WWF/UNESCO/Kew People Plant Initiative)

Posey DA 1990 Intellectual property rights: what is the position of ethnobiology? J Ethnobiol 10:93–98

Rambo AT 1985 Primitive polluters, Semang impact on the Malaysian tropical rain forest ecosystem. University of Michigan, Ann Arbor, MI (Anthropol Pap 76)

Redford KH, Padoch C (eds) 1992 Conservation of neotropical forests, working from traditional resource use. Columbia University Press, New York

Schultes RE 1992 Ethnobotany and technology in the Northwest Amazon: a partnership. In: Plotkin M, Famolare L (eds) Sustainable harvest and marketing of rain forest products. Island Press, Washington, DC, p 7–13

Shiva V, Anderson P, Schücking H, Gray A, Lohmann L, Cooper D 1991 Biodiversity, social and ecological perspectives. Zed Books London/World Rainforest Movement, Penang

Toledo VM, Batis AI, Becerra R, Martinez E, Ramos CH 1992 Products from the tropical rainforests of Mexico: an ethnoecological approach. In: Plotkin M, Famolare L (eds) Sustainable harvest and marketing of rain forest products. Island Press, Washington, DC, p 99–109

von Hildebrand M 1992 Colombia: setting a precedent for the world. In: The rainforest harvest, sustainable strategies for saving the tropical forests? Friends of the Earth, London, p 61–63

Wells M, Brandon K 1992 People and parks, linking protected area management with local communities. The World Bank, Washington, DC/WWF/USAID

Williams NM, Baines G 1993 Traditional ecological knowledge: wisdom for sustainable development. Centre for Resource and Environmental Studies, Australian National University, Canberra, ACT

World Resources Institute 1992 Global biodiversity strategy. WRI/IUCN/UNEP, Washington, DC

World Resources Institute 1993 Biodiversity prospecting. WRI, Washington, DC/Instituto Nacional de Biodiversidad/Rainforest Alliance/African Centre for Technology Studies

DISCUSSION

Balick: I have the impression that perhaps 70–90% of all ethnobotanical research is never published—the plant lists and the material. Partly because people are looking at sterile material and the taxonomists do not want to identify sterile material any more. I've recently heard an ethnobotanical specimen that vouched an important medicinal use called 'a useless piece of fuel' by a taxonomist—at least it had a use!

You are turning your material over to the Malaysians. Are they then taking charge of identifying it? If so, is there a time frame? Will it be distributed to specialists? A lot of people have gladly turned over their material to us. In some cases, with large collections, I've had to ask our staff to put it in boxes and wait until we have more support for processing the collections, if people aren't paying to have that done.

Martin: In our work with local ethnobotanists, one of the stipulations is that all plant collections must be fertile. One of the benefits of working with people who live in the community all the time is that they're in a position to wait until things come into flower and fruit in order to get really optimal material. Our collections from Malaysia are all fertile, except for some of the rattan palms, which John Dransfield of the Royal Botanic Gardens, Kew has said he will identify even if they are sterile.

The *Universiti Malaysia Sarawak* is one of the three partners in this venture; its botanists will receive the botanical collections and identify them. Other sets go to herbaria in Kinabalu Park and the Sabak Forestry Department, in Sandakan; they are capable of identifying a large number of these plants. Other sets will be sent to Kew, Leiden and to specialists. So a mechanism has been set up to identify these plants. This is not dumping them on somebody who has no stake in the project, but rather giving them to partners who have agreed to do this jointly.

Balick : So there is a great deal of taxonomic expertise in Malaysia?

Martin: Yes, absolutely.

Prance: Mike, this is a problem of large herbaria. I remember many arguments with the curator of the New York Botanical Garden herbarium about sterile material of my own. I'm constantly discussing with the keeper of the Kew herbarium the value of the large amount of sterile material that we're generating now that we're more into ethnobotany and economic botany. Local herbaria can often be very useful for the treatment of sterile material. I've just been to the INPA herbarium in Manaus to identify 100 sterile collections of Chrysobalanaceae, because it was much easier there where the herbarium contained only Amazonian specimens, without species from the rest of the world mixed in with them.

King: I believe we need a more proactive relationship between conservation organizations and 'industry'. Is Kinabalu Park funded in perpetuity? If it is not, might it be wise to figure out linkages of potential long-term financial support for that area and the people from some sort of sustainable industry around the park or based on small amounts of material?

Martin: Kinabalu Park is funded fairly permanently by the state government and by the national government. It is a recognized state park.

King: My perception to date is that international conservation organizations are taking on the role of arbitrators in terms of biodiversity prospecting, ethics and so on. This is an important aspect of future agreements, but I'm not convinced that all parties have agreed that conservation organizations are the appropriate arbitrators. A goal we might have is that those countries and the people discuss whether this is the case; if they agree it is, arbitrators from the non-governmental organizations (NGOs) would be more formalized and more effective.

Martin: I also have my concerns about the role of NGOs and conservation organizations as brokers in biodiversity prospecting. In the Kinabalu project, there is a direct relationship between the *Universiti Malaysia Sarawak* and the conservation unit in Sabah, Sabah Parks. The People and Plants Initiative is a partner in this research, but we're the odd man out, because the only institutional presence we have in Malaysia is WWF–Malaysia, the affiliate organization to WWF. They are acting as advisers on the question of biodiversity prospecting and making agreements between the park, the university and the local communities. We are interested to see how the relationship evolves. There's a very good relationship between the university personnel, the park personnel and the people from the communities, but good relationships aren't enough, we need written agreements. We need to spell out exactly what benefits are going to come back to which partners in this project.

King: You referred to 'the proper short-term and long-term agreements' with respect to biodiversity prospecting. This is taking an arbitrary role, especially if you have some ideas of what those should be, whether stated explicitly or implicitly. It may be appropriate that, given their world view and experiences, these organizations become arbitrators, but it should become formalized somehow, if both parties are comfortable with that.

Martin: I don't know if arbitrator is the right word, I would use something like adviser. In the case of Sabah Parks, for example, several members of the park staff are concerned about these issues and are looking for advice. We are replying that this is an important issue around the world, various contracts and agreements have been forged in different parts of the world, let's put them on the table. We are suggesting that they bring together head men of the local communities, the people who are responsible in the park and the people who are responsible in the university to work out their own agreements.

Posey: Having an 'honest broker' or 'arbitrator' is becoming increasingly important in this debate. Someone or some institution that has no vested interests in either side, but is sensitive to the cultures and expectations of all parties, is important to facilitate dialogue and agreement. An ombudsman role may also be necessary, because, even if the rights and compensation are guaranteed by law, knowledge of and finances for legal action are beyond the means of local communities.

King: While we talk about the extraordinarily beautiful example of the Kinabalu project you have described, in the same country are the plants and people of indigenous groups such as the Penan, whose rights and resources are being obliterated. I know that you're in a very delicate position, diplomatically, to conduct your research effectively and there are very positive effects of educating people about the value of this in peripheral areas. But we are being dishonest if we do not acknowledge that such situations are occurring at the same moment in the same country.

Martin: There are many things going on in Malaysia. Inside protected areas the policies have a lot of foresight and there is a lot of room to build productive relationships with local people. In areas where there's no protection for the forest and the indigenous groups, the situation is difficult. The role of NGOs in Malaysia is extremely delicate. The usual partnership of an NGO, a university, an indigenous group, and perhaps also a government department, just doesn't exist in Malaysia. We have to find a different path. We are developing one model which is working. We hope that this can be extended and be seen as a viable alternative for other people who are in a much more difficult situation.

Bohlin: You described a training course you gave. Have you considered giving a course on how to find the bioactive principle?

Martin: No. It was a very preliminary course, because we were waiting for word on the MacArthur grant and how this project would evolve in the next years. Now we have obtained funding for three years, Dr Chayan Picheansoonthon will be coming back in summer 1994 to set up a phytochemistry laboratory in Kinabalu Park. The course you suggest may be appropriate in a few more years. There is a very active group of pharmacognosists, phytochemists and ethnopharmacologists in Malaysia and surrounding countries. The evolution of the programme will be very much in their hands.

Cragg: The NCI agreement with Sarawak hasn't been finalised yet, but I'm sure it will incorporate inviting one of the people from the university to come and work in our labs on precisely that sort of project.

Bohlin: It is very important to train the local students in both phytochemistry and bioassays. This combination is most important for future independence.

Lozoya: This project sounds as though you are discovering ethnobotany 20 years later. You have a project in Mexico—my country. You are talking about the collection of medicinal plants in the same terms as was proposed 20 years ago. Are you aware of the work that has been done in the last 10 years?

You are performing some phytochemical testing in the field—what do you mean by that exactly?

Martin: As an introductory step, we are speaking with the people about the various things that scientists do with plants—they collect them, they make voucher specimens, they study how they are classified by local people, they do phytochemical analyses and they carry out economic assessments. How do you explain phytochemical analysis? One preliminary way is to show some of the steps that are taken in screening for alkaloids, screening for very simple compounds. The idea is not to introduce sophisticated research, but to give them a foothold in the techniques that are important in ethnobotany, which have rarely been brought into the community and explained to them.

Lozoya: So Mexicans, probably indigenous practitioners, are now working in the Sierra Norte for you, to detect alkaloids in the plants you are collecting?

Martin: No. Each of the projects has a different plan for supporting the cultural conditions and the particular interests of the people that are involved. We are doing nothing in terms of phytochemistry in Oaxaca.

Lozoya: What is the point of going to a community and saying, 'I want to teach you how to detect alkaloids in the plants surrounding your garden'?

Martin: This is one step in educating local people about ethnobotany and beginning to give them the tools to carry out their own research.

Lozoya: At what level of education? These people don't have primary education.

Martin: The Dusun people who live around Kinabalu Park have diverse levels of education. Although some do not have even a primary education, others are finishing doctorates in fields such as botany and hydrology. By building a phytochemistry laboratory in the Natural History Museum of Kinabalu Park, we are opening the door for members of the Dusun communities to get a first-hand look at how chemical analyses of their medicinal plants are carried out. Selected ethnobotanical promoters (who generally have a high school education), working alongside invited students and professors from Malaysia and other South-East Asian countries, will participate in the initial stages of extraction and screening, enabling them to learn more about ethnopharmacology. In the future, there may well be Dusun people who obtain doctorates in pharmacognosy. Whether or not this goal is achieved, people in the communities will be more aware of why scientists are interested in the chemical properties of plants traditionally used by the Dusun.

Cox: The thing that impresses me about Gary's presentation is that there have been deliberate steps to put as much control as possible of the project in the hands of the indigenous people. If the indigenous people are controlling the purse strings, as in your case, a lot of the problems we've discussed about cultural erosion and cultural imperialism will vanish. I congratulate you on developing a model project.

Lewis: Gary, you are going to start your new work in Bolivia fairly soon. In January 1994 the Rio Convention on Biodiversity will be enacted in most countries. This states: the National Government has rights totally of all genetic material regardless of the desires of indigenous peoples. What plans do you have to satisfy this law in so far as the Bolivian national government is concerned?

Martin: The situation in Bolivia is evolving very rapidly. There is a law planned on biodiversity conservation, which has some very interesting clauses about the rights and responsibilities of local people. There is a fairly good appreciation of the role of indigenous people in this overall scheme. We are going to have to wait to see what the actual law looks like in Bolivia, because that may open the door to certain approaches.

Meanwhile, we are holding an ethnobotany training workshop next week (December 1993) in the Beni region to bring together members of the *Gran Consejo Tsimane* (the Chimane indigenous organization), people from the Beni Biosphere Reserve and people from various universities. This will at least start a process of moving towards an agreement in which these questions are addressed and in which some form of compensation—return of information and control of resources—is contemplated. At this point, it's only a process that we can start and of course it's an internal process in which we are outside observers.

Iwu: I'm really happy that within WWF there is a new generation of people who are concerned about the human aspect of conservation. I have always lumped the World Bank and the WWF together; they are both anti-people, from two different perspectives. The World Bank is more concerned with money, the WWF is more concerned with plants. The two share the common denominator that they don't have any concern for the people who live in the conserved areas. The People and Plants Initiative is a good programme. But most of the WWF programmes fall into the category I call 'safari science'. Scientists come from developed countries for two weeks, bring in very good projects, set up infrastructure that cannot be sustained with no process for internalizing it, then they leave. WWF has been in Africa for 30 years, but there is no structure to allow us to continue their projects. In most cases, these are started with no sensitivity towards the dynamics of the situation on the ground.

I shall give two examples. In Sierra Leone, they created a sanctuary for gorillas. There is an historical animosity between the local tribe and the gorillas. All the money was wasted because as soon as the foreigners left, the people ransacked the sanctuary.

Another example is in East Africa. A tribe was completely wiped out because a forest reserve was created and they were forced to evacuate their traditional homelands.

So it is gratifying that there is this change in the approach of the WWF.

Martin: Change comes slowly but change is coming.

Albers-Schönberg: In all our concerns about conserving cultures and ecologies, what role do the immense migrations of people all over the world, from country to country, from overcrowded cities back to the countryside, have?

Martin: In some cases, such as in the Beni Biosphere Reserve, there is a group of people who have lived there for many centuries and there are recent immigrants. At times there is an exchange of information between these groups, so the ecological knowledge of each group is increased rather than decreased. They learn from each other and develop a corpus of information much bigger than any one group initially had. In other cases, the arrival of new people puts a lot of pressure on forest resources, causing widespread ecological devastation. It may also introduce an element of cultural change that decreases the breadth and depth of local knowledge. We are now trying to determine the dynamics of this process in places such as the Beni Biosphere Reserve.

McChesney: As indigenous populations have migrated, for whatever reason, have they carried with them important plants—medicinal plants or food plants?

Balée: In a number of the systems of indigenous knowledge of plants that we observe in the Amazon, for example, there are introduced plants. However, these are always domesticated plants, used typically as food plants. This is in indigenous societies that have had usually a considerable amount of contact with non-Indians. We don't find so many recently borrowed plants in indigenous societies that have had very little contact or are just beginning to come into contact with Western society, because these societies tend to be associated with a trekking lifestyle. The kinds of borrowings are things like sugar cane, lime trees and other citrus plants. We don't see medicinal plants having been borrowed in those societies. Usually, the uses of the plants were also borrowed from the contact society.

Posey: My experience is completely different from that. When you see any Indian travelling, wherever they go, they always take plants with them. Plants make some of the best presents.

We frequently haven't distinguished between migrant communities and refugee communities. We also know very little about historical patterns of migration. Many of these people in fact are going back and re-inhabiting areas where they had lived before. Some Kayapó groups, for example, are only now returning to sites that their elders or ancestors prepared by planting fruit trees and semi-domesticated species two generations ago. This is a very complicated issue that needs to be looked at in detail.

Prance: I agree. If you meet a Yanomani Indian on a trail in a remote place and you ask what they have in their backpack, it is nearly always germplasm— not necessarily medicinal plants, it may be food plants. I have studied the distribution of some of the economic plants, such as the papaya, which originated in the Amazon, and the peach palm (*Bactris gasipaes*), which originated in the Western Amazon. They were carried all over the place, even in pre-Columbian times; this process is certainly still going on amongst the Indians who travel over long distances. Darrell Posey has pointed out that the Kayapó area of trade is larger than Western Europe; this gives some idea of the possible scale of plant movement by people.

Balée: By borrowings, I mean borrowings from outside the general region within which the society exists. For example, although maize came into South America, probably, from Middle America, in practical terms today it is not a borrowing because it came in so long ago, well before post-Columbian times. I'm not denying that there are borrowings but within the many indigenous societies that are isolated (and are either foragers or trekkers), the majority of their domesicated plants (if they have any) are not borrowed but have been retained in their cultures for very long periods of time.

Iwu: In modern times, there have been significant movements of biogenetic material from Africa to the so-called New World. Two examples are the introduction of oil palm (*Olea guinensis*) to Malaysia and Indonesia from Africa, and the transfer of coffee from East Africa (Ethiopia) to Latin America.

Prance: There are also refugees—boat people often take germplasm with them when leaving their country for political reasons.

Anthropological issues in medical ethnobotany

B. Berlin and E. A. Berlin

Department of Anthropology, University of Georgia, Athens, GA 30602 USA

Abstract. While ethnobotany has emerged as an important discipline in the search for new drugs, this economic impetus should in no way distract from a more ethnobiological and equally critical goal—the codification and promotion of indigenous medical systems as a major factor in the conservation of biocultural diversity. *Codification* of indigenous medical systems requires a holistic view which entails (1) in-depth understanding of the recognized health conditions in the native system and how they might be described in terms of Western biomedicine; (2) comprehensive inventories of medicinal species employed in the native system, descriptions of their modes of preparation and administration and giving priority to those species most likely to merit pharmacological testing; and (3) identification of the pharmacological properties of these species with the goal of discovering how they might be effective in the treatment of the health conditions for which they are employed. *Promotion* of indigenous medical systems requires the development of local training programmes aimed at the active conservation and enhancement of traditional herbal medicinal therapies that have been shown to be pharmacologically effective in the treatment of symptoms of recognized health conditions. The establishment of such programmes is critical at a time when traditional medical systems are often disparaged as worthless by the national societies in which indigenous peoples live, as well as by younger members of the native populations themselves.

1994 Ethnobotany and the search for new drugs. Wiley, Chichester (Ciba Foundation Symposium 185) p 246–265

Over the last several years, it has become increasingly apparent, both to government funding agencies and to private pharmaceutical firms, that ethnobotanists can play a major role as research partners in the discovery of new pharmaceutical products. Within the USA alone, significant new financial support for ethnobotanically informed drug exploration has been forthcoming from the National Cancer Institute and most recently from a consortium of the National Institutes of Health, the National Science Foundation and the US Agency for International Development with the funding of five long-term International Cooperative Biodiversity Groups. In addition, a growing recognition of the importance of ethnobotany in drug discovery is seen in the private sector, the most notable example being the remarkable financial success story of Shaman Pharmaceuticals Inc..

As important as these ethnobotanical partnerships with government and business may be, it is unarguable that most of the new pharmaceuticals currently under development or likely to be produced as a result of ethnobotanical surveys will have little direct importance for traditional societies residing in the developing countries of the world. Furthermore, the present emphasis by funding agencies on the discovery of new drugs principally for the treatment of cancer and AIDS (acquired immune deficiency syndrome) has unfortunately led to a decreased concern for the scientific validation of traditional herbal therapies employed against the serious illnesses prevalent in developing countries such as diarrhoeal disorders, respiratory diseases, skin infections and other chronic tropical diseases. In the small-scale societies with which ethnobiologists normally work, remedies used to treat these common but often life-threatening conditions are clearly more important than a medication that might be useful in the treatment of leukaemia.

As an ethnobiologist and a medical anthropologist with strong interests in medical ethnobotany, we are concerned that the economically motivated search for new drugs may distract from the equally important goals of *codification* and *promotion* of traditional medical systems as a major aspect of the conservation of biocultural diversity and the improvement of human health in developing countries. Indigenous knowledge systems, especially those associated with the health of so-called indigenous fourth-world peoples, represent an irreplaceable human resource, the documentation and active enhancement of which are crucial for the long-term well being of small-scale societies everywhere.

The tasks facing ethnobiologists and their collaborators in first describing and then moving to conserve indigenous medical systems are enormous but not insurmountable. This paper outlines some of the ways that our project on medical ethnobiology among the Highland Maya of Chiapas, Mexico, has striven to address the questions of codification and promotion of ethnomedical knowledge. While recognizing that the local conditions facing traditional peoples will vary, we hope that some of the general features of our work will be widely applicable.

Since 1987, a multidisciplinary, US–Mexican research project has been underway that deals with the medical ethnobotany, ethnomedicine and ethnopharmacology of the Tzotzil and Tzeltal, two Maya-speaking peoples who reside in the Highlands of southern Mexico. This project, referred to locally as the Programa de Colaboración Sobre Medicina Indígena Tradicional y Herbolaria (PROCOMITH), is being carried out as part of a collaborative research agreement between the University of California at Berkeley and a number of Mexican institutions.

Our work has aimed to be fully regional in scope, including comprehensive ethnobotanical and ethnomedical surveys of 14 central Tzeltal and Tzotzil municipalities that constitute the Central Maya Highlands, an area containing more than 300 000 people. From the beginning, our collaborative research has been motivated by three major sets of questions that lead to a basic understanding of the Highland Maya ethnomedical system:

(1) *Ethnomedicine*: What are the recognized health conditions in the native system and how can they be understood in terms of Western biomedicine?

(2) *Medical ethnobotany*: What species of plants are recognized as medicinal by the Maya, how are they prepared and administered and which of these species are the most likely candidates for pharmacological testing?

(3) *Ethnopharmacology*: What are the pharmacological properties of the ethnobotanically most significant medicinal plants and how might these properties be effective in the treatment of the health conditions for which these species are employed?

As our research progressed and as the general outlines of the ethnomedical system began to emerge, we moved to develop educational programmes aimed at promoting those herbal remedies that appear to be effective in the treatment of symptoms associated with the most common health conditions suffered by the Highland Maya. These programmes include:

(1) Native language publications that describe the results of our research in a series of multilingual, illustrated ethnomedical handbooks or manuals in a form readily available to and easily understood by the local populations of the area. There is also extensive use of computers and CD-ROM materials that can be employed as educational tools and to demonstrate the benefits of Highland Maya herbal remedies.

(2) The establishment of a permanent ethnobotanical herbarium of the medicinal flora of Highland Chiapas, organized and maintained in such a way as to advance the use of traditional herbal medicine throughout the Chiapas Highlands.

(3) The development of permanent ethnobotanical gardens of medicinal plants to be used for training, education and experimentation in the propagation of medicinal plants and the preparation and administration of Highland Maya herbal remedies that have been shown experimentally to be pharmacologically beneficial.

Highland Maya ethnomedicine

In the traditional medicine of the Highland Maya, as in many other ethnomedical systems, the maintenance or re-establishment of a state of health is dependent on events and interactions in two separate realities, the natural, usually visible, reality that follows predictable physical norms, and another, frequently invisible, reality that relates to extranatural phenomena. We adopt Foster's division of medical systems into *naturalistic* and *personalistic* to characterize these two cognitive frameworks (Foster & Anderson 1978).

A health condition in the naturalistic system is empirically determined and is based primarily on immediately apparent signs and symptoms. For example, the naturalistic gastrointestinal condition *ch'ich' tza'nel* 'bloody diarrhoea' (lit. 'blood' + 'faeces') is unambiguously recognized by the presence of blood in the stool and severe pain in the lower abdomen. For a naturalistic condition such as bloody diarrhoea, it is the norm that one treats oneself with medicinal plants or, lacking this knowledge, one consults individuals who are themselves knowledgeable about medicinal plants.

In contrast, diagnosis of the personalistic condition *jme'tik jtatik* (lit. 'our ancestral mothers and fathers') is based on 'retrospective presumption of aetiologic agent', such as an inadvertent encounter with ancestral spirits. In this case, which might demonstrate gastrointestinal symptoms that are unresponsive to herbal remedies, diagnosis and treatment frequently involve the intervention of healers with special powers. While personalistic conditions may at times also be treated with herbal medications, Maya curers normally employ remedies that require ceremonial healing rituals, special prayers and shamanic divination.

Almost all earlier work dealing with Highland Maya ethnomedicine has focused on its personalistic bases. The general outlines of this aspect of Maya ethnomedicine are well known (for some of the more important examples, see Guiteras-Holmes 1961, Holland 1963, Metzger & Williams 1963, Holland & Tharp 1964, Silver 1966a,b, Vogt 1966, 1970, 1976, Fabrega 1970, Fabrega & Silver 1970, 1973, Fabrega et al 1970, Harman 1974, Foster & Anderson 1978). This preoccupation with the cosmological aspects of Maya healing has tended to shift emphasis away from the wealth of medical ethnobiological knowledge possessed by the Tzeltal and Tzotzil. Consequently, it is generally concluded that Highland Maya medicine incorporates a poor understanding of human anatomy, has but a weak relationship to physiological processes, and that healing primarily satisfies psychosocial needs through magical principles (Arias 1991, p 40–43; Fabrega & Silver 1973, p 86, 211, 218; Holland 1963, p 155, 170–171; Vogt 1969, p 611). Because magical principles have little to do with science, a reading of these works supports a view that Maya ethnomedicine is anything but scientific and that the Maya themselves lack a scientific understanding of health and disease.

Our findings on the naturalistic aspects of Maya ethnomedicine lead to quite the opposite conclusions. The extensive medical ethnobiological materials we have collected demonstrate that the Highland Maya have a remarkably complex ethnomedical understanding of the anatomy, physiology and symptomatology of particular health conditions. In addition, they have identified a wide range of medicinal plant species that treat the symptoms associated with these conditions. Our findings show that almost all of the medicinal plant species that make up the Highland Maya pharmacopoeia specifically target individual health conditions. In this sense, Highland Maya traditional medicine is an ethnoscientific knowledge system based on astute and accurate observation that

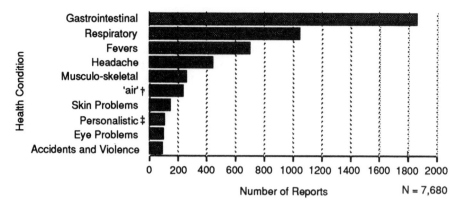

FIG. 1. Ethnoepidemiological reports of the health problems of the Highland Maya grouped by major ethnomedical category.
†'Air' is roughly equivalent to the Spanish concept of '*aire*'; it usually refers to a condition where cold air is said to enter the body and cause localized pain and discomfort.
‡Personalistic is a general category of conditions that derive from extranatural phenomena and sensate beings, such as souls, deities, demons, ancestors and sorcerers.

could only have been elaborated over many years of explicit empirical experimentation with the effects of herbal remedies on bodily function.

The Highland Maya recognize approximately 250 individual health conditions that are grouped into around 15 major illness classes. Several of these classes account for the large majority of health conditions deemed by the Maya to be serious illnesses, as shown by the results of our extensive epidemiological surveys throughout the Highland region. Gastrointestinal and respiratory conditions are by far the most prevalent, as can be seen from the frequency of reports of primary health problems in the region (Fig. 1). These data are drawn from retrospective epidemiological surveys conducted in 14 Maya municipalities from 1989–1992. Because these are retrospective self-reports of health problems, these data are indicators of the cultural saliency as well as the perceived frequency of illness.

Highland Maya ethnobotany

More than 7000 botanical collections have been made as part of a general ethnobotanical collecting programme, resulting in an inventory of 1600 species of purported medicinal value. This represents approximately one third of the total flora of Highland Chiapas (Breedlove 1981). This high proportion of medicinal plants in relation to the total flora of the state demonstrates that the Maya continue to employ a broad range of herbal medications in spite of almost 500 years of Hispanic cultural influence.

TABLE 1   Regression analysis of the 10 most significant plant families in the medicinal flora of Highland Chiapas

Botanical family	Total species	No. used	No. predicted	Residual
Asteraceae[a]	602	264	139	124
Lamiaceae[a]	129	59	30	29
Solanaceae	164	60	38	22
Rosaceae[a]	71	34	17	17
Apiaceae	48	25	11	14
Iridaceae	24	18	6	12
Araliaceae	18	15	4	11
Boraginaceae	51	21	12	9
Rhamnaceae	21	14	5	9
Onagraceae	31	16	7	9

[a]The three most important families of medicinal plants for North American Indians according to Moerman (1986).
From Berlin et al (1994).

Standard regression analysis allows one to recognize the most significant families of medicinal plants in the Chiapas Highlands (Table 1). Three of the top-ranked families (Asteraceae, Lamiaceae and Rosaceae) are also the three most medicinally important families in the ethnopharmacopoeias of North American Indians (Moerman 1986). This finding indicates continent-wide patterns of medicinal plant selection by aboriginal peoples of the New World. It underscores the importance of ethnobotanical studies in the search for pharmacologically active phytochemicals.

The medicinal uses the Maya attribute to these botanical families vary significantly. Species of the sunflower family (Asteraceae) are selected predominantly as sources of herbal medications for gastrointestinal problems, especially diarrhoeas and abdominal pains. The Lamiaceae and Rosaceae are favoured as sources of herbal remedies for the treatment of respiratory problems. Species of the Solanaceae are widely sought as medications for treating major skin infections, infections of the eyes, mouth and throat, and, in some cases, gastrointestinal problems. Species of the Apiaceae are preferred for remedies for fevers.

The distribution of medicinal species reflects the great ecological diversity of the central plateau of the Chiapas Highlands. Numerous species occur as restricted endemics; many species of the same genus exhibit allopatric distributions. Others are widespread, not only in their distribution but also in their patterns of use in traditional healing. These patterns have formed the basis of the selection of species for detailed pharmacological and phytochemical assays.

While almost all medicinal plant species are employed for several health conditions, our research indicates clearly that the Maya select medicinal plant species that are often condition specific, as can be ascertained from the high

TABLE 2 Distribution of medical use derived from reports by the Tzeltal Maya for the principal species employed in the treatment of major gastrointestinal diseases

Botanical species	Diarrhoea	Bloody diarrhoea	Mucoid diarrhoea	Abdominal pain	Epigastric pain	Abdominal distension	'Mother of man'	Worms	Other uses
Verbena carolina	33	8	6	11	3	—	1	—	5
Verbena litoralis	29	8	10	17	6	—	—	—	7
Psidium guineense	26	15	9	2	2	—	—	—	12
Baccharis serraefolia	25	—	9	3	—	—	—	3	12
Baccharis trinervis	23	3	4	8	1	—	—	—	7
Byrsonima crassifolia	22	9	5	2	—	—	—	—	8
Tagetes filifolia	20	1	—	4	—	—	—	—	27
Crataegus pubescens	7	30	8	—	1	—	—	—	9
Calliandra houstoniana	2	26	8	1	1	—	—	—	19
Baccharis vaccinioides	15	21	4	4	4	—	—	—	22
Acacia angustissima	3	20	5	1	1	—	—	—	13
Calliandra grandiflora	5	17	8	—	—	—	—	—	22
Rubus coriifolus	4	16	6	—	—	1	—	—	22
Cissampelos pareira	10	12	13	8	3	—	—	—	12
Lepidium virginicum	7	2	14	11	10	—	4	—	6
Nicotiana tabacum	—	—	—	25	3	11	1	—	34
Lantana camara	13	8	5	16	6	—	—	—	20
Ocimum selloi	3	—	—	15	4	12	—	—	11
Tagetes lucida	3	1	—	14	13	1	7	—	13
Ageratina ligustrina	10	1	2	13	11	—	—	—	13
Perymenium ghiesbreghtii	3	2	1	13	6	—	1	—	15
Vernonia leiocarpa	6	6	2	13	6	1	—	—	10
Bidens squarrosa	5	5	2	12	—	—	1	6	17
Chromolaena collina	1	8	2	12	10	—	1	—	11
Lobelia laxiflora	—	2	—	2	1	7	16	—	13
Fuchsia splendens	2	—	—	1	1	—	14	—	24
Chenopodium ambrosioides	4	1	—	5	4	—	—	42	8
Helianthemum glomeratum	7	—	—	1	1	—	—	24	13

Contingency table analysis: total Chi-square: 1273.28, $P = 0.001$. 'Mother of man' probably represents cholecystitis (gall bladder disease).

levels of agreement or consensus among our collaborators on the particular medicinal uses of individual species or sets of species. If we take the gastrointestinal conditions as an example, our preliminary findings strongly suggest that the use of a limited set of species is restricted to particular types of gastrointestinal diseases.

The Tzeltal recognize 28 principal species in the treatment of the major gastrointestinal diseases (Table 2). The Tzotzil recognize 26 principal species in the treatment of such diseases, as determined from data from the Tzotzil Traveling Herbarium (Table 3).

Our findings show that three major genera comprising six species are favoured by the Highland Maya in the treatment of unspecified or, most generally, watery diarrhoea. These species are *Baccharis serraefolia* and *Baccharis trinervis*, *Ageratina ligustrina* and *Ageratina pringlei*, and the two most common species of *Verbena* in the area, *Verbena carolina* and *Verbena litoralis*.

A second series of species is strongly favoured in the treatment of bloody and mucoid diarrhoeas, namely, *Psidium* spp., especially *Psidium guineense*, *Crataegus pubescens*, *Calliandra houstoniana*, *Acacia angustissima*, *Sonchus oleraceus*, *Rubus coriifolius*, *Cissampelos pareira* and *Lepidium virginicum*.

The favoured species for epigastric pain are *Tagetes lucida* (by both groups) and *Foeniculum vulgare* (by the Tzotzil). Abdominal pain is almost universally treated with admixtures of tobacco and garlic among the Tzotzil and with *Stevia ovata* by the Tzeltal.

Finally, in the treatment of worms, the universal herbal remedy of choice is *Chenopodium ambrosioides,* as is the case in most of Latin America, although the Tzeltal also commonly use *Helianthemum glomeratum* for the same purpose.

Highland Maya ethnopharmacology

We mentioned at the outset that some 1650 species are recognized by the Highland Maya as having medicinal value. It is unlikely that each of these species will show important biological activity and we have given priority to species for pharmacological analysis on the basis of widespread informant consensus as to the use of a particular species in the treatment of particular conditions.

It is now widely recognized by ethnobotanists that the use of indigenous knowledge can be critical in deciding which species of a local flora may show important pharmacological activity (Balick 1990, Cox 1989, 1990, Elisabetsky 1986, Farnsworth 1984, Heinrich et al 1992, Svendensen & Scheffer 1982). However, the use of intragroup patterning of ethnobotanical knowledge as a guide to biologically active plants, following the proposals of Trotter & Logan (1986), has not been widely adopted. It is possible that one of the major reasons patterns of group consensus have not been generally employed ethnographically is the still common bias among ethnobotanists that detailed information on the medicinal properties of plants is, in the main, secret or specialized knowledge,

TABLE 3 Distribution of medical use derived from reports by the Tzotzil Maya for the principal species employed in the treatment of major gastrointestinal diseases

Botanical species	Diarrhoea	Bloody diarrhoea	Mucoid diarrhoea	Abdominal pain	Epigastric pain	Abdominal distension	'Mother of man'	Worms	Other uses
Baccharis trinervis	54	9	8	3	4	2	1	2	12
Ageratina pringlei	48	4	11	13	3	2	1	—	10
Baccharis serraefolia	47	6	5	8	1	4	1	1	14
Ageratina ligustrina	45	4	6	14	4	—	7	—	13
Lantana camara	43	24	15	12	2	—	2	1	20
Verbena carolina	35	25	5	6	7	—	3	—	30
Lepechinia schiedeana	30	5	1	3	—	3	—	—	24
Helianthemum glomeratum	29	2	4	6	3	—	—	—	18
Lantana hispida	27	10	11	9	2	—	3	—	25
Borreria laevis	27	9	5	9	4	2	2	—	19
Psidium guineense	26	38	18	1	6	—	2	—	31
Sonchus oleraceus	9	38	1	3	4	—	1	2	27
Crataegus pubescens	5	37	1	4	3	—	—	—	30
Byrsonima crassifolia	12	31	5	4	5	2	2	—	29
Calliandra houstoniana	21	26	—	1	7	1	—	—	28
Calliandra grandiflora	25	23	2	1	6	1	—	1	24
Verbena litoralis	22	23	4	5	16	1	4	1	51
Nicotiana tabacum	6	—	—	41	7	27	2	1	41
Stevia ovata	9	1	2	38	1	—	6	—	8
Baccharis vaccinioides	25	1	2	35	9	—	2	—	41
Tithonia diversifolia	15	—	—	22	—	—	8	—	16
Smallanthus maculatus	8	2	—	21	7	1	9	—	12
Foeniculum vulgare	—	—	—	3	137	—	—	—	31
Tagetes lucida	12	3	1	8	34	7	—	1	14
Dahlia imperialis	2	—	—	—	1	—	13	—	34
Chenopodium ambrosioides	2	—	—	—	1	—	—	58	44

Contingency table analysis: total Chi-square: 2980.31, $P = 0.0001$. 'Mother of man' probably represents cholecystitis (gall bladder disease).

known only to those individuals who hold the specialized roles of curers or shamans. As a consequence, ethnobotanists have tended to work with a few, select individuals and broad-ranging surveys aimed at tapping commonly held knowledge of large numbers of persons of the community have not been general practice.

While it is, of course, possible that 'Doña María' or 'Don Trino' has discovered particularly effective medicinal plant species that control the symptoms of diarrhoea, respiratory conditions or skin infections and that they have kept this knowledge secret from everyone else, we find such a possibility unlikely. Consequently, our collecting efforts among the Highland Maya have been motivated by the hypothesis, first stated by Trotter & Logan, that 'The greater the degree of consensus [among native collaborators] regarding the use of a plant based therapy, the greater the likelihood that the remedy in question [will be] physiologically active or effective' (Trotter & Logan 1986, p 95).

The species presented in Tables 2 and 3 are those for which a high level of consensus is found among the Maya for the treatment of gastrointestinal conditions. These are among 125 species from which 500 pharmacological samples have been evaluated experimentally for their antimicrobial, phototoxic, cytotoxic and spasmolytic activities by our collaborators at the Center for Investigation of Medicinal Plants at Xochitepec, Morelos, Mexico. Remarkably, 73% of the plant material examined shows moderate-to-high antimicrobial activity, sufficient to justify further studies on their pharmacological efficacy (see Table 4). Furthermore, approximately 47% exhibit medium-to-strong antispasmodic properties that would be especially important for the relief of the symptoms of gastrointestinal disorders. Three of these species are reported as having sedative or anticonvulsive properties or effects on the central nervous system that would affect bleeding through haemagglutination and haemostatic activity. Even at this early stage in the work, it is apparent to us that these species should be marked as worthy of promotion for continued use by the Maya populations as part of their traditional pharmacopoeia. Similar patterns have been reported by Michael Heinrich and his colleages in the ethnopharmacopoeia of the Lowland Mixe in the neighbouring Mexican state of Oaxaca (Heinrich et al 1992).

Efforts at promoting traditional herbal medicine among the Highland Maya

As work on the codification of Highland Maya ethnomedicine proceeds, our research team has been actively exploring ways to make the results of our work available to the Maya people themselves in a form that would both conserve and promote cultural knowledge of herbal therapies. As part of these efforts, we have recently published the first of a series of manuals on Highland Maya medicine in three dialects of the Tzeltal and Tzotzil language, as well as in Spanish (Berlin et al 1990). These native language manuals have been made

TABLE 4 Results of pharmacological screening of principal species used in the treatment of gastrointestinal conditions by the Highland Maya

Species	Family	S. aureus	E. coli	Candida albicans	Spasmolytic	UVA phototoxicity	KB	P388
Ageratina ligustrina	Asteraceae	+	−	−	+	+	+	+
Baccharis serraefolia	Asteraceae	+	−	−	+	+	−	+
Baccharis trinervis	Asteraceae	−	−	−	+	−	−	+
Baccharis vaccinioides	Asteraceae	+	−	−	+	−	−	−
Bidens squarrosa	Asteraceae	+	−	−	−	−	−	+
Smallanthus maculatus	Asteraceae	+	−	−	+	−	−	−
Sonchus oleraceus	Asteraceae	−	?	−	+	−	−	+
Stevia ovata	Asteraceae	?	?	−	+	−	−	−
Stevia serrata	Asteraceae	+	?	−	+	−	−	+
Tagetes erecta	Asteraceae	?	?	+	+	+	−	−
Tagetes lucida	Asteraceae	+	+	−	+	+	+	+
Tithonia longistrata	Asteraceae	+	+	−	+	+	+	+
Lobelia laxiflora	Campanulaceae	−	−	−	−	−	−	−
Chenopodium ambrosioides	Chenopodiaceae	+	+	+	+	−	−	+
Helianthemum glomeratum	Cistaceae	+	+	+	+	+	−	−
Acacia angustissima	Fabaceae	+	+	−	+	+	−	−
Calliandra grandiflora	Fabaceae	+	+	−	−	+	−	−
Ocimum selloi	Labiatae	+	−	−	+	−	−	+
Allium sativum	Lilliaceae	−	−	+	+	−	−	−
Cissampelos pareira	Menispermaceae	+	+	+	+	−	−	+
Byrsonima crassifolia	Malpighiaceae	+	+	−	+	−	−	+
Psidium guajava	Myrtaceae	+	+	−	+	−	−	+
Psidium guineense	Myrtaceae	+	+	+	−	+	−	+
Crataegus pubescens	Rosaceae	+	+	−	−	−	−	−
Rubus coriifolius	Rosaceae	+	+	+	−	+	−	+
Borreria laevis	Rubiaceae	+	−	−	−	−	−	+
Nicotiana tabacum	Solanaceae	−	−	−	−	+	+	−
Foeniculum vulgare	Umbelliferae	−	−	−	−	−	+	+
Lantana camara	Verbenaceae	+	−	−	+	+	+	+
Lantana hispida	Verbenaceae	+	−	−	+	−	+	+
Verbena carolina	Verbenaceae	+	−	−	−	+	−	+
Verbena litoralis	Verbenaceae	+	−	−	−	+	−	+

E. coli, Escherichia coli; KB, human nasopharyngeal cell line; S. aureus, Staphylococcus aureus; P388, mouse leukaemia cell line.

widely available throughout the Highland region and have been warmly welcomed by the Mayan people. In addition, with the strong support of the Chiapas state government, arrangements have been made for the publication of the first two volumes of the *Enciclopedia Etnomédica Maya: bases científicas de la medicina tradicional en los Altos de Chiapas, Mexico*, a comprehensive comparative atlas on herbal medicine. The major body of the work is in Spanish; there is also ample textual information in Tzeltal and Tzotzil. Each medicinal plant species discussed is illustrated by a technical drawing of high quality produced by one of our Mayan botanical illustrators. Copies of the encyclopaedia will be placed in the govermental centres of each Maya municipality, as well as in rural health clinics throughout the region.

In addition to the publication of research results in the native languages of our indigenous collaborators, we have formed a medicinal plants herbarium (Herbario Etnobotánico de Chiapas), which currently houses more than 10 000 mounted collections of the medicinal flora of Chiapas. This research and teaching institution was formed as a result of agreements between our research organization and the Chiapas state Institute of Natural History. The Herbarium has recently added a multimedia laboratory for the production of interactive, CD-ROM programs that can be used by both literate and non-literate Maya speakers, allowing them to study and compare the preparation and use of the major medicinal plant species documented in our work thus far.

As an outgrowth of the herbarium project, we have begun to develop a series of ethnobotanical gardens that will be located at several Community Development Centers throughout the state and will promote *ex situ* and *in situ* conservation practices. At present, these centres provide public educational programmes for groups of Maya school children and other visitors. In the future, our trained Maya collaborators will offer workshops in the community centres on the preparation and administration of herbal remedies and on the horticulture of medicinal plants that have been shown experimentally to be pharmacologically beneficial.

Each of these efforts—a native language publication programme, a medicinal plants herbarium with multimedia capabilities and a series of educational/experimental medicinal plants gardens—emphasizes the scientific basis of traditional herbal knowledge. They aim to present the codification of the Highland Maya ethnomedical system in such a way as to promote more enlightened attitudes about the Mayan people among the local Mexican populations of Chiapas. As importantly, these efforts at promotion are aimed at educating younger Maya, many of whom have incomplete and partial knowledge of herbal medicine. It is our hope that our work will lead these young people to the recognition that the traditional medical knowledge of their forebears is an invaluable and irreplaceable cultural heritage that should be mastered and transmitted to future generations as an integral part of Maya society and culture.

Acknowledgements

The work reported here describes research carried out with numerous colleagues in pharmacology, botany and linguistics. We wish to acknowledge specifically the important contributions of Xavier Lozoya, Mariana Meckes, María Luisa Villarreal and Jaime Tortoriello (pharmacology), Dennis E. Breedlove and Guadalupe Rodríguez (botany) and Luisa Maffi and Robert Laughlin (linguistics). Our principal Maya collaborators, who have assisted us at all stages of the research on traditional Maya medicine, include Nicolás Hernández Ruíz, Domingo Sánchez Hernández, Carmelino Sántis Ruíz, Feliciano Gómez Sántis, Antonio Ruíz Hernández, Magdalena Girón López, Alonso Méndez Girón, Juliana López Pérez, Martín Gómez López, Esteban Sántiz Cruz, Marcos Pérez Pérez, Eleuterio Pérez López, Lorenzo González González and Bartolo Hidalgo Vázquez. We are grateful for the generous financial support provided our work by the National Science Foundation, the US Agency for International Development, the Chiapas State Government, the Centro de Investigaciones Ecológical del Sudeste and the University of California Consortium on Mexico and the United States. From mid-1994, work will continue as part of the ecological, ethnobiological and environmental anthropology programme of the University of Georgia, Athens.

References

Arias J 1991 El mundo numinoso de los Mayas. Talleres Gráficos del Estado, Tuxtla Gutiérrez, Chiapas

Balick MJ 1990 Ethnobotany and the identification of therapeutic agents from the rainforest. In: Bioactive compounds from plants. Wiley, Chichester (Ciba Found Symp 154) p 22–39

Berlin B, Berlin EA, Breedlove DE, Duncan T, Jara V, Velasco T 1990 La herbolaria médica Tzeltal–Tzoltzil en los Altos de Chiapas. Gobierno del Estado de Chiapas: Tuxtla Gutiérrez, Chiapas, México

Berlin B, Berlin EA, Lozoya X, Meckes M, Villarreal ML, Tortoriello J 1994 The scientific basis of diagnosis and treatment of gastrointestinal diseases by the Tzeltal and Tzotzil of Chiapas, Mexico. In: Nader L (ed) The anthropology of science. American Association for the Advancement of Science, Washington, DC, p xx–xx, in press

Breedlove DE (ed) 1981 Introduction to the flora of Chiapas. California Academy of Sciences, San Francisco, CA, p 1–35

Cox PA 1989 Pharmacological activity of the Samoan ethnopharmacopoeia. Econ Bot 43:487–497

Cox PA 1990 Ethnopharmacology and the search for new drugs. In: Bioactive compounds from plants. Wiley, Chichester (Ciba Found Symp 154) p 40–55

Elisabetsky E 1986 New directions in ethnopharmacology. J Ethnobiol 6:121–128

Fabrega H 1970 Dynamics of medical practice in a folk community. Milbank Mem Fund Q 48:391–412

Fabrega H, Silver DB 1970 Some social and psychological properties of Zinacanteco shamans. Behav Sci 15:471–486

Fabrega H, Silver DB 1973 Illness and shamanistic curing in Zinacantán. Stanford University Press, Stanford, CA

Fabrega H, Metzger D, William GE 1970 Psychiatric implications of health and illness in a Maya indian group: a preliminary statement. Soc Sci & Med 3:609–626

Farnsworth NR 1984 How can the well be dry when it is filled with water? Econ Bot 38:4–13

Foster GM, Anderson BG 1978 Medical anthropology. Wiley, New York

Guiteras-Holmes C 1961 Perils of the soul. The Free Press of Glencoe, New York

Harman RC 1974 Cambios medicos y sociales en una comunidad Maya-Tzeltal. Instituto Nacional Indigenista, México, DF

Heinrich MH, Rimpler H, Barrera NA 1992 Indigenous phytotherapy of gastrointestinal disorders in a lowland Mixe community (Oaxaca, Mexico): ethnopharmacologic evaluation. J Ethnopharmacol 36:63–80

Holland WR 1963 Medicina Maya en Los Altos de Chiapas. Instituto Nacional Indigenista, México City

Holland WR, Tharp RG 1964 Highland Maya psychotherapy. Am Anthrop 66:41–66

Metzger D, Williams G 1963 Tenejapa medicine: the curer SW. J Anthropol 19:216–234

Moerman DE 1986 Medicinal plants of native America. University of Michigan Museum of Anthropology, Ann Arbor, MI (vols 1 & 2)

Silver DB 1966a Enfermedad y curación en Zinacantán: esquema provisional. In: Vogt E (ed) Los Zinacantecos: un pueblo Tzotzil de los altos de Chiapas. Instituto Nacional Indigenista, México, DF

Silver DB 1966b Zinacanteco shamanism. PhD thesis, Harvard University, Cambridge, MA, USA

Svendensen AB, Scheffer JJC 1982 Natural products in therapy: prospects, goals and means in modern reseach. Pharmacol Wkly Bull (Sci Ed) 4:93–103

Trotter RT, Logan MH 1986 Informant consensus: a new approach for identifying potentially effective medicinal plants. In: Etkin N (ed) Plants in indigenous medicine and diet. Bedford Publishing Company, New York, p 91–112

Vogt EZ (ed) 1966 Los Zinacantecos. Instituto Nacional Indigenista, México, DF

Vogt EZ 1969 Zinacantan: a Maya community in the highlands of Chiapas. Belknap Press of Harvard University Press, Cambridge, MA

Vogt EZ 1970 The Zinacantecos of Mexico. Holt, Rinehart & Winston, New York

Vogt EZ 1976 Tortillas for the gods. Harvard University Press, Cambridge, MA

DISCUSSION

Dagne: You made a distinction between different types of diarrhoea, abdominal pain and worms (Tables 2 and 3). Was that based on a diagnosis?

Berlin: These are not Western categories even though we have Western names for them. The category called 'mother of man', now shown to be gall bladder disease, wouldn't have been noticed if we weren't interested in discovering the correspondance between Western and Maya systems of disease classification.

The clumpings that we found are based on Maya notions of similarity. They are groupings of the most closely related conditions according to the Maya themselves. Confirmation that they do represent clusters of widely recognized folk conditions is seen by the association of the plant species used to treat them. This is independent confirmation that illness groupings form valid sets.

Martin: Your extensive set of plant collections and interviews with people from different communities opens the door to assessing cultural change. Among ethnobotanists, there is general concern about cultural loss and how systems of ecological knowledge are disappearing. Have you any way of quantifying, through the data that you have collected, the rate of loss? If you interviewed older informers in their 50s and 60s and people in their 20s and 30s, would you be able to quantify loss and understand how it affects knowledge about certain plants or certain disease conditions?

Berlin: We would certainly be able to quantify it. I don't think we can explain why it happens. We have a set of major species that everybody ought to know. We could simply work with people and find out what proportion of those plants they do know. This is one of the ways in which we initially screen our collaborators. We are interested in people who know about medicinal plants. If they can't pass a simple test on common medicinal plants, we don't use them as collaborators.

Martin: Is knowledge about certain plants being lost even while other species continue to be widely used? That is, do older people know and use certain plants that the younger generation ignores?

Berlin: No. It's a generalized phenomenon. If you're taken out of the context of walking down the trails and looking at these plants and knowing what they're used for, that whole piece of knowledge is something you never get. It's pretty discouraging when you find people who are 20 years old who can't identify a *Verbena litoralis*.

Posey: I'm willing to believe that there's this generalized knowledge. Would you say there are no differences in traditional mixtures or preferences for administration of herbal remedies?

Berlin: There are differences. Your mother makes an apple pie different than my mother makes an apple pie. But the generalized knowledge about what the constituents are is very widespread. You are not going to find somebody using *Baccharis vaccinioides* in combination with *Psidium guajava* for the treatment of some skin condition.

Think of it this way. How could you maintain a secret about the treatment of such common conditions? If you have serious diarrhoea, once you have found a plant that stops it, it is unlikely that nobody else is going to know about it.

Cox: Generally, there are two realms in any society. There is a generalist realm that everybody knows; for example in Polynesia they all know that you use *Mikania* leaves to stop bleeding. Then there is a specialist realm. Michael Balick refers to plants in the specialist realms sometimes as being powerful plants. Frequently, in the specialist realm there is not a lot of consensus because this is almost proprietary information of specialist healers. In Polynesia, this information is transmitted matrilineally. So if you are dealing with a specialist realm, do you believe that consensus is still a good indicator of pharmacological activity?

Berlin: The realms of the specialist are generally for the kinds of conditions that have not responded to herbal treatment. The treatment for personalistic conditions includes prayer, divination and other aspects of non-naturalist intervention. Those do vary a great deal.

Balick: I'm having trouble with the hypothesis that the consensus value equals the efficacy. There is a set of plants that everybody knows. We all agree that they work and they are useful. But in your scenario, when you have a condition that is not a common condition, why would you go to the healer? People go to the healer because they're going to a specialist to deal with a problem. They don't know what the plants are for that condition, so they go to the healer. Based on my experience with the people in Belize, there are two very different groups of plants.

I would like to test the hypothesis. One could look at the powerful plants that the healers employ, some of which are dangerous, versus the most common plants that everybody knows.

Berlin: Let's imagine testing it with these data. There are 1600 species. Many of these species have been collected only once; they are not endemics, they are not geographically restricted in any way. Doña María says that plant A is useful for the treatment of something, but nobody else in 7000 collections said that plant A was useful for that. Now if you had a dollar to bet on biological activity, would you put your money on the single species that Doña María had said was useful, or one that's been collected 50 times and everybody says is useful for the treatment of skin conditions?

Balick: I look at it a different way. We have sat in the house of 'Doña María' and watched her for seven years. Plant A is consistently used for this condition, perhaps something more serious than a skin condition, and people over time seem to be treated with it and feel fine. It's not asking someone, but just watching them over time.

Berlin: We watch too. What would you bet?

Balick: I agree that the plants that *everyone* knows are mostly going to be useful, but there is also this different group of plants that the specialists know. Some may be highly toxic, such as *Abrus precatorius*, which is used in Ayurvedic medicine in India. There are analogies in Western medicine—we don't go to a general practitioner with an oncological problem.

This has been framed in terms of diarrhoea for the most part; this falls within the area of general knowledge. What happens when one has leishmaniasis, which maybe 1% of the people do? If I asked around in the community about what to do, why would I get a better answer from a young boy or his father than from the person who has treated it for 60 years?

Berlin: You ask folks how to treat leishmaniasis. If you talk to 100 people and they give you the names of 100 different species, then this is *prima facie* evidence that nobody really knows anything about how to treat leishmaniasis.

Balick: I disagree.

Berlin: We can test the hypothesis, if we can get somebody to run the assays.

Elvin-Lewis: Do as I did in Ghana and you will get the answer. I had 100 species of plants used as chewing sticks to consider. We interviewed our study population of over 800 people to determine their preferences according to a variety of parameters associated with the nature of the plant and how they viewed it as efficacious. We compared these data with other parameters like their age, sex, profession and religion. We found that many factors influenced their choices but overall the most favoured species were the most efficacious. Ultimately, these proved to be bioreactive against odontoperopathogens.

Cox: Chewing sticks, by definition, belong to the generalist realm, not the specialist realm.

Elvin-Lewis: There are certain peoples who rely on generalist things; they don't have shamans, they don't have specialist healers. When we are in the Amazon, we occasionally find people who know more about plants than others, but usually it's general knowledge in the village, except for the divination, etc.

Rarely, we find somebody who has a 'magic potion'. The proof of the pudding is when the use of these magic potions spreads out of the region. In Africa, a lot of the information is proprietary. It is the bread and butter of the healer in a particular village and it's always very difficult to get information from these people, because it's rare and exotic. In other areas, for example Belize, there are individuals who are half-way between the Western world and their own indigenous system. They will rely upon somebody who personifies a medical practitioner. This is all related to the social idea of how healing takes place.

Our greatest joy was finding that after many years of working with the Jívaro villages the younger people were beginning to see the value of what we were doing.

Lozoya: I would like to support Brent Berlin's hypothesis. In Mexico, the use of medicinal plants is a common practice in the rural and semirural societies. In 1989, we performed a national ethnobotanical study. We talked with 25 000 healers. We collected 40 000–50 000 plants and saw that there are two levels of knowledge; one of the specialist or healer and the other of the common people, the cultural medical knowledge of every person of that rural area. The role of the healers is particularly important in the diagnosis and interpretation of the disease. The people go to the healer to find out what type of disease they have. Then comes the use of medicinal plants, in some cases not necessarily in collaboration with the healer. For example, in Chiapas, gastrointestinal disorders are very important. Everybody knows in general terms the most important plants used to treat these. Everybody knows that guava is very important for the treatment of certain types of diarrhoea.

Elisabetsky: Even common symptoms might be associated with a personalistic explanation. A person might go to a healer for treatment of these symptoms, even if the plant they are given is one that they might already know for use in diseases that do not have a personalistic interpretation.

Iwu: It is important to separate healing attributed to personalistic causes from healing based on naturalistic concepts. If the cures for naturalistic disorders are selected for biological evaluation, a totally different picture emerges. We have tested this hypothesis in the selection of plants that we are currently investigating at the Walter Reed Institute of Research. All the plants in our inventory were derived from the ethnomedical evaluation of plants used for the treatment of diseases that the local people can identify very clearly as being of naturalistic causation. There is a distinct agreement on clinical symptoms for a disease known to both the modern and traditional medical systems.

It is important to differentiate folklore from ethnomedical information. Anecdotal information has little room in science. It is our duty to tailor the bioassay to the stated activity without using restrictive, mechanism-based assays that will give a false correlation between the traditional medical use of a given remedy and the observed activity. In our studies on antiparasitic agents, a correlation of 70–80% was observed.

We have to differentiate the role of the healer. Healers don't arise spontaneously, they are trained and information is passed on. Information on the use of the powerful drugs is often not supposed to be given up at all. Some are really secret and may be sacred or necessary for the protection of the entire village.

An aspect of your presentation which interests me is the training programme. I am delighted to note that the younger generations are learning directly from their elders. This approach is a lot better than the 'shaman's apprentice' programme advocated by some ethnobotanists. In that approach, the fundamental concept is to use notes generated in ethnobotanical studies (compiled in a foreign medium) for the training of a new generation of shamans. The approach is based on the erroneous assumption that indigenous cultures do not have methods for passing on information between generations. Just because such transfer of information occurs in a format that does not fit into the Western mode of cognition and education, it does not necessarily mean that there is no apprenticeship. Forcing indigenous tradition to pass through the reductionist Western notebook could lead to serious distortion of the original culture; it might amount to blatant cultural imperialism. In effect, what is transmitted is no longer authentic to the host culture. Any externally motivated training programme should be limited to facilitating a wholly indigenous process of knowledge transfer. Once oral information has passed into the written system of the Western world, it loses so much. It has to be translated, recorded, then translated back into the native language and distortion invariably occurs.

Dagne: If an ethnobotanist goes to a community and makes an inventory of their medicinal plants, then divulges the information in a publication, what effect would that have on the herbalists of that community? Some healers make their living by dispensing medicinal plants.

Berlin: Since the knowledge about herbal remedies is public knowledge, the Maya have no problems about publishing it. They are delighted to see this. They are very anxious to find out which plants are thought to be useful in terms of pharmacological activity. It is almost impossible to imagine the private information associated with the personalistic realm being something that anyone would be nervous about being published. This information is also very idiosyncratic; each individual healer would have some special 'cure' that he or she knows.

Martin: Publishing knowledge derived from oral traditions is something that many of us are proposing as part of the benefit to the community—to bring back information in written form. I think it's a very good policy, but we have to exercise a considerable amount of caution.

We have talked about Ayurveda becoming a written tradition, moving into communities that had an oral tradition and displacing local knowledge. We know this happened in Europe with Dioscorides' *'De Materia Medica'*, which over a period of 1600 years moved into parts of Europe outside the Mediterranean region and displaced or at least stopped empirical study of medicinal plants being used in those regions of northern Europe. I think we are doing something similar by publishing information and bringing it back to communities. What impact will this have on the diversity of knowledge that exists today? Will a certain set of plants become the standardized plants just because they are the ones cited in written works?

The way to tackle this problem, particularly in situations of rapid culture loss, is to follow Brent Berlin's example: to make consensus surveys of many communities and identify the plants that are most widely used. It is best to go about our studies in a systematic way, always aware that we may be reducing the diversity of knowledge that exists. We must avoid publishing and distributing knowledge that we have gained from just one or two local people.

Lozoya: One of the most important contributions of the projects of Dr Berlin is the direct participation of the local people. In this project, over many years, we have seen a very important aspect, this is the Indian people learning to do some of the classifications by themselves.

This is the first time in the study of traditional medicine in Mexico that documents have been published in the native language. Historically, all the information on traditional medicine in Mexico has been controlled by the dominant culture—first the Spanish and then Mexican institutions. It is particularly welcome that for the first time we are printing information in a local Indian language.

Jain: I have a a general observation that relates to all the issues we have discussed today and a word of caution. Every one of us wants to see the indigenous people benefit by their knowledge. We have all emphasized that ethnomedical research needs to be done soon for various reasons. We have emphasized the role of non-governmental organizations (NGOs) in defining what is beneficial to the indigenous people, etc. Many NGOs are very sincere and very good. But many NGOs can politicize the situation and hinder genuine research, or bargain for personal gains. It is only after 20 or 30 years of sincere work, research and collaborative effort by ethnobotanists that contact and confidence are built up in many indigenous socieities of the world. We need to see that this movement of indigenous peoples' rights doesn't get out of hand.

We do not want to study only ethnomedicine; we also want to study linguistics, faith, culture, etc. There are associations of traditional healers; these can do very constructive work. But unless carefully guided or monitored this can lead to 'unionism' and get into dangerous extremes, which can block even genuine academic research.

Conclusions

Ghillean T. Prance

Royal Botanic Gardens, Kew, Richmond TW9 3AB, UK

Ethnobotany is a truly interdisciplinary science that has come of age through both the new emphasis on ethnomedicine and the pressures of the environmental crises to produce solutions for conservation.

Ethnobotany is dependent upon close collaboration with other disciplines. In the case of ethnomedicine, we have repeatedly heard here, e.g. from Elaine Elisabetsky and Norman Farnsworth, of the need for the involvement of physicians in the field work to assist with the diagnosis of symptoms treated by medicinal herbs. Farnsworth recommended that budding ethnobotanists take some medical courses. Other essential collaborators who have already been much more involved are, of course, chemists and ethnopharmacologists. Perhaps this collaboration could be stimulated through the creation of centres of excellence. Gary Martin's reminder that the search for new products is only a small part of ethnobotany should also be remembered.

Walter Lewis showed that ethnobotany is becoming much more a scientific discipline of its own because it has moved from basic recording to a more quantitative and experimental approach. Michael Balick showed us his adaptation of species–area curves of ecologists to the gathering of information about use of plants. Martin suggested adding applied ethnobotany to Lewis's list.

There is an obvious concern amongst us about the loss of culture and of information about plant medicines that is occurring because of pressure to Westernize and to become acculturated; for example, the various types of healers in Belize who have no apprentices. Our goal must be to encourage the continuity of cultural information whilst at the same time taking care that we are not creating living museums or zoos of indigenous peoples. Cultural preservation must allow the natural evolution of a society at the pace it wants to develop. The examples we have heard from Mexico, China and India show that medicinal plant culture can survive in modern literate societies. This also means that ethnomedical information should be collected in such a way that cultural disturbance is minimal.

It is now generally accepted that all bioprospecting should be carried out with the consent of the local people involved. We must be sure to see that this is always the case. There is a need further to develop guidelines for the distribution of compensation to local peoples and the nation states in which they are found. This must obviously be done with the advice of the local peoples themselves.

266

As often repeated, there has been the shortage of ethnobotanists and ethnomedical experts. Another challenge put to us is to improve and expand training. I would emphasize, as did Gary Martin, that training should not be confined to the ivory towers of our universities but rather concentrate also on training of local people, paraethnobotanists and the children who can continue to retain and perpetuate knowledge.

We have seen that there are many approaches to the search for new medicines. The hit rate seems very low in the approach used by large pharmaceutical companies or the National Cancer Institute. However, government sponsorship of the Institute or the funding of the Walter Reed Hospital allows them to investigate leads that are not attractive to commercial companies; thus we now have taxol on the market. Time will tell whether the newly developed approach of Shaman Pharmaceuticals Inc. will lead to a greater hit rate or not.

How are we going to respond to the challenge put to us by Dr Lozoya? Are the Western developed countries losing out because of their emphasis on pure compounds? Will the Japanese, Chinese and Koreans produce the medicines of the future because of their emphasis on the clinical trial approach that does not necessarily involve pure compounds?

We have had a clear message that not enough is being done to address the common diseases of the developing world, such as malaria and leishmaniasis. (I notice that no one has mentioned Chagas' disease, the plague of this region.) Can we be more sensitive to the great needs of the Third World and focus our efforts on diseases for which the medicines will not generate a huge profit for pharmaceutical companies and their investors, but rather will solve some of the great social problems of our world? We have a moral obligation to rise to this challenge and to redouble our efforts to improve the health of the local people with whom we work.

Related to the last point, we have heard several times that the reasons for the focus on pure molecules is often that they can be patented and profitable, hence an overemphasis on alkaloids. Several speakers have suggested that local peoples are using some simple, well known compounds, such as tannins, also mixtures of different substances that interact in such a way as to create a more powerful therapeutic effect.

Following on from this is the challenge put to us about the involvement of indigenous peoples' opinions. It is not just in Samoa that the local medical practitioners feel that healing is something spiritual for which you could not possibly charge. Here is an area that we need to take much further in the future. How about a Ciba Foundation Symposium with indigenous participants? Darrell Posey certainly feels there are many suitable people for such a meeting.

I was pleased to hear that most programmes, whether financed by the National Cancer Institute, Conservation International or pharmaceutical companies, are including some up-front funds in their projects for capacity building and for the direct benefit of the people who are providing ethnobotanical information.

It is important for us to follow the example of Shaman Pharmaceuticals both in providing for some of the immediate needs of the community because it will be a long wait for royalties and in a more general distribution of royalties so that all people who provide their information benefit.

We have seen the possible pitfalls if an ethnomedicine is developed and patented by one company or institute. Someone else can then synthesize and modify the molecule, often for the motive of improved effect of the drug but, at the same time, getting round the original patent based on ethnobotanical information. We obviously have an important role to monitor such cases and to try to persuade the companies involved to share their profits with the original source of information.

This leads on to another area we have not explored adequately: the role of conservation non-governmental organizations as arbitrators/advisers/brokers/ intermediaries in the issues of intellectual property rights. We are helping to define a new function for these organizations.

Thank you all for coming and I especially thank *all* the speakers for their collaboration with the format of a Ciba Symposium. I am most grateful to you for making my job so easy by ending your talks on time. I also thank you all for making the discussions so lively that we could have done with twice the time.

I hope this meeting has challenged us to work more closely together to pool our limited resources and number of people to further both the search for medicines and the rights and social needs of local peoples around the world.

We have heard the urgency of the task; let us go into the field and apply what we have learned here.

Index of contributors

Non-participating co-authors are indicated by asterisks. Entries in bold type indicate papers; other entries refer to discussion contributions.

Indexes compiled by Liza Weinkove

Albers-Schönberg, G., 145, 148, 193, 244
*Alencar, J. W., **95**

Balée, W., 41, 55, 56, 57, 58, 73, 113, 126, 165, 175, 207, 208, 244, 245
Balick, M. J., **4**, 18, 19, 21, 22, 23, 24, 140, 141, 144, 145, 168, 212, 225, 239, 240, 260, 261
Barton, J. H., 74, 103, 165, 191, 195, 196, **214**, 222, 223, 224, 225, 226, 227
Berlin, B., 141, 142, **246**, 259, 260, 261, 263
*Berlin, E. A., **246**
Bohlin, L., 36, 54, 93, 103, 173, 191, 242
*Boyd, M. R., **178**

*Cardellina, J. H., **178**
Cox, P. A., **25**, 37, 38, 39, 40, 41, 127, 147, 166, 168, 208, 211, 225, 243, 260, 262
Cragg, G. M., 52, 53, 59, 75, 102, 114, 128, 145, 165, 176, **178**, 190, 191, 192, 193, 195, 196, 242
Craveiro, A. A., 55, **95**, 102, 103, 104, 176, 222, 223

Dagne, E., 18, 40, 73, 140, 144, 150, 165, 195, 259, 263

Elisabetsky, E., 20, 38, 40, **77**, 91, 92, 93, 104, 143, 148, 168, 213, 224, 225, 262
Elvin-Lewis, M. P., 22, 56, 58, **60**, 74, 75, 102, 166, 176, 261, 262

Farnsworth, N. R., 19, 20, 37, 40, 41, 42, 52, 55, 56, 57, 74, 90, 128, 145, 165, 166, 168, 174, 192, 213

Iwu, M. M., 20, 24, 39, 40, 56, 93, 103, 104, 113, 114, **116**, 126, 127, 128, 145, 147, 151, 173, 194, 209, 210, 221, 222, 223, 224, 243, 245, 262

Jain, S. K., 23, 57, 72, 74, 112, 113, 149, **153**, 165, 166, 167, 168, 212, 264

King, S. R., 20, 37, 51, 54, 59, 75, 103, 112, 126, 190, 195, **197**, 206, 207, 208, 209, 210, 211, 225, 226, 240, 241

Lewis, W. H., 21, 38, 56, **60**, 72, 73, 74, 75, 126, 167, 174, 176, 191, 193, 206, 211, 243
Lozoya, X., 91, 92, 113, 128, **130**, 140, 141, 142, 143, 144, 146, 149, 174, 175, 207, 242, 262, 264

McChesney, J. D., 23, 52, 53, 104, 115, 126, 127, 129, 148, 165, 175, 176, 191, 194, 244
*McCloud, T. G., **178**
*Machado, M. I. L., **95**
Martin, G. J., 21, 39, 57, 73, 93, 103, 114, 115, 141, 142, 167, 208, 211, **228**, 239, 240, 241, 242, 243, 244, 259, 260, 264
Matos, F. J. A., **95**, 102

*Newman, D. J., **178**

de Oliveira, A. B.,   92

Peter, H. H.,   53, 113, 174, 190, 194
Posey, D.,   21, 37, 58, 77, 115, 164, 193,
    196, 206, 207, 222, 226, 241, 245, 260
Prance, G. T.,   1, 19, 57, 59, 72, 151,
    194, 208, 240, 245, 266

Rubin, S. M.,   195, 209, 210, 212

Schultes, R. E.,   106
Schwartsmann, G.,   37, 38, 54, 75, 92,
    147, 151
*Snader, K. M.,   178

*Tempesta, M. S.,   197

Xiao, P. G.,   169, 173, 174, 175, 176

Subject index

Acacia angustissima, 252, 253, 256
Acalypha arvensis, 5
Aconitum, 171, 172, 173–174
Aframomum melegueta, 124
Africa, 116–129, 244
  biodiversity prospecting, 123–125
  Bioresources Development and
    Conservation Programme,
    120–121
  Biotechnology Development Agency,
    120
  medicinal plants and drug development,
    119
  Salvage Ethnography Project, 119–120
  Shaman Pharmaceuticals, 122, 126
  traditional medicine, 118
  Tropical Diseases Chemotherapy Project,
    122–123, 126–127
Ageratina ligustrina, 252, 253, 254, 256
Ageratina pringlei, 253, 254
Ageratum conyzoides, 124
Agonandra racemosa, 13
agromophol, 172
Allium sativum (garlic), 253, 256
Amazon, 56, 59, 106–115, 140, 207–208
  ethnobotanical drug discovery, 108–110
  plants worthy of study, 110–111
  see also individual countries
9-aminocamptothecin, 182
Ammi majus, 161
analogues, 194, 218
Ancistrocladus abbreviatus, 186
Ancistrocladus korupensis, 185–186, 190
Ancistrocladus tectorius, 69–70
Anemia oblongifolia, 82
Annona, 82, 85–86
trans-annonene, 98–99
anti-HIV compounds, 48, 179, 188
  ethnobotanical approach, 5, 66,
    183–184, 200
  potential, 185–187
  see also michellamine B; prostratin

anticancer drugs, 54, 92–93, 179, 188
  discovered 1960-1982, 48–49, 180–183
  ethnobotanical approach, 5–6, 183–184
  toxicity, 70
  see also individual drugs
antifungal plants, 200–201, 202
antiinflammatory agents, 67, 74
antiviral compounds, 68, 70
  ethnopharmacological search, 5, 65–66,
    77–94
  interactions between, 69
  Shaman Pharmaceuticals and, 198,
    200–202
  see also anti-HIV compounds
ants, 56, 57, 58–59, 110–111
apprenticeships, 21–22, 263
Arachis, 82
Aralia, 171
Aristolochia, 19, 82, 84, 86, 171
Aristolochia trilobata, 13, 86
aristolochic acid, 19
Arrabidaea cf. cinnamomea, 84
Arrabidea chica, 62, 86
Arrabidea xanthophylla, 86
Artemisia, 93–94, 171
artemisinin, 94, 172
arthritis, 67, 75
Asclepias curassavica, 161
aspirin, 68
Ayurvedic medicine, 133, 148, 159, 160,
  165–168
Azadirachta indica, 124

Baccharis serraefolia, 252, 253, 254, 256
Baccharis trinervis, 252, 253, 254, 256
Baccharis vaccinioides, 252, 254, 256, 260
Bacopa monnieri, 161
Balanites aegyptica, 124
Balansia cyperi, 67
Banisteriopsis (formerly *Banisteria*) *caapi*,
  108–109
Banner, Horace, 79

bao-gong-teng A, 172
Barjonia, 84
Belize, 5–6, 9, 21, 141
  ethno-biomedical forest reserve, 14–17
  ethnobotany project, 9
  forest valuation studies, 12–14, 23–24
Belize Association of Traditional
  Healers, 16, 21
Berberis, 171
Bidens squarrosa, 252, 256
bioassays, 2, 43, 136–137
biodiversity conservation, *see* conservation
biodiversity prospecting, 123–125,
  194–196
  conservation organizations and, 236,
  240–241
Biomedical Research Center in Traditional
  Medicine and Natural Products, 133
Bioresources Development and Conservation
  Programme, 120–121
Biotechnology Development Agency, 120
biphenyl dimethoxy dicarboxylate, 172
Bleekeria vitensis, 183
Blepharodon, 84
Bolivia, 63–64, 108, 234–235, 243
Borreria, 85
Borreria laevis, 254, 256
Boswellia serrata, 161
botanical gardens, 162–163, 248
Brazil, 1, 58, 63–64, 78, 91, 108
  north-eastern, natural product chemistry,
  95–105
  patents system, 104, 222, 223
  Pilocarpus overexploitation, 3, 8
  studies of herbal remedies, 143
Bridelia ferruginea, 124
Bursera, 19
Bursera simaruba, 13, 14
Butyrospermum paradoxum, 124
Byrsonima, 86
Byrsonima aerugo, 84
Byrsonima crassifolia, 252, 254, 256

caatinga, 95, 97
Caesalpina crista, 161
caffeine-rich plants, 66, 108
Cajanus cajan, 124
calanolide A, 186, 187, 190, 192–193
Calliandra grandiflora, 252, 254, 256
Calliandra houstoniana, 252, 253, 254
Calophyllum, 187

Cameroon, 121, 190, 195
Camptotheca acuminata, 182–183
camptothecin, 49, 182–183
cancer
  concepts, 5–6, 38, 54, 76
  prophylaxis, 128
  see also anticancer drugs
candle-wood oil, 97
canelinha oil
  anethole type, 96
  estragole type, 97
Carica papaya, 124
Caryocar microcarpum, 110–111
trans-cascarillone, 98–99
Cassia, 84, 124
Catharanthus roseus, 179, 180–181
Centella asiatica, 161
Cephalotaxus harringtonia, 183
chalcone dimers, 99–100
changroline, 172
chemistry, plant, *see* phytochemistry
chemotaxonomy, 49
Chenopodium ambrosioides, 252, 254,
  256
Chenopodium anthelminticum, 46
chewing sticks, 261–262
Chimane, 235, 243
Chinese medicine, 93–94, 130–131, 133,
  169–177
  ethnopharmacological studies, 170–172
  medicinal plant use, 169–170, 171
  new drug development, 91, 92, 172–173
Chromolaena collina, 252
Cimicifuga, 172
Cissampelos pareira, 86, 252, 253, 256
Cissampelos tropaeolifolia, 84
Clematis, 171
Cleome guianensis, 82
clinical studies, 91–92, 137–138, 142–147,
  267
codification, traditional medicine, 247,
  248–255, 257
Codonopsis, 171
Cola nitida, 124
collaborative arrangements, 195–196,
  218–219
  biodiversity conservation, 230
  legal situation, 219–221
  National Cancer Institute, 187–188,
  196, 218
  Nigeria, 122–123, 126

Shaman Pharmaceuticals, 122, 126,
  203, 204–205, 218
  see also compensation
collections, plant, 195–196
  ethno-directed, see ethnobotanical drug
    discovery
  multiple use curves, 10–12
  random, 48–49, 179, 180, 191
  Shaman Pharmaceuticals, 199, 200
Colombia, 108, 109, 110
colonialism, new, 34
Commiphora mukul, 161
community, local, see local community
compensation, 192–196, 209–213,
  223–224, 266
  Africa, 123–125, 126
  indigenous/local peoples, 34–35,
    73–74, 122, 209–213, 236–237,
    267–268
  legal situation, 219–221
  National Cancer Institute, 34–35,
    187–188, 212
conocurvone, 186, 187, 190, 191
consent, 34–35, 37, 79
conservation, 228–245
  biodiversity, 35, 129, 203–204, 205
  Africa, 120–121
  ethnobotany/drug development links,
    4–24
  India, 161–163, 165
  urgency, 107–108, 115
  cultural, 1, 17, 203–204, 205
  ethnobotany and, 229–232
  future, 236–237
  organizations, 233, 236, 240–241, 268
Conservation International, 209–210, 211
Convention on Biodiversity, 203,
  215–216, 219, 243
Cordyceps sinensis, 173
Corydalis, 171
costatolide, 187, 192–193
Coutoubea ramosa, 82, 84, 86
CPT-11, 182
Crataegus pubescens, 252, 253, 254, 256
Croton aff. agraphilius, 82
Croton sonderianus, 55, 98–99
Croton zehntneri, 96, 97
crude plant extracts, see extracts, plant
cucurbitacins, 99
cultivation, 9, 74, 128–129
  Chinese medicinal plants, 176

Indian medicinal plants, 161–162, 165
cultural change, 231–232, 259–260, 266
cultural conservation, 1, 17, 203–204, 205
cultural handicap, 137
Curcuma longa, 161, 162
Cymbopogon citratus, 124
Cynanchum, 171
Cyperus, 67

Dahlia imperialis, 254
data, ethnobotanical, see information,
  ethnobotanical
databases, 9, 46–47, 134–135
  see also NAPRALERT database
Delphinium, 171
dementia, senile, 110
Dendrobium, 171
dental health, 62, 63, 166–167
dereplication, 5, 19–20, 66, 185
derived knowledge, 217–219, 220
Desmodium adscandens, 84
Desmoncus, 85
developing countries
  diseases, 31, 38–40, 267
  science funding, 90–91
diarrhoea/dysentery, 69–70, 150
  Kayapó treatment, 78, 81–87
  Maya treatments, 252, 253, 254
  plants used in India, 156
Dichroa febrifuga, 46
diet, 118, 127–128
digitalis, 26, 29, 147
Dioscorea, 13, 14, 140–141
Dioscorea composita, 140, 141
Dioscorea floribunda, 141
Dioscorea mexicana, 133, 134
diosgenin, 133, 140
Diplopterys cabrerana, 109
diseases
  categories, 30–31, 41, 259
  concepts, 2, 6, 38–39, 70, 75–76,
    127
  descriptions, 69–70, 199, 206
  developing countries, 31, 38–40, 267
  Highland Maya, 250
  India, major, 160
  introduced, 73, 79–80
diterpenes, 174–175
doctrine of signatures, 47
Dorstenia multiradiata, 124
Dracaena mannii, 124

drug development
  Africa, 119
  China, 91, 92, 172–173
  ethnobotanical approach, *see* ethno-
    botanical drug discovery
  ethnobotany/conservation links, 4–24
  ethnopharmacology and, 42–59
  methodological issues, 91–92, 136–138,
    142–151, 267
drugs
  administration methods, 36–37, 83
  derived, legal issues, 217–219, 220
  plant-derived, 25, 27, 40–41, 43
    categories, 149
    economic value, 6–7
Dusun, 234, 242–243
Dyplopterys pauciflora, 84

Eclipta alba, 100
economic incentives, drug development,
  32–33
economic values
  forest products, 12–14, 23–24, 231
  plant-derived drugs, 6–7
educational programmes, 248, 255–257
Eli Lilly and Company, 203
elliptinium/ellipticine, 182, 183
ergot alkaloids, 67
essential oils, 96–97, 176
ethics, 1–2, 123–125
  see also compensation
ethnobotanical drug discovery, 17, 25–41,
  45, 65–67, 191
  Amazon, 108–110
  analysis of indigenous plant uses,
    30–33
  anticancer drugs, 5–6, 183–184
  antiviral compounds, 5, 65–66, 77–94
  efficacy, 213, 267
  historical success, 26–28
  limitations, 45–46
  National Cancer Institute, 180, 184,
    188
  obligations to indigenous peoples, *see*
    compensation
  plant preferences and, 74
  recent success, 28–29
  Shaman Pharmaceuticals, 198–200
  strengths, 29–30
ethnobotanists, 18–19, 201
  training, 46, 111, 112–113, 114, 121, 267

ethnobotany, 266
  basic, 61–63, 73
  conservation and, 229–232
  definitions, 43, 106, 113–114, 153
  drug development/conservation links,
    4–24
  experimental, 65–70, 73
  medical
    anthropological issues, 246–265
    Highland Maya, 248, 250–253
  quantitative, 63–65, 73
  research phases of future, 60–76
ethnomedicine, 43, 266
  Highland Maya, 248–250
ethnopharmacology, 42–59, 107
  Chinese medicinal plants, 170–173
  Highland Maya, 248, 253–255
etoposide, 70, 179, 181
Eucalyptus globulus, 124
Eugenia, 86, 87
extraction industries, 203, 204
extracts, plant, 2, 20–21, 138
  antiviral activity, 202
  Chinese medicinal plants, 173, 174, 176
  preparation methods, 75, 83, 165–166
  screening, 136
  standardization, 147, 148, 167, 174
  testing for bioactivity, 184
  see also herbal remedies

Faramea, 85, 87
field trials, 144
flavones, 68, 85
Foeniculum vulgare, 253, 254, 256
Food and Drug Administration (FDA),
  143, 145–146
forests
  Amazon, 107
  conservation, 231
  economic value, 12–14, 23–24, 231
  ethno-biomedical reserve, 14–17
  quantitative ethnobotany, 63–65
formulations, drug, 54, 175
frightening syndrome, 38, 70, 76
Fritillaria, 171
Fuchsia splendens, 252

Ganoderma lucidum, 173, 174
Garcinia, 63, 124
gastrointestinal (GI) disorders, 31, 38, 40,
  250

Kayapó treatments, 78, 81–87
Maya treatments, 252, 253, 254, 256
General Agreement on Tariffs and Trade
(GATT), Uruguay Round, 215, 222
Gentiana, 171
Geophila gracilis, 85
Ghana, 63, 261–262
ginseng, 176–177
Gnetum nodiflorum, 110
guava leaf tea, 150, 262

hallucinogens, 108–109, 161
hardwickic acid, 98–99
harringtonine, 182, 183
Healing Forest Conservancy, 204, 212
health care, 122, 202, 235
Helianthemum glomeratum, 252, 254, 256
Heliconia psittacorum, 84, 86
Helosis, 82, 86
herbal industry, international, 6–7
herbal products, 149
herbal remedies, 132, 138, 149
    methodology of evaluation, 91–92,
        137–138, 142–151, 267
    preparation methods, 75, 83, 165–166
    standardization, 147, 148, 167, 174
    synergistic interactions, 69, 138
    see also extracts, plant; medicinal plants
herbaria, 9, 134–135, 140, 240, 248, 257
homoharringtonine, 182
Homolanthus acuminatus, 69–70
Homolanthus nutans, 28, 187
Houttuynia, 172
Human Genome Diversity Project, 196
human immunodeficiency virus (HIV), *see*
    anti-HIV compounds
huperzine A, 172
hydroxy-bisabolol, 98
Hypericum, 171

Igbo, 120
Ilex, 171
Ilex guayusa, 66, 72, 108
INBio, 7–8, 193, 195, 218
Index Kewensis, 47
India, 57–58, 74, 148, 153–168, 212, 222
    biological screening of plants, 161
    conservation, 161–163, 165
    ethnobotanical information, 154–156
    medicinal plants research, 156–160

indigenous peoples, 267
    analysis of plant uses, 30–33
    communication of information among,
        56–59
    compensation, *see* compensation,
        indigenous/local peoples
    definition, 225
    legal rights, 216–217, 220–222, 223,
        225–227, 264–265
    migrations, 22, 244–245
    see also Jívaro; Kayapó; local commu-
        nity; Maya; *other specific groups*
indirubin, 172
informants, 64, 78–79, 199, 259–260
information, ethnobotanical, 22–23
    additional applications, 46–47
    collection, 23, 61–64, 78–79, 199–200
    deficiencies, 5–6, 45–46, 55–56
    India, 154–156
    legal protection, 215
    Mexico, 133–136, 141–142
    precautions in interpretation, 47–48
    primary, 61, 63
    return to local community, 19, 21,
        208–209, 234–235, 248, 255–257,
        263–264
    secondary, 61
    see also knowledge, local ethnobotanical
Institute for the Study of Medicinal Plants
    (IMEPLAM), 133, 134
intellectual credit, 210, 211
intellectual property rights, 34, 123–125,
    210–211, 214–227
    current legal situation, 219–221
    derived pharmaceutical products,
        217–219
    National Cancer Institute and, 187–188
    source material and knowledge, 214–217
International Labour Organization (ILO),
    216, 225
inventories, ethnobotanical, 9, 121, 234,
    263
Iris, 161

Jacaranda, 62
Jívaro, 22, 56–57, 58, 61, 262
    compensation, 73–74
    frightening syndrome, 38, 70, 76
    medicinal plants, 66–67
    plant names, 62–63
Justicia pectoralis, 100

Ka'apor, 56, 58–59
Kayapó, 77–94, 245
    gastrointestinal diseases, 81–83
    medical system, 79–80
    pharmacy, 83
    sociological considerations, 80–81
KIBORD project, 120
knowledge, local ethnobotanical, 231
    generalist versus specialist, 260–262
    loss, 231–232, 260, 266
    patterns of group consensus, 253–255,
        260–261
    spread, 56–59
    see also information, ethnobotanical

language, 199–200, 206, 248, 255–257, 264
Lantana camara, 252, 254, 256
Lantana hispida, 254, 256
lappaconitine, 172
Larren tridentata, 37–38
Latin names, 46–47
leishmaniasis, 122, 261, 267
Lepechinia schiedeana, 254
Lepidium virginicum, 252, 253
leukaemia, 70, 92
ligustrazine, 172
Linalool, 91
Lippia dulcis, 46
Lippia sidoides, 97, 100, 102
liver complaints, 157
Living Pharmacy programme, 100,
    102–104
Lobelia laxiflora, 252, 256
local community, 266
    biodiversity conservation, 229–232
    compensation, 34–35, 73–74, 122,
        209–213, 236–237, 267–268
    Living Pharmacy programme, 100,
        102–104
    People and Plants Initiative, 233–236,
        240–244
    reciprocity, 198, 203–205, 207
    return of information to, 19, 21,
        208–209, 234–235, 248, 255–257,
        263–264
    see also indigenous peoples
Lysimachia, 171

Madagascar, 235
malaria, 40, 122, 126, 147, 267
Malaysia, 211, 234, 239–243

Malpighia glabra, 100
Mandevilla cf. seabra, 84
Mandevilla steyermarkii, 110
Mandevilla tenuifolia, 82, 84
Maya, 21, 22, 141, 246–265
    ethnobotany, 250–253
    ethnopharmacology, 253–255
    promoting traditional herbal medicine,
        255–257
    traditional medicine, 142, 248–250
Mectisan, 148–149
MEDFLOR database, 46
medical doctors, 70, 75–76, 136, 201, 266
medicinal plant knowers (mẽkuté pidjá
        mari), 78, 80, 83
medicinal plants
    Africa, 119
    Amazon, 110–111
    analysis of indigenous uses, 30–33
    antiviral activity, 5, 200–202
    China, 169–170, 171
    common uses in different areas, 56–59,
        155
    economic value, 6–7, 12
    Highland Maya, 250–253
    India, 156–160, 165–166
    international law, 215–216
    Living Pharmacy programme, 100,
        102–104
    north-eastern Brazil, chemistry, 97–100
    rationale for studying, 44–45
    see also herbal remedies
mẽkuté pidjá mari (medicinal plant
        knowers), 78, 80, 83
Mentha spicata, 97
Mentha X villosa, 97, 100
Merck, Sharp and Dohme, 7–8, 42–43,
        193, 218
mestizo, 75, 234–235
Mexico, 91–92, 128, 130–152, 222, 242
    ethnobotanical information, 133–136,
        141–142
    Highland Maya, 246–265
    screening process, 136–138
michellamine B, 54, 70, 185–186, 190, 195
Miconia barbigera, 84, 86
migrations, 22, 244–245
Morinda lucida, 124
mouthwashes, 102
multidisciplinary approach, 55, 75, 118,
        136, 201, 266

multiple use curves, 10–12
Myracroduon urundeuva, 99–100

NAPRALERT database, 19, 31–32, 41, 46, 49–51, 57
National Cancer Institute (NCI), 4–5, 52, 178–196, 207
anticancer drug discovery 1960-1982, 48–49, 180–183
collaboration/compensation issues, 34–35, 187–188, 196, 212, 218
current plant acquisition/screening programme, 183–184
Developmental Therapeutics Program, 8, 178–179
plant acquisition 1960-1982, 179
recent successes, 28–29, 53, 66, 184–187
naturalistic systems, 118, 248–249, 262–263
nature reserves, 230
Neurolaena lobata, 10
New York Botanical Garden, 1, 4, 9, 114
Nicotiana tabacum (tobacco), 109, 252, 253, 254, 256
Nigeria, 119–120, 121, 122–123, 126
non-governmental organizations (NGOs), 221, 222, 226–227, 240–241, 264, 268

Ocimum gratissimum, 124
Ocimum selloi, 252, 256
Olyra latifolia, 84
Opuntia, 150
Organization of African Unity, 119
Ouratea hexasperma, 85
overexploitation/overharvesting, 3, 8–9
prevention, 16, 128–129
see also sustainable harvesting
Oxalis barrelieri, 85
oxytocic agents, 67

PADETEC, 100–101, 103
Palicourea, 85, 87
Passiflora, 82, 86
patents
circumvention, 192–194, 207, 268
human genetic material, 196
laws, 104, 214, 218, 222–223, 224–225
Paullinia yoco, 108
People and Plants Initiative, 100, 233–236, 240–244

pepper, black, 166
pepper-alecrim oil, 96–97
Periandra heterophylla, 84
personalistic systems, 118, 248, 249, 262
Peru, 64–65, 67, 75, 108, 209
Perymenium ghisbreghtii, 252
pharmaceutical industry, 52–53, 131, 133–134
collaborative agreements, 218–221
compensation issues, 6, 7, 123–125
role, 197–213
see also Shaman Pharmaceuticals Inc.
pharmacognosy, 54–55, 160
pharmacopoeias, 47, 119, 169, 207–208
Phaseolus, 85
Phlebodium decumanum, 5, 13
Phyllanthus amarus, 58
physicians, 70, 75–76, 136, 201, 266
Physostigma venenosum, 127, 128–129
phytochemistry
ecological variations, 2, 59, 175
India, 160
north-eastern Brazil, 95–105
Picralima nitida, 122, 124, 126
Picrorrhiza kurrooa, 161
Pilocarpus, 3, 8
'ping-pong' method, 137
Piper, 87
Piper guineense, 124
Piper jaquemontianum, 11
Piper schultesii, 110
Piper snethlagei, 82
Piperaceae, 72, 86–87
placebo effect, 47–48, 151
plants
chemistry, *see* phytochemistry
crude dried, 149
drug development approaches, 48–51
introduced, 244–245
Latin names, 46–47
vernacular names, 62–63, 72–73
see also extracts, plant; medicinal plants; powerful plants
podophyllotoxin, 38, 179, 181
Podophyllum, 38, 57, 70, 179, 181
Polygala, 171
Polygonum, 171
Polypodium phyllitidis, 82
polysaccharides, 19–20, 185
powerful plants, 201–202, 260, 261

Programa de Colaboración Sobre Medicina
    Indígena Tradicional y Herbolaria
    (PROCOMITH), 247
promotion, traditional medicine, 247,
    255–257
proscillaridin, 26
prostratin, 28–29, 37, 70, 186, 187
    compensation issues, 34–35, 37
Protium, 84, 86
Provir™, 198, 206–207
Prunus africana, 3, 233–234
Psidium guajava, 100, 124, 256, 260
Psidium guineense, 252, 253, 254, 256
Psittacanthus biternatus, 87
psychoactive plants, 108–109, 161
Psychotria, 85, 87
Psychotria lupulina, 82, 87
Psychotria viridis, 109
Pueraria lobata, 173

Qualea, 87
quality control, 167, 174
Quassia amara, 46

Rabdosia, 171
Radix puerariae, 174
random screening, 48–49, 179, 180, 191
Rauvolfia, 158, 159
respiratory disorders, 70, 250
Rourea induta, 84
Royal Botanic Gardens, Kew, 100, 104,
    194, 233
royalties, *see* compensation
Rubus, 171
Rubus coriifolius, 252, 253, 256
Ruellia, 82, 85
Ruta graveolens, 161

Sabiaceae calycina, 124
sacrality, 117
salicin derivatives, 68
Salvage Ethnography Project, 119–120
Salvia, 170, 171
Salvia miltiorrhiza, 173
Samoa, 28–29, 34–35, 36–37, 39, 226
sample size, 10–12
Sapindus saponaria, 99
Sapium poeppigii, 84
Saponaria officinalis, 47
saponins, 99, 141
Sauvagesia, 82, 86

schistosomiasis, 39
Schultezia guianensis, 98
schultezin, 98
Schwenkia guineensis, 124
Sclerocarya birrea, 124
screening methods, 2, 54, 179, 183
screening programmes, 43, 44–45, 52,
    65–66
    India, 161
    Mexico, 136–138
    National Cancer Institute, 179, 183–184
    random, 48–49, 179, 180, 191
    targeted, *see* ethnobotanical drug
        discovery
    toxicity and, 68–69
Scutellaria, 171
scutellarin, 172
Sebastiania corniculata, 84
Selaginella penniformis, 87
Shaman Pharmaceuticals Inc., 45–46, 52,
    197–213, 226, 268
    antiviral activity of medicinal plants,
        200–202
    biocultural conservation and, 203–204
    collaborative relationships, 122, 126,
        203, 204–205, 218
    history and methods, 198–200
shamans, 78, 80–81, 200
Siddha medicine, 157, 159, 160, 167–168
Simaruba glauca, 13
skin conditions, 31–32, 37–38, 39–40,
    158, 207
slashing, tree, 19
Smallanthus maculatus, 254, 256
Smilax lanceolata, 13
snake bite remedies, 56–57, 58
snuffs, 109
sodium houttuyfonate, 172
Sonchus oleraceus, 253, 254, 256
Souroubea, 110
South America, 60–76
spearmint oil, 97
species–area curve, 10, 11–12
spices, 164–165
Spigelia anthelmia, 84
Spiranthera odoratissima, 85
spiritual healers, 208
standardization, plant extracts, 147, 148,
    167, 174
Stevia ovata, 253, 254, 256
Stevia serrata, 256

street vendors, 165
Struthanthus marginatus, 87
supplies, 203–204
  by country of origin, 8, 74, 204
  problems, 52–54, 75
sustainable harvesting, 3, 203–204, 232
  forest valuation studies, 12–14
  People and Plants Initiative, 233–234, 236
  research, 16
  see also overexploitation/overharvesting
sweeteners, non-sugar, 223–224
Swertia, 171
synergistic interactions, 69, 138

Tabernaemontana heterophylla, 110
Tagetes erecta, 256
Tagetes filifolia, 252
Tagetes lucida, 252, 253, 254, 256
Tamarindus indica, 124
tannins, 5, 19–20, 86, 185
tax yields, 7
taxol, 49, 92, 137, 180, 181–182
  supply problems, 3, 53–54
taxonomy, 2–3, 58
taxotere, 182
Taxus, 53, 181–182
technological development park, 100–101, 103–104, 105
teniposide, 179, 181
Terminalia, 84
Terra Nova Rain Forest Reserve, 15, 16
Tetrapleura tetraptera, 124
Thalictrum, 171
Third World countries, *see* developing countries
Tithonia, 254
tobacco, *see Nicotiana tabacum*
topical administration, 36–37
Topotecan, 182
tourism, ecological, 16
Tournefortia cuspidata, 110
toxic plants
  Africa, 127
  Amazonian, 110–111
  north-eastern Brazil, chemistry, 97–100
toxicity
  anticancer compounds, 70
  Ayurvedic medicines, 166–167
  broad screening programmes and, 68–69

medicinal plants, 19, 32, 173–174
trade secrets, 215, 218
traditional healers, 22, 95, 263
  African, 118
  apprentices, 21–22, 263
  collection of information from, 23, 135, 200
  as ethnobotanists, 18–19
  management of ethno-biomedical forest reserve, 16
  recognition of diseases, 70, 75–76
  see also shamans
traditional medicine, 43–44, 131–133, 202
  Africa, 118
  codification, 247, 248–255, 257
  current attitudes, 150–151
  development of local industry, 8–9
  India, 156–158, 165–168
  Mexico, 132–133, 135, 142
  promotion, 247, 255–257
  see also Chinese medicine; herbal remedies; medicinal plants
training
  ethnobotanists, 46, 111, 112–113, 114, 121, 267
  local people, 236, 242–243, 255–257, 263
trichosanthin, 69
Tripterygium wilfordii, 173, 174–175
tropical diseases, 38, 39
  chemotherapy project, 122–123, 126–127
trypanosomiasis, 122
Typha angustata, 173
Tzeltal, 247–257
Tzotzil, 247–257

ubiquitous compounds, 67–68
United Nations Commission on Human Rights, 217
United Nations Convention on Environment Development (1992), 194, 203
United Nations Educational, Scientific and Cultural Organization (UNESCO), 216, 233
United States of America (USA), 114, 143, 144–146, 168
  see also National Cancer Institute
United States Army Medical Command, 122–123

United States Department of Agriculture
    (USDA),   179, 180, 183, 188
universities, industrial links,   100–101,
    103, 104–105
Utricularia oliverana,   84
Uvaria chamae,   124

valuation studies, forests,   12–14, 23–24
Vanillosmopsis arborea,   97, 98
Venezuela,   63–64, 108, 109
Verbena carolina,   252, 253, 254, 256
Verbena litoralis,   252, 253, 254, 256,
    260
vernacular names,   62–63, 72–73
Vernonia amygdalina,   124
Vernonia leiocarpa,   252
Veronia herbacea,   82
vinca alkaloids,   137, 147, 179, 180–181
Virend™,   198
Viriola,   109–110, 112
Vismia tomentosa,   110
Vitex gaumeri,   10

Waika,   109–110
Walter Reed Army Institute of Research,
    122–123, 127, 262–263
wayanga (shamans),   78, 80–81
Wilbrandia,   99
Withering, William,   26, 29, 147
World Health Organization (WHO),   90,
    119, 130–131
World Resources Institute,   218, 219, 221
World Wide Fund for Nature (WWF),
    233, 240–241, 243–244
Wulffia baccata,   84, 86

Xylopia aethiopica,   124
Xyris,   82

Yanomani,   208, 245
yellow fever,   28, 37
Yunani medicine,   157, 159, 160, 167–168

Zanthoxylum xanthoxyloides,   124
Zingiber officinale,   82, 124

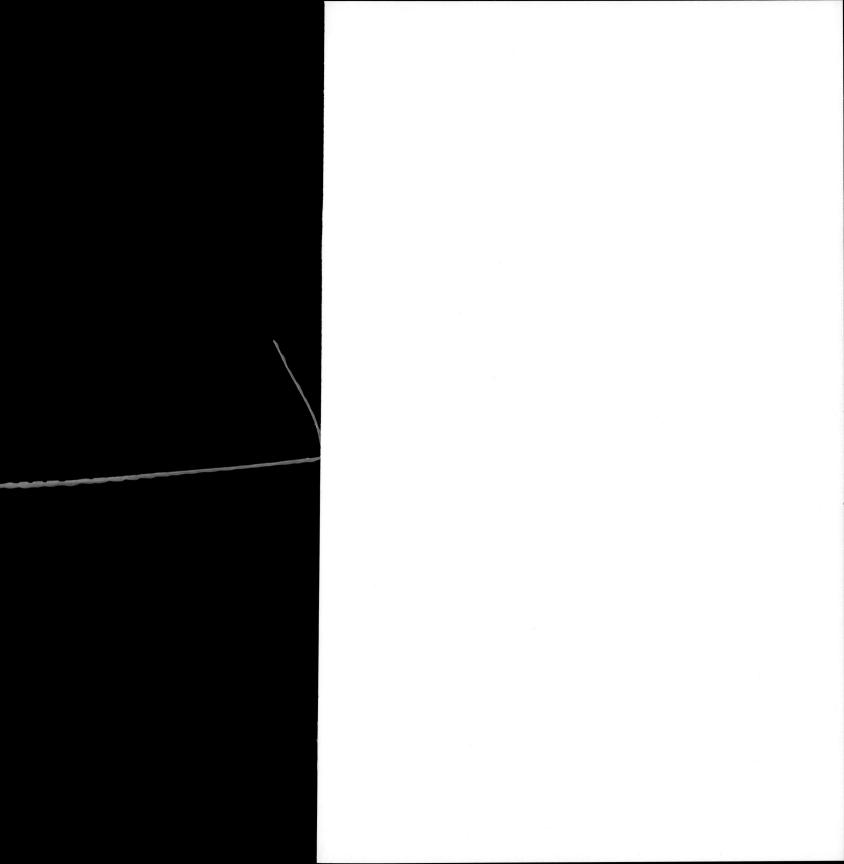

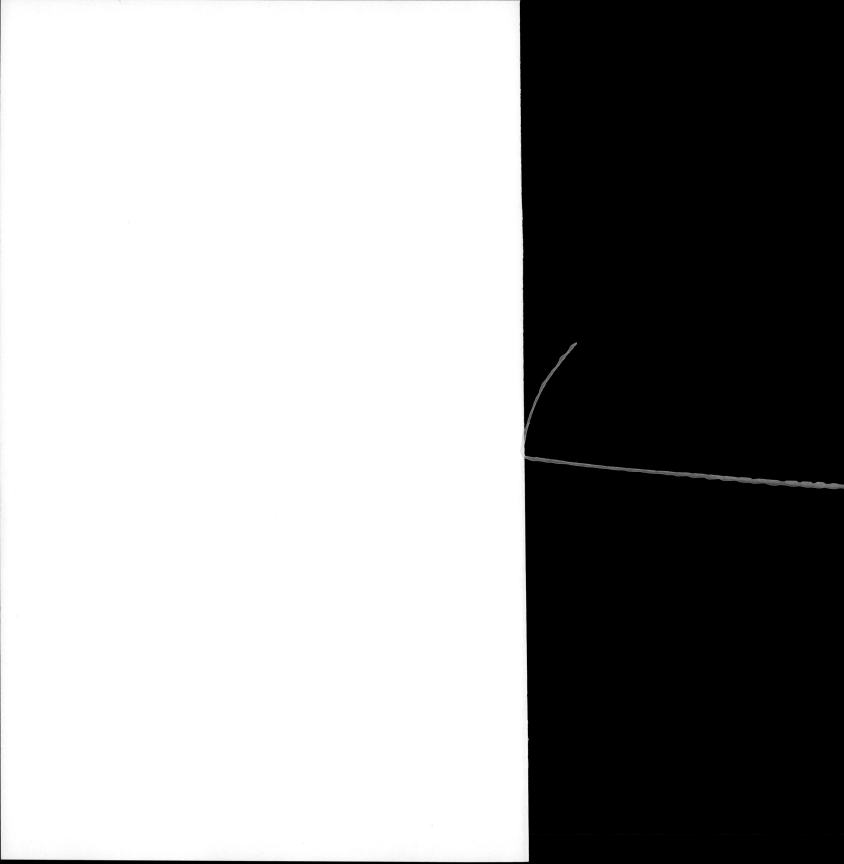